# Systems & Control: Foundations & Applications

Founding Editor

Christopher I. Byrnes, Washington University

Arie Feuer
Graham C. Goodwin

# Sampling in
# Digital Signal Processing
# and Control

1996
Birkhäuser
Boston • Basel • Berlin

Arie Feuer
Dept. of Electrical Engineering
Technion-Israel Institute of
Technology
Haifa 32000
Israel

Graham C. Goodwin
Faculty of Engineering
The University of Newcastle
Callaghan NSW 2308
Australia

**Library of Congress Cataloging-in-Publication Data**

Feuer, Arie, 1943-
    Sampling, in digital signal processing and control / Arie Feuer,
Graham C. Goodwin.
        p.  cm. -- (Systems & control)
    Includes bibliographical references.
    ISBN-13:978-1-4612-7546-6       e-ISBN-13:978-1-4612-2460-0
    DOI: 10.1007/978-1-4612-2460-0
    1. Signal processing--Digital techniques.    2. Signal processing-
-Statistical methods.  3. Sampling (Statistics)  4. Control theory.
I. Goodwin, Graham C. (Graham Clifford), 1945-    II. Title.
III. Series.
    TK5102.9.F48  1996                   96-24284
    621.382'23--dc20                        CIP

Printed on acid-free paper

© 1996 Birkhäuser Boston        *Birkhäuser*  ®
Softcover reprint of the hardcover 1st edition 1996

ISBN-13:978-1-4612-7546-6

Typeset by the Authors in LaTeX.

9 8 7 6 5 4 3 2 1

# Contents

# Chapter 2   Sampling and Reconstruction

# Chapter 3   Analysis of Discrete-Time Systems

# Chapter 4   Discrete-Time Models of Continuous Deterministic Systems

# Chapter 5   Optimal Linear Estimation with Finite Impulse Response Filters

## Chapter 6   Optimal Linear Estimation with State-Space Filters

## Chapter 7   Periodic and Multirate Filtering

## Chapter 8   Discrete-Time Control

# Chapter 9   Sampled Data Control

# Chapter 10   Generalized Sample-Hold Functions

# Preface

Undoubtably one of the key factors influencing recent technology has been the advent of high speed computational tools. Virtually every advanced engineering system we come in contact with these days depends upon some form of sampling and digital signal processing. Well known examples are digital telephone systems, digital recording of audio signals and computer control.

These developments have been matched by the appearance of a plethora of books which explain a variety of analysis, synthesis and design tools applicable to sampled-data systems. The reader might therefore wonder what is distinctive about the current book. Our observation of the existing literature is that the underlying continuous-time system is usually forgotten once the samples are taken. The alternative point of view, adopted in this book, is to formulate the analysis in such a way that the user is constantly reminded of the presence of the underlying continuous-time signals. We thus give emphasis to two aspects of sampled-data analysis:

Firstly, we formulate the various algorithms so that the appropriate continuous-time case is approached as the sampling rate increases.

Secondly we place emphasis on the continuous-time output response rather than simply focusing on the sampled response.

This philosophy has several advantages including:

(i) making the sampling period explicit in all calculations and hence allowing one to evaluate the effect of different sampling strategies,

(ii) allowing one to evaluate the performance loss due to sampling,

(iii) enhancing the numerical properties,

(iv) showing that some operations are highly undesirable when viewed in the limiting continuous-time case, and hence equally undesirable when implemented with fast sampling,

(v) indicating that some designs lead to large intersample behaviour, and

(vi) highlighting the origins of the differences between continuous-time and discrete-time models.

The issue of intersample response in sampled-data control has become a central focus of recent research work. We believe this book will provide a useful starting point for accessing this literature.

A distinctive feature of our treatment is that we combine both time and frequency domain methods. The frequency domain setting is a particularly natural one especially when periodic sampling or control strategies are employed. Another distinctive feature of the book is that we present a unified view of both signal processing and control. Many of the basic techniques are common to these two areas and hence a unified treatment allows one to combine ideas from the two fields of study.

Many of the ideas in the book were tested in a teaching environment at the University of Newcastle, Australia. In part, this encouraged us to write the book since the students taking these courses reported to us that their previous exposure to traditional methods of discrete analysis had often seemed unsatisfactory; little or no attempt had been made to connect the methods of analysis used for sampled-data systems to those used for continuous-time systems. By way of contrast, our approach shows that sampled-data analysis and continuous-time signal analysis can be viewed as special cases of a broader picture.

The book has been written as a text book for students or as a reference source for private study. All key results are proved in detail and each chapter includes numerous examples and exercises. Part I, i.e. Chapters 1 to 7, could form the basis of a senior undergraduate course on digital signal processing. Part II, i.e. Chapters 8 to 13 could form the basis of a senior undergraduate course on sampled data control. The whole book would be suitable for a course on sampled-data systems. The prerequisites for the book are elementary algebra and a junior level undergraduate course on control or signals and systems. We emphasise engineering insights rather than mathematical completeness.

Chapter 1 gives a comprehensive treatment of Fourier theory. Our treatment here is distinctive since we study both continuous and discrete methods in a unified fashion. Each transform pair is shown to be a special case of the continuous Fourier integral pair. The basic concepts are reinforced by applying them to non-traditional problems. For example, we present Fourier transform methods applicable to data sampled in periodic patterns but not necessarily with a uniform sampling interval.

Chapter 2 is concerned with the process of sampling and signal reconstruction. The Shannon reconstruction theorem is developed as well as the properties of various hold circuits. We also develop reconstruction methods applicable to more general problems such as data sampled in general periodic patterns.

Chapter 3 gives a review of analysis methods for discrete-time systems. $Z$ and delta transform methods are described and used to analyse the at-sample response of discrete-time systems. Some brief remarks are also made on the implementation of digital filters in shift and delta form.

Chapter 4 shows how discrete-time models can be developed for continuous linear time-invariant systems. Both time and frequency domain methods are treated as well as their inter-relationship.

Chapter 5 treats signal processing issues and develops results on finite impulse response optimal filtering. A distinctive feature of the treatment is that continuous-time and discrete-time results are presented in a unified framework.

Chapter 6 develops discrete-time state-space models for stochastic linear systems. These models are then used to derive optimal filtering results in state-space form. Again emphasis is placed on the connection between discrete-time and continuous-time results.

Chapter 7 considers more general filtering problems which apply to cases where the system is inherently time-varying. Both time domain and frequency domain methods of analysis are presented.

Chapter 8 gives a survey of discrete-time control methods. State space and polynomial techniques are considered. Design methods based on the parameterization of all stabilizing controllers and pole assignment are covered in detail. Also, a brief introduction to linear quadratic (LQ) optimal control is presented together with dual relationships to optimal linear filtering.

Chapter 9 takes the methods of Chapter 7 further by focusing attention on the intersample response. Frequency domain design methods are described which allow one to predict the continuous-time output response of a system under the action of sampled-data control. The sensitivity of the resultant closed loop system to changes in the underlying continuous-time system are studied. Also, optimization methods aimed at the design of sampled-data controllers for continuous-time plants are developed. Several methods are studied including direct design of sampled-data controllers in both the time and frequency domain as well as indirect methods which approximate the performance of a given continuous-time controller. Optimal Linear Quadratic sampled data controllers are described in detail.

Chapter 10 describes non-traditional hold circuits. It is shown that by use of these 'generalized' holds, one can design controllers achieving remarkable properties for the *at-sample* response. It is also shown that these properties are generally achieved at the expense of the intersample behaviour. Both time and frequency domain descriptions are given. Also, it is shown how the at-sample response relates to the intersample performance. Sensitivity issues are again treated.

Chapter 11 covers periodic control of linear time-invariant plants. It is shown that this class of control laws leads to some interesting properties which cannot be achieved with linear time-invariant controllers. Also, emphasis is placed on the cost of achieving these benefits.

Chapter 12 deals with multi-rate control. In particular, various schemes applicable to either fast-output/slow-input sampling or slow-output/fast-input sampling are reviewed. Also, general techniques applicable to multivariable, multi-rate control problems are presented.

Chapter 13 develops optimal control and filtering results for periodic linear systems. Raising techniques are used to relate the periodic control and filtering problems to associated stationary problems.

Finally a word on terminology. Throughout this book we use the adjective 'discrete' to imply sequences and operations on sequences. Of course, we always have in the back of our minds that these sequences may well be related to continuous signals in some way. We use the adjective 'sampled data' when the continuous signals are themselves an integral part of the analysis. These systems are also sometimes called 'hybrid' to emphasize the mixture of continuous and discrete concepts.

The authors wish to acknowledge the inspiration provided by the Australian bushland since significant parts of the book were written at rugged mountain retreats. The book was expertly typed by Denise Taft whose skill and friendliness combined to make the author's job much easier. Rob Newton gave valuable assistance with the diagrams and extremely helpful feedback on a preliminary draft of the book was obtained from Steve Weller and David Mayne.

Finally, the authors want to thank their respective families for their unstinting support. In the case of the Feuer family they allowed Arie time to travel across the world to visit Australia. In the case of the Goodwin family they tolerated Graham spending long hours at the office and they put up with the vague looks that inevitably came from a preoccupied book writer.

*Arie Feuer*
*Graham Goodwin*
*April, 1996*

# Notation

*The symbols are in approximately the same order as they appear in the text. The section numbers refer to the section in which the notation is first used.*

## Section 1.1

$\{y(t) \; ; \; t \epsilon R\}$                 Continuous-time signals.

$\{y(t_k) \; ; \; k \epsilon Z\}$             Sampled sequence.

$\Delta$                               Sampling period.

$\{y[k] = y(k\Delta) \; ; \; k \epsilon Z\}$     Sampled sequence with period $\Delta$ .

## Section 1.2.1

$\tilde{x}(t)$                           Periodic time function (period $T$).

$$\overline{X}_k = \int_{\frac{-T}{2}}^{\frac{T}{2}} \tilde{x}(t) e^{-j\left(\frac{k2\pi}{T}\right)t} dt \qquad \text{Fourier series.}$$

## Section 1.2.2

$$X(\omega) = \int_{-\infty}^{\infty} x(t) e^{-j\omega t} dt \qquad \text{Continuous Fourier transform of } x(t).$$

$$x(t) \overset{F}{\leftrightarrow} X(\omega)$$

Fourier transform pair.

$$\mu(t) = \begin{cases} 1 & \text{for } t > 0 \\ 1/2 & \text{for } t = 0 \\ 0 & \text{for } t < 0 \end{cases}$$

Unit step.

$$\delta(t)$$

Dirac delta function or unit impulse.

## Section 1.2.3

$$\Delta$$

Sampling period.

$$X^d(\omega) = \Delta \sum_{k=-\infty}^{\infty} x[k]e^{-j\omega\Delta k}$$

Discrete-time Fourier transform (DTFT).

$$x[k] \overset{DF}{\leftrightarrow} X^d(\omega)$$

Discrete-time Fourier transform pair.

$$\mu[k] = \begin{cases} 1 & \text{for } k \geq 0 \\ 0 & \text{for } k < 0 \end{cases}$$

Discrete unit step.

## Section 1.2.4

$$\tilde{x}[k]$$

Periodic sequence (period $N$).

$$X'_m = \sum_{k=0}^{N-1} \Delta\tilde{x}[k]e^{-j2\pi k\frac{m}{N}}$$

Discrete Fourier transform of sequence $\{\tilde{x}[k]\}$ of period $N$.

$$W_N = \frac{1}{\sqrt{N}} \begin{bmatrix} 1 & 1 & 1 & \cdots & 1 \\ 1 & W & W^2 & & W^{N-1} \\ 1 & & & & \\ \cdot & & & & \\ \cdot & & & & \\ \cdot & & & & \\ 1 & W^{N-1} & W^{2(N-1)} & & W^{(N-1)(N-1)} \end{bmatrix} \quad ; \quad W = e^{j\frac{2\pi}{N}}$$

DFT matrix.

## Section 1.3

$x^*$                                          Complex conjugate of $x$.

$\angle X(\omega)$                             Angle of $X(\omega)$.

$h(t) \otimes x(t) := \displaystyle\int_{-\infty}^{\infty} h(t)x(t-\tau)d\tau$          Convolution (continuous-time).

## Section 1.4

$h[k] \otimes x[k] := \Delta \displaystyle\sum_{m=-\infty}^{\infty} h[m]x[k-m]$   Convolution (discrete-time).

## Section 1.5

$s(t, \Delta) = \displaystyle\sum_{k=-\infty}^{\infty} \Delta\delta(t-k\Delta)$                        Delta-impulse sampler.

$S(\omega, \Delta) = 2\pi \displaystyle\sum_{k=-\infty}^{\infty} \delta(\omega - k\frac{2\pi}{\Delta}) \overset{F}{\leftrightarrow} s(t, \Delta)$   Transform of delta-impulse

sampler.

$\tilde{f}(t) = \displaystyle\sum_{k=-\infty}^{\infty} f(t-kT)$                        Periodic time domain function

formed by repetition.

## Section 1.6.2

$x^s(t) = \Delta \displaystyle\sum_{k=-\infty}^{\infty} x[k]\delta(t-k\Delta)$   Dirac pulse signal associated with the

sequence $\{x[k]\}$.

$X^s(\omega)$                                  Continuous-time Fourier transform of

$x^s(t)$.

## Section 1.6.4

$$f_1(u) \overset{\text{sample}}{\underset{a}{\rightarrow}} f_2(u) \Rightarrow f_2(u) = a \sum_{k=-\infty}^{\infty} f_1(ka)\delta(u - ka) \qquad \text{Sampling.}$$

$$g_1(u) \overset{\text{fold}}{\underset{a}{\rightarrow}} g_2(u) \Rightarrow g_2(u) = \sum_{k=-\infty}^{\infty} g_1(u - ka) \qquad \text{Folding.}$$

## Section 1.7.1

$$x_{\text{dec}}[k] = x[kn] \qquad\qquad\qquad \text{Decimation.}$$

$$x_{\text{fill}}[k] = \begin{cases} x_{\text{dec}}[k/N] & k \text{ a multiple of } N \\ 0 & \text{elsewhere} \end{cases} \qquad \text{Filling.}$$

$$s_d[k, N] = \Delta \sum_{m=-\infty}^{\infty} \delta_d[k - mN] \qquad \text{Discrete impulse sequence.}$$

$$\delta_d[k] = \begin{cases} \dfrac{1}{\Delta} & \text{for } k = 0 \\ 0 & \text{elsewhere} \end{cases} \qquad \text{Discrete } \dfrac{1}{\Delta} \text{ pulse.}$$

$$S^d(\omega, N) = \frac{2\pi}{N} \sum_{m=-\infty}^{\infty} \delta(\omega - m2\pi/N\Delta) \qquad \begin{array}{l}\text{Discrete transform of periodic}\\ \text{sampling sequence.}\end{array}$$

## Section 1.7.2

$$s(t, \{\Delta_i\}) = \sum_{k=-\infty}^{\infty} \Delta_k \delta(t - t_k) \qquad \begin{array}{l}\text{Non-uniform } \Delta \text{ impulse stream}\\ \text{where } t_k - t_{k-1} = \Delta_k.\end{array}$$

## Section 2.4

$$\omega_s = \frac{2\pi}{\Delta} \qquad\qquad\qquad \text{Sampling frequency.}$$

$$H_s(\omega) = \begin{cases} 1 & \text{for } |\omega| \leq \dfrac{\omega_s}{2} \\ 0 & \text{otherwise} \end{cases}$$   Ideal low pass filter.

## Section 2.5

$$h_0(t) = \begin{cases} \dfrac{1}{\Delta} & \text{for } 0 \leq t < \Delta \\ 0 & \text{elsewhere} \end{cases}$$   Impulse response of zero-order-hold.

$$H_0(s) = \frac{1 - e^{-s\Delta}}{s\Delta}$$   Continuous transform function of ZOH.

## Section 3.2

$q : qx[k] := x[k+1]$   Shift operator.

$\tilde{A}(q)$   Polynomial in $q$ : $\tilde{a}_n q^n + \tilde{a}_{n-1} q^{n-1} + \dots + \tilde{a}_0$.

$\left[ A_q,\ B_q,\ C_q,\ D_q \right]$   Shift domain state-space matrices.

## Section 3.3

$X_q(z) = Z\{x[k]\} \overset{\Delta}{=} \displaystyle\sum_{k=0}^{\infty} z^{-k} x[k]$   (One sided) $z$-transform.

$\tilde{G}(z)$   Discrete (shift domain) transfer function.

## Section 3.4

$\delta : \delta x[k] := \dfrac{x[k+1] - x[k]}{\Delta}$   The delta operator.

$$\left[A_\delta, \ B_\delta, \ C_\delta, \ D_\delta\right]$$    Delta domain state space model matrices.

$\overline{A}(\delta)$    Polynomial in $\delta$ :

$$\overline{a}_n\delta^n + \overline{a}_{n-1}\delta^{n-1} + \ ... \ + \overline{a}_0 \ .$$

$\delta^{-1}$    Lower Riemann sum $\delta^{-1}u = \displaystyle\sum_{m=0}^{k-1} \Delta u[m]$ .

## Section 3.6

$$Y(s) = L\{y(t)\} = \int_0^\infty e^{-st}y(t)dt$$    Continuous (one-sided)
Laplace transform.

$L^{-1}\{Y(s)\}$    Inverse Laplace transform of
$Y(s)$ .

$$Y_\delta(\gamma) = D\{y[k]\} = \sum_{k=0}^\infty \Delta(1 + \Delta\gamma)^{-k}\{y[k]\}$$    (One-sided) delta transform
of $y[k]$ .

## Section 3.7

$\overline{G}(\gamma)$    Discrete (delta domain) transfer function.

## Section 3.11

$$\gamma_\omega = \frac{e^{j\omega\Delta} - 1}{\Delta}$$    Discrete frequency variable for delta domain
models.

## Section 3.13

$\tilde{G}(z)$    Discrete (shift-domain) transfer function.

## Section 4.1

$G_p(s)$                                Continuous plant transfer function.

$F(s)$                                  Anti-aliasing filter.

## Section 4.3

$U^s(s)$                                Laplace transform of the Dirac pulse signal
$$u^s(t).$$

$S_\Delta\{x(t)\}$                      Sampling operator with period $\Delta$ producing
sequence $\{x[k]\}$.

## Section 4.6

$$\overline{G}_p(\gamma) = D\left\{S_\Delta\left\{L^{-1}\{F(s)G_p(s)H_0(s)\}\right\}\right\}\bigg|_{\gamma_\omega}$$    Discrete plant transfer
function.

$$\overline{G}_p^s(s) = \overline{G}_p(\gamma)\bigg|_{\gamma = \dfrac{e^{s\Delta}-1}{\Delta}}$$    Change of variable in discrete
transfer function.

## Section 5.2

$s'$                                    Continuous signal to be estimated.

$\hat{s}'$                              Estimate of $s'$.

$y'$                                    Continuous measurement.

## Section 5.3

$S_s(\omega)$                           Signal spectral density.

| | |
|---|---|
| $S_v'(\omega)$ | Noise spectral density. |
| $\bar{y}\,[k]$ | Filtered and sampled output. |
| $s$ | Discrete signal to be estimated. |

## Section 5.4

| | |
|---|---|
| $\hat{s}$ | Estimate of $s$. |
| $h[\cdot,\cdot]$ | Weighting function in forward signal estimator of lattice filter. |
| $b[\cdot,\cdot]$ | Weighting function in backwards signal estimator of lattice filter. |
| $e_f\,[\cdot,\cdot]$ | Forward estimation error of lattice filter. |
| $e_b[\cdot,\cdot]$ | Backward estimation error of lattice filter. |
| $K[\cdot]$ | Gain sequence in discrete lattice filter. |

## Section 5.5

| | |
|---|---|
| $h'(\cdot,\cdot)$ | Impulse response in continuous forward lattice filter. |
| $b'(\cdot,\cdot)$ | Impulse response in continuous backward lattice filter. |
| $e_f'\,(\cdot,\cdot)$ | Forwarded error in continuous lattice filter. |
| $e_b'(\cdot,\cdot)$ | Backward error in continuous lattice filter. |

## Section 6.2

$\dot{v}$                          Continuous-time white process noise.

$\dot{\omega}$                          Continuous-time white measurement noise.

$\Omega$                          Spectral density of continuous process noise.

$\Gamma$                          Spectral density of continuous measurement noise.

$y' = \dfrac{dy}{dt}$          Output of continuous stochastic system.

$\bar{y}$                          Filtered output of continuous stochastic system.

## Section 6.3

$\bar{y}[k+1] = \delta y[k]$          Sampled filtered output of stochastic system.

## Section 6.4

$v_\delta[k]$                          Process noise in discrete stochastic system.

$\omega_\delta[k]$                          Measurement noise in discrete stochastic system.

$\begin{bmatrix} \Omega_\delta & S_\delta \\ S_\delta^T & \Gamma_\delta \end{bmatrix}$          Discrete *spectral density* for the joint process $\begin{bmatrix} v_\delta[k] \\ \omega_\delta[k] \end{bmatrix}$

## Section 6.5

$H_\delta$                          Discrete Kalman filter gain.

| | |
|---|---|
| $P$ | Discrete state estimator error covariances. |
| $P_\infty$ | Steady state solution of discrete Riccati equation of optimal filtering. |

## Section 6.6

| | |
|---|---|
| $H(t)$ | Continuous Kalman filter gain. |

## Section 7.3

| | |
|---|---|
| $e_R$ | Raised vector $\left[e[mN], \quad \dots, \quad e[mN+N-1]\right]^T$. |
| $A_R, \; B_R, \; C_R, \; D_R$ | Matrices associated with raised system. |
| $\tilde{C}_R(z)$ | Multi-input multi-output discrete transfer function (shift domain) associated with a raised system. |

## Section 7.6

| | |
|---|---|
| $\begin{bmatrix} \bar{\Omega}_s & \bar{S}_s \\ \bar{S}_s^T & \bar{\Gamma}_s \end{bmatrix}$ | Discrete *variances* for the joint (shift domain) process $\begin{bmatrix} v_s[k] \\ \omega_s[k] \end{bmatrix}$. |
| $\begin{bmatrix} \bar{\Omega}_R & \bar{S}_R \\ \bar{S}_R^T & \bar{\Gamma}_R \end{bmatrix}$ | Discrete *variances* for the raised system. |

## Section 8.3

| | |
|---|---|
| $\bar{C}(\gamma)$ | Discrete (delta domain) controller transfer function. |

$\overline{G}_c(\gamma)$ Closed loop transfer function.

## Section 8.4

$\overline{S}(\gamma)$ Discrete sensitivity function.

$\overline{T}(\gamma)$ Discrete complementary sensitivity function.

$\Delta \overline{G}$ Small change in $\overline{G}$.

## Section 8.5

$\overline{Q}(\gamma)$ Transfer function parameterising all stabilizing control laws.

$\overline{Q}_1(\gamma)$ Injection transfer function in simplified form of all stabilizing control laws.

## Section 8.6

$Q_\delta^c$ Controllability matrix.

$K_\delta$ State feedback gain.

$J_\delta$ Observer gain.

## Section 8.8

$Q_d$ State weighting matrix in discrete-time linear regulator.

$R_d$ Control weighting matrix in discrete-time linear regulator.

$\Sigma_g[k]$                                          Solution of discrete Riccati equation of optimal
                                                       control.

$L_g[k]$                                               Optimal discrete feedback gain.

## Section 9.1

$\omega_k = \omega - k\omega_s$                        Shifted frequency variable.

$[X]^s \stackrel{\Delta}{=} \displaystyle\sum_{k=-\infty}^{\infty} X\big(j(\omega - k\omega_s)\big) = \displaystyle\sum_{k=-\infty}^{\infty} X(\omega_k)$    Folded version of continuous
                                                       frequency response $X(j\omega)$.

## Section 9.2

$P(\omega) = \overline{C}(\gamma_\omega)G_p(j\omega)H_0(j\omega)\overline{S}(\gamma_\omega)$    Reference gain function.

$D(\omega) = 1 - P(\omega)F(j\omega)$                  Disturbance gain function.

$P_f(\omega) = P(\omega)F(j\omega)$                    Filtered reference gain function.

## Section 9.4

$P_\epsilon^s(\omega) = \displaystyle\sum_{k=-\infty}^{\infty} P_f(\omega_k)e^{-j\omega_k\epsilon}$    Modified z-transform frequency response.

$[H_0G_pF]^s$                                          Discrete (i.e. folded) equivalent of the
                                                       continuous-time frequency domain
                                                       function $H_0G_pF$.

## Section 9.8

$Q_c$                                                  State weighting matrix in continuous-time linear
                                                       regulator.

| | |
|---|---|
| $R_c$ | Control weighting matrix in continuous-time linear regulator. |
| $L$ | Gain matrix in optimal linear state feedback. |
| $\Sigma$ | Riccati variable in optimal linear regulator. |

## Section 9.9

$$\bar{S}(s(\cdot))t = \sum_{k=-\infty}^{\infty} \Delta s[k]\delta(t-k\Delta)$$   Sampling operator converting continuous time signal into $\Delta$ pulse samples.

## Section 10.2

| | |
|---|---|
| $h_f(t),\ h_g(t)$ | Impulse responses of generalized sample holds. |
| $H_f(s),\ H_g(s)$ | Laplace transforms of $h_f(t),\ h_g(t)$. |
| $\bar{h}_g(t) = \sum_{k=-\infty}^{\infty} h_g(t-k\Delta)$ | Periodic extension of generalized hold function. |

## Section 10.4.4

| | |
|---|---|
| $\bar{H}_f(\omega),\ \bar{H}_g(\omega)$ | Fourier transforms of $\bar{h}(t),\ \bar{g}(t)$. |
| $H^s(\omega)$ | Frequency response of closed-loop using generalized holds. |
| $H_f(j\omega),\ H_g(j\omega)$ | Frequency responses of generalized holds corresponding to $h_f(t),\ h_g(t)$. |
| $a_f[p],\ a_g[p]$ | Fourier series coefficients of $\bar{h}_f(t),\ \bar{h}_g(t)$. |

## Section 12.2

$\Delta_0$                                    Output sampling period.

$\Delta$                                      Basic (fast) sampling period.

$\Delta_i$                                    Input sampling period.

## Section 13.2

$$\begin{bmatrix} Q[k] & S^T[k] \\ S[k] & R[k] \end{bmatrix}$$

Time varying weighting matrices in periodic optimal control problem.

$$\begin{bmatrix} Q_R & S_R^T \\ S_R & R_R \end{bmatrix}$$

Weighting matrix in equivalent 'raised' time-invariant optimal control problem.

# Chapter 1

# Fourier Analysis

## 1.1 Introduction

The frequency domain plays a central role in our discussion and analysis throughout this book. It is thus appropriate to begin with a treatment of Fourier theory. There are four basic Fourier transforms. These are:

(i) Fourier series that map *periodic continuous-time* functions into *sequences* in the frequency domain;

(ii) continuous-time Fourier transforms that map *continuous-time* functions into *continuous* frequency domain functions;

(iii) discrete-time Fourier transforms (DTFT) that map time domain *sequences* into *periodic continuous* frequency domain functions;

(iv) discrete Fourier transforms (DFT) that map *periodic sequences* in the time domain to *periodic sequences* in the frequency domain.

We see that the DFT combines the results of Fourier series with those of the DTFT.

Each of the above four transforms can be defined independently. However, by replacing the sequences in (i), (iii) and (iv) by impulses, it can be shown that each of the transforms can be conceived as a special case of the continuous-time Fourier transforms. This chapter will present these two alternative views of the various transform methods.

We suggest that all readers, whether or not they are familiar with Fourier methods, should review this chapter since we introduce notation that will be used subsequently. Also we define the various transforms (Fourier series, DTFT, DFT etc.) so that they are compatible. This is not typically done elsewhere in the literature where various scaling factors (e.g. $\Delta$, $T$, $2\pi$ etc.) apply. Our reason for using compatible definitions is so that we can mix the various transforms together when we study sampled-data signal processing and control. Also, we take our study of Fourier methods somewhat beyond that which is normally found elsewhere. For example, we develop discrete Fourier transforms for data sampled in a periodic pattern. This will allow us to treat more sophisticated sampling strategies later in the book but also serves the purpose of reinforcing the understanding of Fourier methods by applying them to more difficult problems.

Fourier transforms are alternative representations of continuous signals or discrete sequences. These representations bring forth many signal properties which are of central importance in the understanding of the signals, and their behaviour.

In this and subsequent chapters, we shall need to refer to both continuous-time signals and discrete-time sequences. We shall use the notation $\{y(t); t \in R\}$ for continuous-time signals. Given a continuous signal $\{y(t); t \in R\}$, sampling at a set of time instances $\{t_k \in R; k \in Z\}$ will result in a sequence $\{y(t_k); k \in Z\}$. We will frequently discuss the case where $t_k = k\Delta$. Then $\Delta$ is referred to as the *sampling period*. This sampling is commonly referred to as a constant sampling rate and the resulting sampled data sequence is $\{y[k] = y(k\Delta); k \in Z\}$. We use a distinct notation with square brackets, i.e. $y[k]$, to emphasize the fact that $\{y[k]\}$ is a sequence.

## 1.2   The Basic Transforms

In this section we will define the various transforms that will be used in the remainder of the book. Later, in Section 1.6 we will show that the various transforms can be viewed as special cases of the continuous-time Fourier transform.

## 1.2.1 Fourier Series

The reader has probably previously met the idea that a continuous-time function of period $T$ can be expanded as a sum of sinusoidal components via a Fourier series. The form of the Fourier series that we will employ is defined below:

**Definition 1.2.1** *Fourier series*

$$\overline{X}_k = \int_{\frac{-T}{2}}^{\frac{T}{2}} \tilde{x}(t) e^{-j\left(\frac{k2\pi}{T}\right)t} dt \tag{1.2.1}$$

**Definition 1.2.2** *Inverse Fourier series*

$$\tilde{x}(t) = \frac{1}{T} \sum_{k=-\infty}^{\infty} \overline{X}_k e^{j\left(\frac{k2\pi}{T}\right)t} \tag{1.2.2}$$

where $\tilde{x}(t)$ denotes a periodic continuous-time function of period $T$ and $\overline{X}_k$ are the Fourier coefficients. Note that the time domain function is periodic and $\overline{X}_k$ is a sequence.

The reader may recall that the coefficients $\overline{X}_k$ can be interpreted as representing the frequency content of the signal $\tilde{x}(t)$. This allows one to obtain useful insights into the nature of the signal. Also, the orthogonality of the terms $e^{j\left(\frac{k2\pi}{T}\right)t}$ for different values of $k$ proves extremely useful in analyzing the response of systems to the signal $\tilde{x}(t)$.

Note that it is usual in most books on Fourier analysis to include the factor 1/T in (1.2.1) rather than in (1.2.2) as we have done. However, we prefer the form given in (1.2.1), (1.2.2) to achieve compatability with the continuous-time Fourier transform as will become clear in the sequel.

## 1.2.2   The Continuous-Time Fourier Transform

If we let the period, $T$, in (1.2.1), (1.2.2) tend to infinity then we can (formally) convert a periodic signal into an a-periodic one. Also, the frequency spacing between the discrete frequency domain components in (1.2.1) tends to zero as the period goes to infinity. Letting $\omega = k2\pi/T$ and $\Delta\omega = 2\pi/T$ and then taking the limit $T \rightarrow \infty$ , we see that, the Riemann sum in (1.2.2) becomes an integral. This motivates us to introduce the following continuous-time Fourier transform pair for an a-periodic time function $x(t)$.

**Definition 1.2.3** The *Fourier transform* of the function $x(t), x : R \rightarrow C$, is given by (c.f. (1.2.1))

$$X(\omega) = \int_{-\infty}^{\infty} x(t)e^{-j\omega t}dt \qquad\qquad (1.2.3)$$

**Definition 1.2.4** The *inverse Fourier transform* of the function $X(\omega)$, $X : R \rightarrow C$ is given by (c.f. (1.2.2))

$$x(t) = \frac{1}{2\pi} \int_{-\infty}^{\infty} X(\omega)e^{j\omega t}d\omega \qquad\qquad (1.2.4)$$

Equation (1.2.3) is sometimes referred to as the 'analysis equation' and equation (1.2.4) as the 'synthesis equation' and the two equations as the 'Fourier transform pair'.

In the sequel we will typically use lower case letters for the time domain functions (signals) and uppercase letters for their frequency domain representations. We will use the notation $X(\omega) = F\{x(t)\}$ to indicate that $X(\omega)$ is the

Fourier transform of $x(t)$ and also refer to $x(t)$ and $X(\omega)$ as being a Fourier transform pair by writing

$$x(t) \overset{F}{\leftrightarrow} X(\omega)$$

Note that, in this case, the time and frequency domain functions are both
*continuous* functions.

Various restrictions apply to the function $x(t)$ so that $X(\omega)$ is well defined.
However, we will not distract the reader with these issues at this stage.

Some elementary continuous-time Fourier transform pairs are summarized below:

$$\text{sgn}(t) = \begin{cases} 1 & \text{for } t > 0 \\ 0 & \text{for } t = 0 \\ -1 & \text{for } t < 0 \end{cases} \overset{F}{\leftrightarrow} \frac{2}{j\omega} \tag{1.2.5}$$

$$e^{j\omega_0 t} \overset{F}{\leftrightarrow} 2\pi\delta(\omega - \omega_0) \tag{1.2.6}$$

$$\mu(t) = \begin{cases} 1 & \text{for } t > 0 \\ \frac{1}{2} & \text{for } t = 0 = \frac{1}{2}(\text{sgn}(t) + 1) \\ 0 & \text{for } t < 0 \end{cases} \overset{F}{\leftrightarrow} \frac{1}{j\omega} + \pi\delta(\omega) \tag{1.2.7}$$

Note that $\mu(t)$ is the unit step function and $\delta(t)$ is the Dirac delta function
or unit impulse.

$$\mu(t)e^{at} \overset{F}{\leftrightarrow} \frac{1}{j\omega - a} \tag{1.2.8}$$

## 1.2.3   The Discrete-Time Fourier Transform (DTFT)

We next consider a sequence $\{x[k]\}$. The discrete-time Fourier transform
(DTFT) is then defined as follows:

**Definition 1.2.5**   The *discrete-time Fourier transform* (DTFT) of the sequence $\{x[k]\}$, is given by

$$X^d(\omega) = \Delta \sum_{k=-\infty}^{\infty} x[k] e^{-j\omega \Delta k} \qquad (1.2.9)$$

Note that the transform in (1.2.9) can be viewed as a discrete approximation to the integral in (1.2.3). Actually, the connection between the transforms is much deeper as we shall see later. The corresponding inverse transform is given by:

**Definition 1.2.6** The *inverse discrete-time Fourier transform* (IDTFT) of the function $X^d(\omega)$ is given by

$$x[k] = \frac{1}{2\pi} \int_{-\frac{\pi}{\Delta}}^{\frac{\pi}{\Delta}} X^d(\omega) e^{j\omega \Delta k} d\omega \qquad (1.2.10)$$

Note that we prefer to use here $\omega$ rather than the normalized frequency $\Omega = \omega \Delta$ used in many references. Our purpose in doing so is again to emphasize the relationship between the continuous-time and discrete-time representations. Clearly, as $\Delta \rightarrow 0, x[k] = x(k\Delta) \rightarrow x(t), X^d(\omega) \rightarrow X(\omega)$ and (1.2.9), (1.2.10) become (1.2.3), (1.2.4).

In the sequel we will denote a pair of DTFT transforms as:

$$x[k] \overset{DF}{\leftrightarrow} X^d(\omega)$$

Some useful transform pairs are given below

$$\text{sgn } [k] = \begin{cases} 1 & \text{for } k \geq 0 \\ -1 & \text{for } k < 0 \end{cases} \overset{DF}{\leftrightarrow} \frac{2\Delta}{1 - e^{-j\omega \Delta}} \qquad (1.2.11)$$

$$1 \overset{DF}{\leftrightarrow} 2\pi \sum_{m=-\infty}^{\infty} \delta(\omega - m2\pi/\Delta) \qquad (1.2.12)$$

$$\mu[k] = \begin{cases} 1 \text{ for } k \geq 0 \\ 0 \text{ for } k < 0 \end{cases} = \frac{1}{2}\left(\text{sgn } [k] + 1\right)$$

$$\overset{DF}{\leftrightarrow} \frac{\Delta}{1 - e^{-j\omega\Delta}} + \pi \sum_{m=-\infty}^{\infty} \delta(\omega - m2\pi/\Delta) \quad (1.2.13)$$

and

$$e^{ak\Delta}\mu[k] \overset{DF}{\leftrightarrow} \frac{\Delta}{1 - e^{a\Delta}e^{-j\omega\Delta}} \qquad (1.2.14)$$

### 1.2.4   The Discrete Fourier Transform (DFT)

Finally, we consider the case of a *periodic sequence* of period $N$. The appropriate transform of a periodic sequence is

**Definition 1.2.7**   The *discrete Fourier transform* (DFT) of the periodic sequence $\{\tilde{x}[k]\}$ with period $N$ is

$$X'_m = \sum_{k=0}^{N-1} \Delta\tilde{x}[k]e^{-j2\pi k\frac{m}{N}} \qquad (1.2.15)$$

**Definition 1.2.8**   The *inverse discrete Fourier transform* is given by

$$\tilde{x}[k] = \frac{1}{N\Delta} \sum_{m=0}^{N-1} X'_m e^{j2\pi\frac{km}{N}}$$

$$\text{for } k = 0, 1, \ldots , N-1$$

$$(1.2.16)$$

Equations (1.2.15), (1.2.16) can also be expressed in vector form as follows:

$$\underline{X}' = \Delta \sqrt{N} W_N^* \underline{\tilde{x}}$$

$$\underline{\tilde{x}} = \frac{1}{\Delta \sqrt{N}} W_N \underline{X}'$$

where

$$\underline{X}' = \begin{bmatrix} X_0' \\ \cdot \\ \cdot \\ \cdot \\ X_{N-1}' \end{bmatrix} \quad ; \quad \underline{\tilde{x}} = \begin{bmatrix} \tilde{x}[0] \\ \cdot \\ \cdot \\ \cdot \\ \tilde{x}[N-1] \end{bmatrix}$$

and $W_N$ is the following DFT matrix

$$W_N = \frac{1}{\sqrt{N}} \begin{bmatrix} 1 & 1 & 1 & \cdots & 1 \\ 1 & W & W^2 & & W^{N-1} \\ 1 & W^2 & W^4 & & W^{2(N-1)} \\ \cdot & & & & \\ \cdot & & & & \\ \cdot & & & & \\ 1 & W^{N-1} & W^{2(N-1)} & & W^{(N-1)(N-1)} \end{bmatrix}$$

with $W = e^{j\frac{2\pi}{N}}$. Note also that

$$W_N W_N^* = W_N^* W_N = I$$

## 1.3    Properties of Continuous-Time Fourier Transforms

Since much of our subsequent work will be based on the continuous-time Fourier transform we summarize below some of its key properties.

Property 1.3.1 *Linearity*

$$ax_1(t) + bx_2(t) \overset{F}{\leftrightarrow} aX_1(\omega) + bX_2(\omega) \qquad (1.3.1)$$

where

$$x_1(t) \overset{F}{\leftrightarrow} X_1(\omega)$$

$$x_2(t) \overset{F}{\leftrightarrow} X_2(\omega)$$

Property 1.3.2 *Duality*

If

$$f(t) \overset{F}{\leftrightarrow} F(\omega) \qquad (1.3.2)$$

then

$$F(t) \overset{F}{\leftrightarrow} 2\pi\, f(-\omega) \qquad (1.3.3)$$

Proof :  From (1.3.2)

$$f(t) = \frac{1}{2\pi} \int\limits_{-\infty}^{\infty} F(\omega)e^{j\omega t}d\omega$$

so clearly

$$f(-t) = \frac{1}{2\pi} \int\limits_{-\infty}^{\infty} F(\omega)e^{-j\omega t}d\omega$$

$$f(-\omega) = \frac{1}{2\pi} \int\limits_{-\infty}^{\infty} F(t)e^{-j\omega t}dt$$

$$2\pi f(-\omega) = \int_{-\infty}^{\infty} F(t)e^{-j\omega t}dt$$

$$= F\{F(t)\}$$

Comment  The duality property is a central property and enables derivations in the frequency (time) domain to be an immediate consequence of their counterparts in the time (frequency) domain. We will use this extensively in the sequel.

Property 1.3.3 *Symmetry*

  a)  frequency domain

  If $x(t)$ is real valued, then $X(\omega)$ has the following symmetry property

$$X(-\omega) = X^*(\omega) \tag{1.3.4}$$

  b)  time domain

  If $X(\omega)$ is real valued then $x(t)$ has the following symmetry property

$$x(-t) = x^*(t) \tag{1.3.5}$$

Proof :  *We will only establish part a) since part b) follows immediately by duality. We shall also follow this same procedure in subsequent proofs wherever appropriate.*

$$X(-\omega) = \int_{-\infty}^{\infty} x(t)e^{j\omega t}dt$$

$$= \left[ \int_{-\infty}^{\infty} x(t)e^{-j\omega t}dt \right]^*$$

$$= X^*(\omega) \tag{1.3.6}$$

$$\wedge\wedge\wedge$$

Immediate consequences of Property 1.3.3 are

1. For real $x(t)$

$$|X(\omega)| = |X(-\omega)|$$

$$\angle X(\omega) = -\angle X(-\omega) \tag{1.3.7}$$

2. For real and even $x(t)$ (namely, $x(t) = x(-t)$)

$$X(\omega) = X(-\omega) \tag{1.3.8}$$

$X(\omega)$ even and real (by Property 1.3.3)

3. For real and odd $x(t)$ (namely, $x(t) = -x(-t)$)

$$X(\omega) = -X(-\omega)$$

$X(\omega)$ odd and imaginary by (Property 1.3.3).

Property 1.3.4 *Shifting*

a) time domain

$$\boxed{x(t-T) \overset{F}{\leftrightarrow} e^{-j\omega T}X(\omega)} \tag{1.3.9}$$

b) frequency domain

$$\boxed{e^{jWt}x(t) \overset{F}{\leftrightarrow} X(\omega-W)} \tag{1.3.10}$$

Proof :
$$F\{x(t-T)\} = \int_{-\infty}^{\infty} x(t-T)e^{-j\omega t}dt$$

$$= \int_{-\infty}^{\infty} x(\tau)e^{-j\omega(\tau+T)}d\tau$$

$$= e^{-j\omega T} \int_{-\infty}^{\infty} x(\tau)e^{-j\omega\tau}d\tau$$

$\wedge\wedge\wedge$

The above result is also known as the sinusoidal *modulation* property of Fourier transforms.

## Property 1.3.5 *Differentiation*

a) time differentiation

$$\boxed{\frac{dx(t)}{dt} \overset{F}{\leftrightarrow} j\omega X(\omega)}$$
(1.3.11)

b) frequency differentiation

$$\boxed{-jtx(t) \overset{F}{\leftrightarrow} \frac{dX(\omega)}{d\omega}}$$
(1.3.12)

Proof :  Since

$$x(t) = \frac{1}{2\pi} \int_{-\infty}^{\infty} X(\omega)e^{j\omega t}d\omega$$

$$\frac{dx(t)}{dt} = \frac{1}{2\pi} \int_{-\infty}^{\infty} [j\omega X(\omega)]e^{j\omega t}d\omega$$

$\wedge\wedge\wedge$

**Property 1.3.6** *Scaling*

$$\boxed{x(at) \overset{F}{\leftrightarrow} \frac{1}{|a|}X\left(\frac{\omega}{a}\right)}$$

(1.3.13)

where $a$ is a real non-zero constant.

**Proof :**

$$F\{x(at)\} = \int_{-\infty}^{\infty} x(at)e^{-j\omega t}dt$$

$$= \begin{cases} \dfrac{1}{a} \displaystyle\int_{-\infty}^{\infty} x(\tau)e^{-j\frac{\omega}{a}\tau}d\tau & \text{for } a > 0 \\[3ex] -\dfrac{1}{a} \displaystyle\int_{-\infty}^{\infty} x(\tau)e^{-j\frac{\omega}{a}\tau}d\tau & \text{for } a < 0 \end{cases}$$

$$= \frac{1}{|a|}X\left(\frac{\omega}{a}\right)$$

$\wedge\wedge\wedge$

**Property 1.3.7** *Parseval's relation*

$$\int_{-\infty}^{\infty} |x(t)|^2 dt = \frac{1}{2\pi} \int_{-\infty}^{\infty} |X(\omega)|^2 d\omega \qquad (1.3.14)$$

Proof :

$$\int_{-\infty}^{\infty} |x(t)|^2 dt = \int_{-\infty}^{\infty} x(t)x^*(t) dt$$

$$= \int_{-\infty}^{\infty} x(t) \left[ \frac{1}{2\pi} \int_{-\infty}^{\infty} X^*(\omega) e^{-j\omega t} d\omega \right] dt$$

$$= \frac{1}{2\pi} \int_{-\infty}^{\infty} X^*(\omega) \left[ \int_{-\infty}^{\infty} x(t) e^{-j\omega t} dt \right] d\omega$$

$$= \frac{1}{2\pi} \int_{-\infty}^{\infty} X^*(\omega) X(\omega) d\omega$$

$$\wedge\wedge\wedge$$

**Property 1.3.8** *Convolution*

We introduce the symbol $\otimes$ to denote convolution. This is defined as in (1.3.15) below. Then, for

$$x(t) \overset{F}{\leftrightarrow} X(\omega)$$

$$h(t) \overset{F}{\leftrightarrow} H(\omega)$$

a)  time domain

$$h(t) \otimes x(t) : \; = \int_{-\infty}^{\infty} h(\tau)x(t-\tau)d\tau \; \overset{F}{\leftrightarrow} \; H(\omega)X(\omega) \qquad\qquad (1.3.15)$$

b)  frequency domain

$$h(t)x(t) \; \overset{F}{\leftrightarrow} \; \frac{1}{2\pi}H(\omega) \otimes X(\omega) = \frac{1}{2\pi}\int_{-\infty}^{\infty} H(\eta)X(\omega-\eta)d\eta \qquad (1.3.16)$$

**Comment**  This property of frequency domain convolution is commonly referred to as the 'modulation property'. Note that property 1.3.4(b) is a special case where $h(t) = e^{jWt}$ .

**Proof :**

$$F\{h(t) \otimes x(t)\} = \int_{-\infty}^{\infty} \int_{-\infty}^{\infty} h(\tau)x(t-\tau)d\tau e^{-j\omega t}dt$$

$$= \int_{-\infty}^{\infty} h(\tau) \int_{-\infty}^{\infty} x(t-\tau)e^{-j\omega t}dt d\tau$$

$$= \int_{-\infty}^{\infty} h(\tau)e^{-j\omega\tau}d\tau X(\omega)$$

$$= H(\omega)X(\omega)$$

$\wedge\wedge\wedge$

Property 1.3.9 *Integration*

  a) time integration

$$\int_{-\infty}^{t} x(\tau)d\tau \overset{F}{\leftrightarrow} \frac{1}{j\omega}X(\omega) + \pi X(0)\delta(\omega) \tag{1.3.17}$$

  b) frequency integration

$$-\frac{1}{jt}x(t) + \pi x(0)\delta(t) \overset{F}{\leftrightarrow} \int_{-\infty}^{\omega} X(\eta)d\eta \tag{1.3.18}$$

**Proof :**  Follows from the following list of Fourier pairs given in (1.2.5) to (1.2.7).

  In particular, we have

$$\int_{-\infty}^{t} x(\tau)d\tau = \mu(t) \otimes x(t) \overset{F}{\leftrightarrow} \left[\frac{1}{j\omega} + \pi\delta(\omega)\right]X(\omega) = \frac{1}{j\omega}X(\omega) + \pi X(0)\delta(\omega)$$

△△△

Property 1.3.10 *One-sided functions*

  Of special interest are functions of the form

$$y_0(t) = y(t)\mu(t) = \begin{cases} y(t) & \text{for } t > 0 \\ 0 & \text{for } t \le 0 \end{cases} \tag{1.3.19}$$

Clearly

$$y_0(t) \overset{F}{\leftrightarrow} Y_0(\omega) = \int_{-\infty}^{\infty} y(t)\mu(t) \cdot e^{-j\omega t}dt$$

$$= \int_{0^+}^{\infty} y(t)e^{-j\omega t}dt \qquad (1.3.20)$$

The above integral is sometimes referred to as the *unilateral* Fourier transform.

The most important property for these functions is the transform of the one-side derivative.

$$\frac{dy(t)}{dt}\mu(t) \overset{F}{\leftrightarrow} j\omega Y_0(\omega) - y(0^+) \qquad (1.3.21)$$

Proof :  Since

$$y(t)\mu(t) = \frac{1}{2\pi} \int_{-\infty}^{\infty} Y_0(\omega)e^{j\omega t}d\omega$$

and

$$\frac{d}{dt}[y(t)\mu(t)] = \frac{dy(t)}{dt}\cdot\mu(t) + y(t)\delta(t)$$

Then

$$\frac{dy(t)}{dt}\mu(t) = \frac{d}{dt}[y(t)\mu(t)] - y(0^+)\delta(t) \overset{F}{\leftrightarrow} j\omega Y_0(\omega) - y(0^+)$$

$$\wedge\!\wedge\!\wedge$$

Note the difference between (1.3.11) and (1.3.21). We should also point out that when a linear time invariant differential equation is considered typically, it holds from time 0 onwards. In this case it should be multiplied by $\mu(t)$ before applying Fourier transforms and then one uses (1.3.21).

## 1.4    Properties of Discrete-Time Fourier Transforms

Some useful properties of the discrete-time Fourier transform are summarized below. Many of these results are analogous to the properties given in Section 1.3 for the continuous-time Fourier transform.

**Property 1.4.1** *Periodicity of the DTFT*

$$X^d(\omega + 2\pi/\Delta) = X^d(\omega)$$

(1.4.1)

**Proof :**

$$X^d(\omega + 2\pi/\Delta) = \Delta \sum_{k=-\infty}^{\infty} x[k] e^{-j(\omega + \frac{2\pi}{\Delta})\Delta k}$$

$$= \Delta \sum_{k=-\infty}^{\infty} x[k] e^{-j\omega \Delta k}$$

$\triangle\triangle\triangle$

**Property 1.4.2** *Linearity*

For        $x_1[k] \overset{DF}{\leftrightarrow} X_1^d(\omega)$

$x_2[k] \overset{DF}{\leftrightarrow} X_2^d(\omega)$

we have

$$ax_1[k] + bx_2[k] \overset{DF}{\leftrightarrow} aX_1^d(\omega) + bX_2^d(\omega)$$

(1.4.2)

**Property 1.4.3** *Symmetry*

For        $x[k]$   real valued

$$X^d(\omega) = X^d(-\omega)^*$$

(1.4.3)

**Property 1.4.4** *Shifting*

For $\quad x[k] \overset{DF}{\leftrightarrow} X^d(\omega)$

a) time shifting

$$\boxed{x[k-k_0] \overset{DF}{\leftrightarrow} e^{-j\omega \Delta k_0} X^d(\omega)}$$

(1.4.4)

b) frequency shifting

$$\boxed{e^{j\omega_0 \Delta k} x[k] \overset{DF}{\leftrightarrow} X^d(\omega - \omega_0)}$$

(1.4.5)

**Property 1.4.5** *Differencing in the time domain*

For $\quad x[k] \overset{DF}{\leftrightarrow} X^d(\omega)$

$$\boxed{\delta x[k] : \; = \frac{1}{\Delta}\big[x[k+1] - x[k]\big] \overset{DF}{\leftrightarrow} \frac{1}{\Delta}\left(e^{j\omega\Delta} - 1\right) X^d(\omega)}$$

(1.4.6)

**Property 1.4.6** *Differentiation in frequency*

For $\quad x[k] \overset{DF}{\leftrightarrow} X^d[\omega]$

$$\boxed{\Delta k x[k] \overset{DF}{\leftrightarrow} j\frac{dX^d(\omega)}{d\omega}}$$

(1.4.7)

**Property 1.4.7** *Parseval's relation*

For $\quad x[k] \overset{DF}{\leftrightarrow} X^d(\omega)$

$$\boxed{\Delta \sum_{k=-\infty}^{\infty} |x[k]|^2 = \frac{1}{2\pi} \int_{-\frac{\pi}{\Delta}}^{\frac{\pi}{\Delta}} |X^d(\omega)|^2 d\omega}$$

(1.4.8)

**Proof :**   $\Delta \displaystyle\sum_{k=-\infty}^{\infty} |x[k]|^2 = \Delta \displaystyle\sum_{k=-\infty}^{\infty} x[k]x^*[k]$

$$= \Delta \sum_{k=-\infty}^{\infty} x[k]\left[\frac{1}{2\pi}\int_{-\frac{\pi}{\Delta}}^{\frac{\pi}{\Delta}} X^d(\omega)^* e^{-j\omega\Delta k}d\omega\right]$$

$$= \frac{1}{2\pi}\int_{-\frac{\pi}{\Delta}}^{\frac{\pi}{\Delta}} X^d(\omega)^*\left[\Delta \sum_{k=-\infty}^{\infty} x[k]e^{-j\omega\Delta k}\right]d\omega$$

$$= \frac{1}{2\pi}\int_{-\frac{\pi}{\Delta}}^{\frac{\pi}{\Delta}} X^d(\omega)^* X^d(\omega)d\omega$$

$\Delta\Delta\Delta$

**Property 1.4.8** *Convolution*

  a)  time convolution (frequency domain multiplication)

$$\boxed{\; h[k] \otimes x[k] = \Delta \sum_{m=-\infty}^{\infty} h[m]x[k-m] \overset{DF}{\leftrightarrow} H^d(\omega)X^d(\omega) \;}$$   (1.4.9)

  b)  frequency convolution (time domain multiplication)

$$\boxed{\; h[k]x[k] \overset{DF}{\leftrightarrow} \frac{1}{2\pi}H^d(\omega) \otimes X^d(\omega) = \frac{1}{2\pi}\int_{-\frac{\pi}{\Delta}}^{\frac{\pi}{\Delta}} H^d(\eta)X^d(\omega-\eta)d\eta \;}$$   (1.4.10)

Proof :

a) $\quad F\{h[k] \otimes x[k]\} = \Delta^2 \sum\limits_{k=-\infty}^{\infty} \sum\limits_{m=-\infty}^{\infty} h[m]x[k-m]e^{-j\omega\Delta k}$

$$= \Delta \sum\limits_{m=-\infty}^{\infty} h[m] \left[ \Delta \sum\limits_{k=-\infty}^{\infty} x[k-m]e^{-j\omega\Delta k} \right]$$

$$= \Delta \sum\limits_{m=-\infty}^{\infty} h[m]e^{-j\omega\Delta m}X^d(\omega)$$

$$= H^d(\omega)X^d(\omega)$$

b) $\quad F\{h[k]x[k]\} = \Delta \sum\limits_{k=-\infty}^{\infty} h[k]x[k]e^{-j\omega\Delta k}$

$$= \Delta \sum\limits_{k=-\infty}^{\infty} \frac{1}{2\pi} \int\limits_{-\frac{\pi}{\Delta}}^{\frac{\pi}{\Delta}} H^d(\eta)e^{j\eta\Delta k}d\eta \, e^{-j\omega\Delta k}x[k]$$

$$= \frac{1}{2\pi} \int\limits_{-\frac{\pi}{\Delta}}^{\frac{\pi}{\Delta}} H^d(\eta) \left[ \Delta \sum\limits_{k=-\infty}^{\infty} x[k]e^{-j(\omega-\eta)\Delta k} \right] d\eta$$

$$= \frac{1}{2\pi} \int\limits_{-\frac{\pi}{\Delta}}^{\frac{\pi}{\Delta}} H^d(\eta)X^d(\omega-\eta)d\eta$$

$\wedge\wedge\wedge$

**Property 1.4.9** *Summation*

$$\text{For} \qquad x[k] \overset{DF}{\leftrightarrow} X^d(\omega)$$

$$\boxed{\Delta \sum_{m=-\infty}^{k} x[m] \overset{DF}{\leftrightarrow} \frac{\Delta}{1-e^{-j\omega\Delta}} X^d(\omega) + \pi X^d(0) \sum_{m=-\infty}^{\infty} \delta(\omega - m2\pi/\Delta)} \qquad (1.4.11)$$

**Proof :** The result follows from the DTFT pairs given in (1.2.11) to (1.2.13). Thus

$$\Delta \sum_{m=-\infty}^{k} x[m] = \mu(t) * x(t) \overset{DF}{\leftrightarrow} \left[ \frac{\Delta}{1-e^{-j\omega\Delta}} + \pi \sum_{m=-\infty}^{\infty} \delta(\omega - m2\pi/\Delta) \right] X^d(\omega)$$

$$= \frac{\Delta}{1-e^{-j\omega\Delta}} X^d(\omega) + \pi X^d(0) \sum_{m=-\infty}^{\infty} \delta(\omega - m2\pi/\Delta)$$

$$\wedge\wedge\wedge$$

## 1.5 The Δ - Impulse Stream

A function that we will find particularly useful in connecting continuous and discrete transforms is the following ' Δ -impulse stream' or 'time domain picket fence' function

$$\boxed{s(t, \Delta) = \sum_{k=-\infty}^{\infty} \Delta\delta(t - k\Delta)} \qquad (1.5.1)$$

This function is central to many results in Fourier theory and we will use it extensively. Unfortunately however, the derivation of the Fourier transform of this function is difficult. (See the discussion in Section 1.8). We will give some simple insights into the result below:

**Property 1.5.1** Let $D_N(\omega)$ be the Dirichlet kernel defined by

$$D_N(\omega) = \sum_{k=-N}^{N} \Delta e^{-j\omega k\Delta}$$

Then, in the limit as $N \rightarrow \infty$, $D_N(\omega)$ acts as a sum of impulses when integrated against a restricted class (in fact $L_p$) of functions. Thus, for this class of functions, we can write:

$$\lim_{N \rightarrow \infty} \sum_{k=-N}^{N} \Delta e^{-j\omega k \Delta} \equiv 2\pi \sum_{k=-\infty}^{\infty} \delta(\omega - k2\pi/\Delta) \qquad (1.5.2)$$

**Proof :** We will establish the result for a simple class of functions. We first note that for finite $N$ we can use the formula for the sum of a geometric series to obtain

$$\sum_{k=-N}^{N} \Delta e^{-j\omega k \Delta} = \frac{\Delta e^{-jN\Delta\omega} \left[ 1 - e^{j(2N+1)\Delta\omega} \right]}{\left[ 1 - e^{j\omega\Delta} \right]}$$

$$= \Delta \left[ \frac{e^{j\left(N+\frac{1}{2}\right)\Delta\omega} - e^{-j\left(N+\frac{1}{2}\right)\Delta\omega}}{e^{-j\omega\frac{\Delta}{2}} - e^{j\omega\frac{\Delta}{2}}} \right]$$

$$= \frac{\Delta \sin\left[ \left(N + \frac{1}{2}\right)\omega\Delta \right]}{\sin\left[ \frac{\omega\Delta}{2} \right]} \qquad (1.5.3)$$

A plot of this function is shown in Figure 1.5.1 for $N = 2$ and $\Delta = 0.1$.

A simple calculation shows that the area from $-\pi/\Delta$ to $\pi/\Delta$ is exactly $2\pi$ for all $N$. Also, the width of the pulses in Figure 1.5.1 tends to zero as $N$ tends to $\infty$.

To complete the proof of the lemma we will use a suitable test function. We will then apply both sides of equation (1.5.2) to this a test function and show both results to be identical.

Let $f(t) \overset{F}{\leftrightarrow} F(\omega)$ be such that $\sum_{k=-\infty}^{\infty} f(k\Delta) < \infty$ and $\sum_{k=-\infty}^{\infty} F(k\frac{2\pi}{\Delta}) < \infty$.

Integrating each side of equation (1.5.2) against $F(\omega)$ we get the following results for the left and right hand sides respectively:

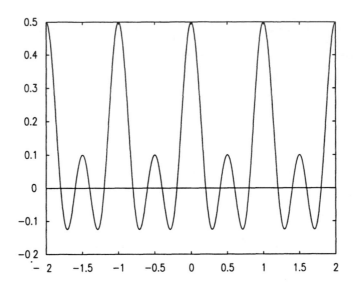

*Figure 1.5.1 : Frequency / sampling frequency*

$$\int_{-\infty}^{\infty} F(\omega) \sum_{k=-\infty}^{\infty} \Delta e^{-j\omega k\Delta} d\omega = \sum_{k=-\infty}^{\infty} 2\pi\Delta \frac{1}{2\pi} \int_{-\infty}^{\infty} F(\omega) e^{-j\omega k\Delta} d\omega$$

$$= 2\pi\Delta \sum_{k=-\infty}^{\infty} f(-k\Delta) \qquad (1.5.4)$$

and

$$\int_{-\infty}^{\infty} F(\omega) 2\pi \sum_{k=-\infty}^{\infty} \delta(\omega - k2\pi/\Delta) d\omega = 2\pi \sum_{k=-\infty}^{\infty} F(k2\pi/\Delta) \qquad (1.5.5)$$

To show that the results in equations (1.5.4) and (1.5.5) are equal define

$$\tilde{f}(t) = \sum_{n=-\infty}^{\infty} f(t - n\Delta) \qquad (1.5.6)$$

Since $\tilde{f}(t + \Delta) = \tilde{f}(t)$, it has a Fourier series representation. Namely, by (1.2.2) we can write

$$\tilde{f}(t) = \frac{1}{\Delta} \sum_{k=-\infty}^{\infty} \overline{F}_k e^{j\left(k\frac{2\pi}{\Delta}\right)t} \tag{1.5.7}$$

Now from (1.2.1)

$$\overline{F}_k = \int_{\frac{-n\Delta}{2}}^{\frac{n\Delta}{2}} \tilde{f}(t)e^{-j\left(\frac{k2\pi}{n\Delta}\right)t}dt = \int_{\frac{-n\Delta}{2}}^{\frac{n\Delta}{2}} \sum_{n=-\infty}^{\infty} f(t-n\Delta)e^{-j\left(\frac{k2\pi}{n\Delta}\right)t}dt$$

$$= \int_{-\infty}^{\infty} f(t)e^{-j\left(\frac{k2\pi}{n\Delta}\right)t}dt = F(k2\pi/\Delta) \tag{1.5.8}$$

Hence from (1.5.6), (1.5.7), (1.5.8)

$$\sum_{n=-\infty}^{\infty} f(t-n\Delta) = \frac{1}{\Delta} \sum_{k=-\infty}^{\infty} F(k2\pi/\Delta)e^{j\left(k\frac{2\pi}{\Delta}\right)t} \tag{1.5.9}$$

Substituting $t = 0$ in (1.5.9) verifies that the right hand sides of (1.5.4), (1.5.5) are equal. This completes the proof.

$\triangle\triangle\triangle$

We then have the following result.

## Property 1.5.2

With the interpretation of $s(t, \Delta)$ given in Property 1.5.1, then its Fourier transform satisfies

$$s(t, \Delta) \overset{F}{\leftrightarrow} S(\omega, \Delta) \tag{1.5.10}$$

where

$$S(\omega, \Delta) = 2\pi \sum_{k=-\infty}^{\infty} \delta(\omega - k2\pi/\Delta) \qquad (1.5.11)$$

**Proof :**

Using (1.2.3) and Property 1.5.1 we have

$$s(t, \Delta) \overset{F}{\leftrightarrow} \int_{-\infty}^{\infty} \Delta \sum_{k=-\infty}^{\infty} \delta(t - k\Delta)e^{-j\omega t} dt$$

$$= \Delta \sum_{k=-\infty}^{\infty} e^{-j\omega k\Delta}$$

$$= 2\pi \sum_{k=-\infty}^{\infty} \delta(\omega - k2\pi/\Delta)$$

$$= S(\omega, \Delta) \qquad (1.5.12)$$

$$\wedge\wedge\wedge$$

The above result has many uses as we shall see below.

## 1.6    Inter-relating the Various Transforms

In this section we will relate the various transforms back to the basic continuous-time Fourier transform.

### 1.6.1    Fourier Series Revisited

We first relate the Fourier series coefficients $\{\overline{X}_k\}$ of Section 1.2 to the continuous Fourier transform. To do this we consider an a-periodic function $x(t)$ where $x(t) \overset{F}{\leftrightarrow} X(\omega)$. We then construct a $T$-periodic function $\tilde{x}(t)$ by repeating $x(t)$ at time spacing $T$; i.e. define

$$\tilde{x}(t) = \sum_{k=-\infty}^{\infty} x(t - kT) \qquad (1.6.1)$$

Substituting (1.6.1) in (1.2.1) we obtain

$$\overline{X}_k = \int_{-\frac{T}{2}}^{\frac{T}{2}} \sum_{n=-\infty}^{\infty} x(t - nT) e^{-j\left(k\frac{2\pi}{T}\right)t} dt$$

$$= \sum_{n=-\infty}^{\infty} \int_{-\frac{T}{2}}^{\frac{T}{2}} x(t - nT) e^{-j\left(k\frac{2\pi}{T}\right)t} dt$$

$$= \sum_{n=-\infty}^{\infty} \int_{-\frac{T}{2}-nT}^{\frac{T}{2}-nT} x(t) e^{-j\left(k\frac{2\pi}{T}\right)t} dt$$

$$= \int_{-\infty}^{\infty} x(t) e^{-j\left(k\frac{2\pi}{T}\right)t} dt$$

and using equation (1.2.3) we note that $\overline{X}_k$ is related to $X(\omega)$ as follows:

$$\overline{X}_k = X(k2\pi/T) \qquad (1.6.2)$$

Summarizing the above we have

Property 1.6.1 *Repetition and Fourier Series*

If a $T$-periodic function $\tilde{x}(t)$ is formed by repeating an $a$-periodic function $x(t)$ where $x(t) \overset{F}{\leftrightarrow} X(\omega)$, then the Fourier series coefficients $\{\overline{X}_k\}$ of $\tilde{x}(t)$ are simply obtained by evaluating $X(\omega)$ at $\omega = k2\pi/T$ for $k = -\infty, \ldots, \infty$ .

$\wedge\wedge\wedge$

We can state Property 1.6.1 loosely by saying that *repetition* in the time domain leads to *sampling* in the frequency domain.

We can also express (1.6.1) more compactly by using the $\Delta$ -impulse stream $s(t, \Delta)$ defined in Section 1.5. Specifically, we have

$$\tilde{x}(t) \;=\; \frac{1}{T} \int_{-\infty}^{\infty} s(\sigma, T)x(t-\sigma)d\sigma \tag{1.6.3}$$

or more compactly

$$\tilde{x}(t) \;=\; \frac{1}{T} s(t, \Delta) \otimes x(t)$$

Using Property 1.5.2 and 1.3.8 we obtain the following continuous-time Fourier transform pair :

$$\boxed{\tilde{x}(t) \;\overset{F}{\leftrightarrow}\; \tilde{X}(\omega)} \tag{1.6.4}$$

where

$$\tilde{X}(\omega) \;=\; \frac{1}{T} X(\omega) S(\omega, T) \tag{1.6.5}$$

with $S(\omega, T)$ as in (1.5.11); we see that the continuous-time Fourier transform of the periodic continuous-time signal $\tilde{x}(t)$ is

$$\boxed{\tilde{X}(\omega) \;=\; \frac{2\pi}{T} \sum_{k=-\infty}^{\infty} X(\omega)\delta(\omega - k2\pi/T)} \tag{1.6.6}$$

We thus add extra insight into Property 1.6.1 by stating the following.

**Property 1.6.2** *Repetition and continuous-time Fourier transforms*

If a $T$-periodic function $\tilde{x}(t)$ is formed by repeating an a-periodic function $x(t)$ where $x(t) \overset{F}{\leftrightarrow} X(\omega)$, then the *continuous-time* Fourier transform of $\tilde{x}(t)$ is the *impulse* sampled version of $X(\omega)$ where the impulses have strength $2\pi/T \cdot X(k2\pi/T)$ for $k = -\infty, \ldots, \infty$.

$\wedge\wedge\wedge$

Substituting (1.6.6) into the inverse continuous-time Fourier transform of (1.2.4) allows us to rederive the Fourier series formula as follows:

$$\tilde{x}(t) = \frac{1}{T} \sum_{k=-\infty}^{\infty} X(k2\pi/T) e^{j\left(\frac{k2\pi}{T}\right)t} \tag{1.6.7}$$

Comparing (1.6.7) with (1.2.2) shows that

$$\boxed{\overline{X}_k = X(k2\pi/T)} \tag{1.6.8}$$

as previously stated in Property 1.6.1.

### 1.6.2   The Discrete-Time Fourier Transform Revisited

In this section we will relate the DTFT of a sequence $\{x[k]\}$ to continuous-time Fourier transforms. The sequence $\{x[k]\}$ could be defined without reference to any continous-time signal. When it does come from sampling of some continuous-time signal $x(t)$, we use the notation $x[k] = x(k\Delta)$.

We saw in Section 1.6.1 (Property 1.6.2), that to retain continuous-time Fourier *integral* transforms for frequency domain sampling it was necessary to use frequency domain *impulse* sampling. By duality, one might suspect that to retain continuous-time Fourier integral transforms with time domain sampling it is necessary to use *time domain impulse sampling*.

With this in mind, we associate with a sequence $\{x[k]\}$ at sampling interval $\Delta$, the following Dirac pulse signal:

$$x^s(t) \;=\; \Delta \sum_{k=-\infty}^{\infty} x[k]\delta(t-k\Delta) \qquad\qquad (1.6.9)$$

The scaling factor $\Delta$ in (1.6.9) has been introduced to parallel the factor $(2\pi/T)$ in (1.6.6). Note that we can think of $x[k]$ as holding over a time interval of length $\Delta$ whereas $X(k2\pi/T)$ can be thought of as holding over a period $(2\pi/T)$.

The *continuous-time* Fourier transform $X^s(\omega)$ of $x^s(t)$ is obtained as follows:

$$X^s(\omega) \;=\; \int_{-\infty}^{\infty} x^s(t)e^{-j\omega t}dt \;=\; \int_{-\infty}^{\infty} \Delta \sum_{k=-\infty}^{\infty} x[k]\delta(t-k\Delta)e^{-j\omega t}dt$$

$$= \Delta \sum_{k=-\infty}^{\infty} x[k]e^{-j\omega k\Delta} \qquad\qquad (1.6.10)$$

Thus we have $x^s(t) \overset{F}{\leftrightarrow} X^s(\omega)$. However, we immediately recognize (1.6.10) as the DTFT of the sequence $\{x[k]\}$ – see equation (1.2.9). We thus have the following interesting result:

**Property 1.6.3** The DTFT, $X^d(\omega)$, of a sequence $\{x[k]\}$, is a periodic function with period $2\pi/\Delta$. Also the DTFT is equal to the *continuous-time Fourier transform*, $X^s(\omega)$, of the Dirac-pulse-signal $x^s(t)$ obtained from $\{x[k]\}$ by adding $\Delta$-scaled impulses.

$\wedge\!\wedge\!\wedge$

Note that the above conclusion holds whether or not $\{x[k]\}$ is obtained by sampling an underlying continuous-time signal. However, we next consider the case where $x^s(t)$ is obtained by sampling an underlying continuous-time signal $x(t)$ where $x(t) \overset{F}{\leftrightarrow} X(\omega)$. In this case, we can use the $\Delta$-impulse stream to write

$$\boxed{x^s(t) = x(t)s(t, \Delta)} \qquad (1.6.11)$$

where $s(t, \Delta)$ is the '$\Delta$-impulse stream' defined in (1.5.1).

Now using Property 1.3.8, we have

$$X^s(\omega) = \frac{1}{2\pi} X(\omega) \otimes S(\omega, \Delta)$$

$$= \frac{1}{2\pi} \int_{-\infty}^{\infty} X(\sigma) \sum_{k=-\infty}^{\infty} 2\pi\delta(\omega - \sigma - k2\pi/\Delta) d\sigma$$

or

$$\boxed{X^s(\omega) = \sum_{k=-\infty}^{\infty} X(\omega - k2\pi/\Delta)} \qquad (1.6.12)$$

Summarizing we have

**Property 1.6.4** If an impulse stream $x^s(t)$ is obtained by impulsively sampling a continuous-time signal $x(t)$ where $x(t) \overset{F}{\leftrightarrow} X(\omega)$, then the continuous-time Fourier transform of $x^s(t)$ which we denote by $X^s(\omega)$ is obtained by repeating $X(\omega)$ at intervals separated by multiples of $2\pi/\Delta$.

$\wedge\wedge\wedge$

This is illustrated in Figure 1.6.1.

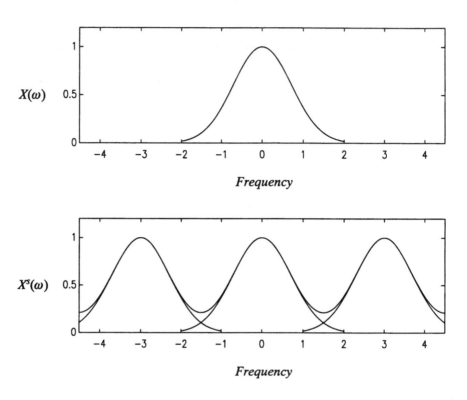

*Figure 1.6.1 :  The spectrum of $x^s(t)$ and its relation to the spectrum of $x(t)$*
        *when  $\Delta = 2$.*

We can state Property 1.6.4 loosely by saying that (impulse) sampling in the time domain leads to repetition in the frequency domain. The dual relationship with Property 1.6.2 is clear.

Finally, we note that, by the continuous-time Fourier transform, the *impulse-sampled* values $x^s(t)$ are recovered by the inverse transform

$$x^s(t) = \frac{1}{2\pi} \int\limits_{-\infty}^{\infty} X^s(\omega)e^{-j\omega t}dt \qquad (1.6.13)$$

However, we are frequently interested in recovering the non-impulsive sequence $\{x[k]\}$. To do this we utilize the transform pair $x(t) \overset{F}{\leftrightarrow} X(\omega)$ to conclude

$$x[k] = x(k\Delta)$$

$$= \frac{1}{2\pi} \int_{-\infty}^{\infty} X(\omega) e^{j\omega k\Delta} d\omega$$

$$= \frac{1}{2\pi} \sum_{n=-\infty}^{\infty} \int_{(2n-1)\frac{\pi}{\Delta}}^{(2n+1)\frac{\pi}{\Delta}} X(\omega) e^{j\omega k\Delta} d\omega$$

$$= \frac{1}{2\pi} \int_{-\frac{\pi}{\Delta}}^{\frac{\pi}{\Delta}} \left\{ \sum_{n=-\infty}^{\infty} X(\omega - n2\pi/\Delta) \right\} e^{j\omega k\Delta} d\omega$$

$$= \frac{1}{2\pi} \int_{-\frac{\pi}{\Delta}}^{\frac{\pi}{\Delta}} X^d(\omega) e^{j\omega k\Delta} d\omega \qquad (1.6.14)$$

where we have used Property 1.6.4.

As the reader would have anticipated equation (1.6.14) is simply a restatement of the inverse discrete-time Fourier transform given in Definition 1.2.6 (equation (1.2.10)).

## 1.6.3   The Discrete Fourier Transform Revisited

We next consider the DFT of Section 1.2.4.

As for Fourier series and DTFT, we can relate the DFT coefficients $X'_k$ directly to the continuous-time transform. To do this consider an $a$-periodic continuous-time function $x(t)$. Let $\tilde{x}(t)$ be formed by repeating $x(t)$ as in (1.6.1) where we assume that the period $T$ is an integer multiple of $\Delta$ i.e. $T = N\Delta$.

Then $\tilde{x}(t)$ has the *continuous-time* transform as given in (1.6.5), i.e.

$$\tilde{X}(\omega) = \frac{1}{T} X(\omega) S(\omega, T)$$

$$= \frac{1}{N\Delta} X(\omega) S(\omega, T) \qquad (1.6.15)$$

Now if we impulse sample $\tilde{x}(t)$ at interval $\Delta$ to produce another signal $\tilde{x}^s(t)$, then by (1.6.11) we have

$$\tilde{X}^s(\omega) = \sum_{k=-\infty}^{\infty} \tilde{X}(\omega - k2\pi/\Delta)$$

$$= \frac{1}{N\Delta} \sum_{k=-\infty}^{\infty} X(\omega - k2\pi/\Delta) \, S(\omega - k2\pi/\Delta, T) \qquad (1.6.16)$$

This is clearly periodic with period $2\pi/\Delta$.

Hence using Property 1.6.3 and equations (1.6.13), (1.6.15), (1.5.10), (1.6.11) the sample values $\{x[k]\}$ can be expressed as

$$\tilde{x}[k] = \frac{1}{2\pi} \int_{0}^{\frac{2\pi}{\Delta}} \tilde{X}^s(\omega) e^{j\omega k\Delta} d\omega$$

$$= \frac{1}{N\Delta} \sum_{m=0}^{N-1} \left\{ \sum_{k=-\infty}^{\infty} X\left[ \frac{m2\pi}{N\Delta} - k2\pi/\Delta \right] \right\} e^{j\frac{m2\pi}{N\Delta} \cdot k\Delta}$$

$$= \frac{1}{N\Delta} \sum_{m=0}^{N-1} X^d(m2\pi/N\Delta) e^{jkm\frac{2\pi}{N}} \qquad (1.6.17)$$

where

$$X^d\left[\frac{m2\pi}{N\Delta}\right] = \sum_{k=-\infty}^{\infty} X\left[\frac{m2\pi}{N\Delta} - \frac{k2\pi}{\Delta}\right]$$

$$(1.6.18)$$

From (1.6.15) and (1.6.17) we have that the DFT coefficients are obviously given by

$$X'_m = X^d \cdot \frac{m2\pi}{N\Delta}$$

$$(1.6.19)$$

Also, by direct calculation using the continuous-time Fourier transform we have

$$\tilde{x}[k] = \frac{1}{2\pi}\int_{-\infty}^{\infty} \tilde{X}(\omega)e^{j\omega k\Delta}d\omega$$

$$= \frac{1}{2\pi}\int_{0}^{\frac{2\pi}{\Delta}} \frac{2\pi}{N\Delta}X(\omega)\sum_{m=-\infty}^{\infty}\delta\left[\omega - \frac{2\pi m}{N\Delta}\right]e^{j\omega k\Delta}d\omega$$

$$= \frac{1}{N\Delta}\sum_{m=-\infty}^{\infty} X\left[\frac{2\pi m}{N\Delta}\right]e^{j2\pi k\frac{m}{N}}$$

$$= \frac{1}{N\Delta}\sum_{m=0}^{N-1} X'_m e^{j2\pi k\frac{m}{N}}$$

$$(1.6.20)$$

where

$$X'_m = \sum_{k=-\infty}^{\infty} X\left[\frac{m2\pi}{N\Delta} - \frac{k2\pi}{\Delta}\right]$$

$$(1.6.21)$$

This re-establishes (1.6.18), (1.6.17).

## 1.6.4  Summary

In the above development we have seen that it is possible to work entirely with continuous-time signals and continuous-time transforms. Fourier series, DTFT's and DFT's then occur as special cases depending upon the range of variables and the conversion of integrals to summations.

The various transforms are summarized in Table 1.6.1.

The inter-relationships between the various operations in Table 1.6.1 enables us to develop a diagram we feel is quite illuminating. This diagram is shown in Figure 1.6.2. Notice that we have used $\omega_0 = \omega_s/N$, with $N > 1$ an integer. We use the notation :

$$f_1(u) \overset{\text{sample}}{\underset{a}{\longrightarrow}} f_2(u) \text{ to imply that } f_2(u) = a \sum_{k=-\infty}^{\infty} f_1(ka)\delta(u - ka) \text{, and}$$

$$g_1(u) \overset{\text{fold}}{\underset{a}{\longrightarrow}} g_2(u) \text{ to imply that } g_2(u) = \sum_{k=-\infty}^{\infty} g_1(u - ka).$$

We recall, as was pointed out in Property 1.6.3, that $X^s(\omega)$ is in fact $X^d(\omega)$, the DTFT of the sequence $x[k]$.

Also, we have shown that $\tilde{X}^s(\omega)$ corresponds to the DFT of a periodic sequence $\tilde{x}[k]$ where the period is $T = N\Delta$.

| Time | Frequency |
|---|---|
| $x(t) = \dfrac{1}{2\pi} \displaystyle\int_{-\infty}^{\infty} X(\omega)e^{j\omega t}d\omega \quad (1.2.4)$ | $X(\omega) = \displaystyle\int_{-\infty}^{\infty} x(t)e^{-j\omega t}dt \quad (1.2.3)$ |

*Table 1.6.1a : Continuous Fourier transform.*

| Time | Frequency |
|---|---|
| $$\tilde{x}(t) = \frac{1}{T} \sum_{k=-\infty}^{\infty} \overline{X}_k e^{j\left(\frac{2\pi}{T}\right)t} \quad (1.2.2)$$ | $$\overline{X}_k = \int_{-\frac{T}{2}}^{\frac{T}{2}} \tilde{x}(t) e^{-j\left(\frac{k2\pi}{T}\right)t} \quad (1.2.1)$$ |
| *(periodic continuous)* | *(sequence)* |
| $$\text{If } \tilde{x}(t) = \sum_{k=-\infty}^{\infty} x(t - kT) \quad (1.6.1) \quad \text{then the continuous transform is}$$ $$\tilde{X}(\omega) = \frac{2\pi}{T} \sum_{k=-\infty}^{\infty} X(\omega)\delta(\omega - k2\pi/T) \qquad (1.6.6)$$ $$\text{and } \overline{X}_k = X(k2\pi/T) \qquad (1.6.8)$$ | |

*Table 1.6.1b : Fourier series.*

The various operations in Figure 1.6.2 and Table 1.6.1 are illustrated by the following example:

Example 1.6.1   Consider the signal

$$x(t) = e^{-|t|} \tag{1.6.22}$$

Walk this signal through the diagram in Figure 1.6.2 with $\Delta = 0.1$ and N=5.

Solution :   In the solution we will make use of the following result (Hansen 1975 p.103)

$$\sum_{k=-\infty}^{\infty} \frac{1}{(ka + b)(kc + d)} = \frac{\pi}{ad - bc} \left[ \cot \frac{\pi b}{a} - \cot \frac{\pi d}{c} \right]$$

The Fourier transform of the signal (1.6.22) is

| Time | Frequency |
|---|---|
| $$x[k] = \frac{1}{2\pi} \int_{-\frac{\pi}{\Delta}}^{\frac{\pi}{\Delta}} X^d(\omega)e^{j\omega\Delta k}d\omega \quad (1.2.10)$$  *(sequence)* | $$X^d(\omega) = \Delta \sum_{k=-\infty}^{\infty} x[k]e^{-j\omega\Delta k} \quad (1.2.9)$$  *(periodic continuous)* |

*If impulse sampling is used* $x^s(t) = \Delta \displaystyle\sum_{k=-\infty}^{\infty} x[k]\delta(t - k\Delta) \quad (1.6.9)$

*then the continuous transform is* $X^s(\omega) = \displaystyle\sum_{k=-\infty}^{\infty} X\big(\omega - k2\pi/\Delta\big) \quad (1.6.12)$

*and* $X^s(\omega) = X^d(\omega)$    *(Property 1.6.3)*

*Table 1.6.1c : Discrete-time Fourier transform.*

$$X(\omega) = \int_{-\infty}^{\infty} x(t)e^{-j\omega t}dt = \int_{-\infty}^{0} e^{t}e^{-j\omega t}dt + \int_{0}^{\infty} e^{-t}e^{-j\omega t}dt$$

$$= \frac{1}{1-j\omega} + \frac{1}{1+j\omega} = \frac{2}{1+\omega^2}$$

We will now follow separately the path in the time domain and the frequency domain and then relate the results. The numbers of the steps are shown in brackets in Figure 1.6.2.

1.  Impulse sampling $x(t)$ with $\Delta = 0.1$ leads to

$$x^s(t) = 0.1 \sum_{k=-\infty}^{\infty} e^{-0.1|k|}\delta(t - 0.1k)$$

| Time | Frequency |
|---|---|
| $$\tilde{x}[k] = \frac{1}{N\Delta}\sum_{m=0}^{N-1}X'_m e^{2\pi k\frac{m}{N}} \quad (1.2.16)$$ (periodic sequence) | $$X'_m = \sum_{k=0}^{N-1}\Delta\tilde{x}(k)e^{-j2\pi k\frac{m}{N}} \quad (1.2.15)$$ (periodic sequence) |
| The coefficients $X'_m$ are given by $X'_m = X^d\left(\dfrac{m2\pi}{N\Delta}\right)$   (1.6.19) $$= X^s\left(\frac{m2\pi}{N\Delta}\right)$$ $$= \sum_{k=-\infty}^{\infty} X\left(\frac{m2\pi}{N\Delta} - \frac{k2\pi}{\Delta}\right) \quad (1.6.18)$$ | |

*Table 1.6.1d : Discrete Fourier transform.*

2. Fold $x^s(t)$ over $N\Delta = 0.5$ leads to the following periodic sampled signal

$$\tilde{x}^s(t) = \sum_{l=-\infty}^{\infty} x^s(t-0.5l) = 0.1\sum_{l=-\infty}^{\infty}\sum_{k=-\infty}^{\infty} e^{-0.1|k|}\delta(t-0.5l-0.1k)$$

$$= 0.1\sum_{m=-\infty}^{\infty}\delta(t-0.1m)\sum_{l=-\infty}^{\infty} e^{-0.5|l-0.2m|}$$

$$= 0.1\sum_{m=-\infty}^{\infty}\delta(t-0.1m)\left[\sum_{l=-\infty}^{\text{int}(0.2m)} e^{-0.5(0.2m-l)} + \sum_{l=\text{int}(0.2m)+1}^{\infty} e^{-0.5(l-0.2m)}\right]$$

where $\text{int}(0.2m)$ denotes the largest integer less than $0.2m$.

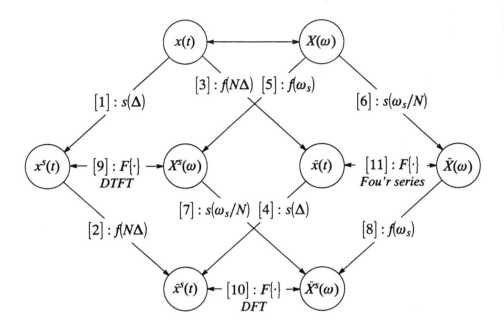

$$s(A) \equiv Sample \ at \ rate \ A \qquad f(B) \equiv Fold \ over \ B \qquad F\{\cdot\} \equiv Fourier \ transform$$

*Figure 1.6.2 :  Relationships between transforms. The numbers in square*
*brackets refer to steps in Example 1.6.1.*

$$\tilde{x}^s(t) = \frac{0.1}{1-e^{-0.5}} \sum_{m=-\infty}^{\infty} \left[ e^{-0.1(m-5\mathrm{int}(0.2m))} + e^{0.1(m-5\mathrm{int}(0.2m)-5)} \right] \delta(t-0.1m)$$

3.  Folding $x(t)$ over $N\Delta = 0.5$ leads to the following periodic
    signal

$$\tilde{x}(t) = \sum_{k=-\infty}^{\infty} x(t-0.5k) = \sum_{k=-\infty}^{\infty} e^{-|t-0.5k|}$$

$$= \sum_{k=-\infty}^{\mathrm{int}(2t)} e^{-(t-0.5k)} + \sum_{k=\mathrm{int}(2t)+1}^{\infty} e^{t-0.5k}$$

$$= \frac{\left(e^{-t}e^{0.5\text{int}(2t)} + e^{t}\cdot e^{-0.5\text{int}(2t)+1)}\right)}{1 - e^{-0.5}}$$

4. Sampling $\breve{x}(t)$ every $\Delta = 0.1$ leads to

$$\frac{0.1}{1 - e^{-0.5}} \sum_{k=-\infty}^{\infty} \left[ e^{-0.1k}e^{0.5\text{int}(0.2k)} + e^{0.1k}\cdot e^{-0.5(\text{int}(0.2k)+1)} \right] \delta(t - 0.1k) = \breve{x}^s(t)$$

which is equal to $x^s(t)$ as in step 2 as expected.

5. Folding $X(\omega)$ over $\omega_s = 20\pi$ leads to

$$X^s(\omega) = \sum_{k=-\infty}^{\infty} X(\omega - 20\pi k) = \sum_{k=-\infty}^{\infty} \frac{2}{1 + (\omega - 20\pi k)^2}$$

$$= 2 \sum_{k=-\infty}^{\infty} \frac{1}{[20\pi k - \omega + j][20\pi k - \omega - j]}$$

$$= -\frac{1}{20j} \left[ \cot\left[\frac{j-\omega}{20}\right] + \cot\left[\frac{j+\omega}{20}\right] \right]$$

$$= 0.1 \left[ 1 + \frac{e^{-0.1(j\omega+1)}}{1 - e^{-0.1(j\omega+1)}} + \frac{e^{-0.1(1-j\omega)}}{1 - e^{-0.1(1-j\omega)}} \right]$$

6. Sampling $X(\omega)$ every $\omega_s/N = 4\pi$ leads to

$$\breve{X}(\omega) = 4\pi \sum_{k=-\infty}^{\infty} \frac{2}{1 + (4\pi k)^2} \delta(\omega - 4\pi k)$$

7. Sampling $X^s(\omega)$ every $\omega_s/N = 4\pi$ leads to

$$\breve{X}^s(\omega) = 0.4\pi \sum_{k=-\infty}^{\infty} \left[ \frac{1 + e^{-0.1(1+4\pi kj)}}{1 - e^{-0.1(1+4\pi kj)}} + \frac{e^{-0.1(1-4\pi kj)}}{1 - e^{-0.1(1-4\pi kj)}} \right] \delta(\omega - 4\pi k)$$

8. Folding $\tilde{X}(\omega)$ over $\omega_s = 20\pi$ leads to the result

$$4\pi \sum_{l=-\infty}^{\infty} \sum_{k=-\infty}^{\infty} \frac{2}{1+(4\pi k)^2} \delta(\omega - 4\pi k - 20\pi l)$$

$$= 8\pi \sum_{m=-\infty}^{\infty} \delta(\omega - 4\pi m) \sum_{l=-\infty}^{\infty} \frac{1}{1+(4\pi)^2(m-5l)^2}$$

$$= 8\pi \sum_{m=-\infty}^{\infty} \delta(\omega - 4\pi m) \sum_{l=-\infty}^{\infty} \frac{1}{\left[20\pi l - 4\pi m + j\right]\left[20\pi l - 4\pi m - j\right]}$$

$$= 0.2\pi j \sum_{m=-\infty}^{\infty} \delta(\omega - 4\pi m) \left[ \cot\left[\frac{j-4\pi m}{20}\right] + \cot\left[\frac{j+4\pi m}{20}\right] \right]$$

$$= 0.4\pi \sum_{m=-\infty}^{\infty} \left[ 1 + \frac{e^{-0.1(1+4\pi mj)}}{1-e^{-0.1(1+4\pi mj)}} + \frac{e^{-0.1(1-4\pi mj)}}{1-e^{-0.1(1-4\pi mj)}} \right] \delta(\omega - 4\pi m)$$

which is equal to $\tilde{X}^s(\omega)$ as in Step 7 as expected.

9. To complete the journey we verify the various connections between the transform pairs as shown by the horizontal lines in Figure 1.6.2.

$$F\{x^s(t)\} = 0.1 \sum_{k=-\infty}^{\infty} e^{-0.1|k|} e^{-j\omega 0.1k}$$

$$= 0.1 \left[ 1 + \sum_{k=-\infty}^{-1} e^{0.1k(1-j\omega)} + \sum_{k=1}^{\infty} e^{-0.1k(1+j\omega)} \right]$$

$$= 0.1 \left[ 1 + \frac{e^{-0.1(1-j\omega)}}{1-e^{-0.1(1-j\omega)}} + \frac{e^{-0.1(1+j\omega)}}{1-e^{-0.1(1+j\omega)}} \right]$$

$$= X^s(\omega)$$

10. To compute the Fourier transform of $\tilde{x}^s(t)$ it is convenient to re-write $\tilde{x}^s(t)$ as

$$\tilde{x}^s(t) = \frac{0.1}{1-e^{-0.5}} \cdot \sum_{l=0}^{4} \sum_{m=-\infty}^{\infty} \left[ e^{-0.1(5m+l)}e^{0.5\text{int}(0.2(5m+l))} \right.$$

$$\left. + e^{0.1(5m+l)}e^{-0.5(\text{int}(0.2(5m+l))+1} \right] \cdot \delta(t-0.5m-0.1l)$$

$$= \frac{0.1}{1-e^{-0.5}} \sum_{l=0}^{4} \left[ e^{-0.1l}e^{0.5\text{int}(0.2l)} + e^{0.1l}e^{-0.5(\text{int}(0.2l)H)} \right]$$

$$\cdot \sum_{m=-\infty}^{\infty} \delta(t-0.1l-0.5m)$$

Then, using the fact that a Fourier transform of an impulse train is an impulse train we get

$$F\{\tilde{x}^s(t)\} = \frac{0.4\pi}{1-e^{-0.5}} \sum_{l=0}^{4} \left[ e^{-0.1l} + e^{0.1l-0.5} \sum_{m=-\infty}^{\infty} \delta(\omega-4\pi m)e^{-j0.4\pi m \cdot l} \right]$$

$$= 0.4\pi \sum_{m=-\infty}^{\infty} \left[ 1 + \frac{e^{-0.1(1+4\pi mj)}}{1-e^{-0.1(1+4\pi mj)}} + \frac{e^{-0.1(1-4\pi mj)}}{1-e^{-0.1(1-4\pi mj)}} \right] \delta(\omega-4\pi m)$$

$$= \tilde{X}^s(\omega)$$

11. To calculate $F\{\tilde{x}(t)\}$ we make use of the fact that $\tilde{x}(t)$ is periodic with periodicity 0.5. Hence $\tilde{x}(t)$ can be written in a Fourier se-ries form as

$$\tilde{x}(t) = 2 \sum_{k=-\infty}^{\infty} \overline{X}_k e^{j4\pi kt}$$

where

$$\bar{X}_k = \frac{1}{1-e^{-0.5}} \int_0^{0.5} \left( e^{-t} + e^{t-0.5} \right) e^{-j4\pi kt} dt$$

$$= \frac{1}{1-e^{-0.5}} \left[ \frac{1-e^{-0.5}}{1+4\pi kj} + \frac{1-e^{-0.5}}{1-4\pi kj} \right]$$

$$= \frac{2}{1+(4\pi k)^2}$$

So we have

$$F\{\bar{x}(t)\} = 4\pi \sum_{k=-\infty}^{\infty} \bar{X}_k \delta(\omega - 4\pi k)$$

$$= 4\pi \sum_{k=-\infty}^{\infty} \frac{2}{1+(4\pi k)^2} \delta(\omega - 4\pi k)$$

$$= \tilde{X}(\omega)$$

Figure 1.6.3 displays the various transform pairs developed above.

## 1.7    Special Topics

In this section we will use the results of the previous sections to study special sampling strategies. In sub-Section 1.7.1 we will study the sampling of a sequence to produce another (decimated) sequence. In sub-Section 1.7.2 we study sampling using non-uniform but periodic sampling patterns. Both of these special topics will serve to reinforce our understanding of Fourier methods by using them in more difficult situations. Note that we will also build on these results later when we study signal reconstruction and other deeper topics.

### 1.7.1    Sampling of Sequences

Here consider the following problem:

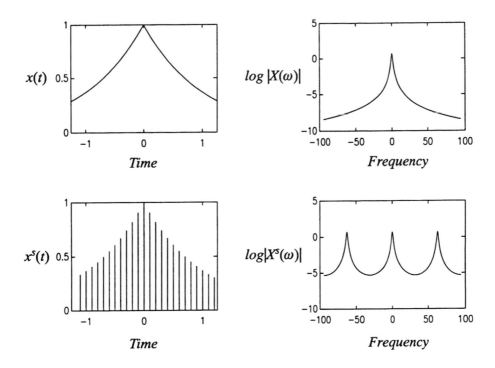

*Figure 1.6.3 : Transform Pairs for Example 1.6.1*

Given the sequence $\{x[k]\} \overset{DF}{\leftrightarrow} X^d(\omega)$ with an associated sampling period $\Delta$, we create a new sequence, $\{x_{dec}[k]\}$, by taking every $N^{th}$ element of $\{x[k]\}$. Namely we form

$$x_{\text{dec}}[k] = x[kN] \qquad (1.7.1)$$

We want to calculate the DTFT for the new sequence and relate it to the DTFT of $x[k]$.

We call the operation given in (1.7.1) *decimation* (sometimes also called *downsampling* in the literature).

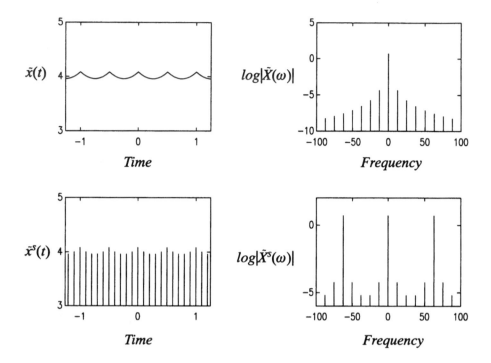

*Figure 1.6.3  (continued)*

A related operation is called *filling* (sometimes also called *upsampling* in the literature). The 'filling' of $x_{dec}[k]$ leads to a new sequence $\{x_{fil}[k]\}$ defined as

$$x_{fil}\ [k]\ =\ \begin{cases} x_{dec}[k/N] \text{ if } k \text{ is a multiple of } N \\ 0 \quad \text{elsewhere} \end{cases} \tag{1.7.2}$$

In the digital signal processing literature, filling is often called *interpolation*. We prefer the word filling because the space between samples is filled with zeros. On the other hand interpolation gives the impression that some attempt is being made to 'estimate' the missing data values.

The relationship between $x[k], x_{dec}[k]$ and $x_{fil}[k]$ is illustrated in Figure 1.7.1 for a particular case with $N = 3$.

We have the following property relating the transforms of $x_{dec}[k]$ and $x_{fil}[k]$.

## Property 1.7.1

$$x_{dec}[k] \overset{DF}{\leftrightarrow} X_{dec}^d(\omega) \text{ with } \Delta_{dec} = N\Delta$$

$$x_{fil}[k] \overset{DF}{\leftrightarrow} X_{fil}^d(\omega) \text{ with } \Delta_{fil} = \Delta$$

and

$$\boxed{X_{fil}^d(\omega) = 1/N \; X_{dec}^d(\omega)} \qquad\qquad (1.7.3)$$

## Proof :

From (1.2.9)

$$X_{fil}^d(\omega) = \Delta \sum_{k=-\infty}^{\infty} x_{fil}[k] e^{-j\omega\Delta k}$$

$$= \Delta \sum_{m=-\infty}^{\infty} x_{fil}[mN] e^{-j\omega\Delta mN}$$

$$= \Delta \sum_{m=-\infty}^{\infty} x_{dec}[m] e^{-j\omega N\Delta m} = \frac{1}{N} X_{dec}^d(\omega)$$

$$\wedge\wedge\wedge$$

Note that decimation of $\{x[k]\}$ is equivalent to sampling $x(t)$ at $N$ times slower rate.

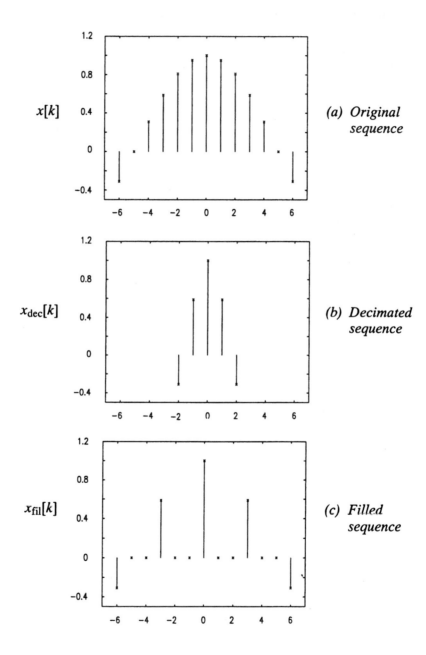

*(a) Original sequence*

*(b) Decimated sequence*

*(c) Filled sequence*

*Figure 1.7.1 : Example of decimation and filling.*

We next want to relate the transforms of $\{x_{dec}[k]\}$ and $\{x_{fil}[k]\}$ back to the transform of $\{x[k]\}$. One possible approach is to imitate what we did in the case of continuous-time sampling. Define the discrete impulse sequence as

$$s_d[k, N] = \Delta \sum_{m=-\infty}^{\infty} \delta_d[k - mN] \qquad (1.7.4)$$

with $\delta_d[k]$ the discrete $1/\Delta$ pulse defined as follows:

$$\delta_d[k] = \begin{cases} 1/\Delta & \text{for } k = 0 \\ 0 & \text{elsewhere} \end{cases} \qquad (1.7.5)$$

Note that $s_d[k, N]$ is analogous to $s(t, \Delta)$ defined in (1.5.1).

Now if we modulate $x[k]$ by $s_d[k, N]$ we obtain

$$x_2[k] = \Delta \sum_{m=-\infty}^{\infty} x[mN]\delta[k - mN]$$

$$= \begin{cases} x[k] & \text{when } k \text{ is a multiple of } N \\ 0 & \text{otherwise} \end{cases}$$

$$= x_{fil}[k] \qquad (1.7.6)$$

To form $\{x_{dec}[k]\}$ out of $\{x_{fil}[k]\}$ we need to take out the $N$ in $\{x_{fil}[k]\}$; i.e. we need to decimate the sequence to form

$$x_{dec}[k] = x_{fil}[kN] \qquad (1.7.7)$$

Repeating the above steps in the frequency domain we have by using (1.5.2) that the DTFT of $s_d[k, N]$ is

$$S^d(\omega, N) = \frac{2\pi}{N} \sum_{m=-\infty}^{\infty} \delta\left[\omega - \frac{m2\pi}{N\Delta}\right] \tag{1.7.8}$$

so, by (1.4.10)

$$X_{\text{fil}}^d(\omega) = \frac{1}{2\pi} X^d(\omega) \otimes S^d(\omega, N)$$

$$= \frac{1}{2\pi} \int_0^{\frac{2\pi}{\Delta}} X^d(\omega - \eta) \frac{2\pi}{N} \sum_{m=-\infty}^{\infty} \delta\left[\eta - m\frac{2\pi}{N\Delta}\right] d\eta$$

$$= \frac{1}{N} \sum_{m=0}^{N-1} X^d\left[\omega - m\frac{2\pi}{N\Delta}\right] \tag{1.7.9}$$

Combining (1.7.9) and (1.7.3) we get

$$X_{dec}^d(\omega) = \sum_{m=0}^{N-1} X^d\left[\omega - m\frac{2\pi}{N\Delta}\right] \tag{1.7.10}$$

The result obtained in (1.7.10) could in fact be reached directly by recognizing that if the sequence $x[k]$ is the result of sampling a signal $x(t)$ at period $\Delta$, $\{x_{\text{dec}}[k]\}$ will be the result of sampling $x(t)$ at period $N\Delta$. Hence, we have from (1.6.12) that

$$X^s(\omega) = \sum_{m=-\infty}^{\infty} X(\omega - m2\pi/\Delta)$$

and

$$X_{dec}^s(\omega) = \sum_{m=-\infty}^{\infty} X\left[\omega - m\frac{2\pi}{N\Delta}\right]$$

$$= \sum_{k=-\infty}^{\infty} \sum_{l=0}^{N-1} X\left[\omega - (kN + l)\frac{2\pi}{N\Delta}\right]$$

$$= \sum_{l=0}^{N-1}\left[\sum_{k=-\infty}^{\infty} X\left[\omega - l\frac{2\pi}{N\Delta} - k\frac{2\pi}{\Delta}\right]\right]$$

$$= \sum_{l=0}^{N-1} X^s\left[\omega - l\frac{2\pi}{N\Delta}\right] \tag{1.7.11}$$

We note that (1.7.11) is analogous to (1.7.10) where we have used impulse sampling and continuous-time Fourier transforms in place of ordinary sampling and discrete-time Fourier transforms.

## 1.7.2   Irregular Sampling in Periodic Patterns

So far we have considered only the case of uniform sampling with period $\Delta$. In this section we develop Fourier methods applicable to the case of irregular sampling in periodic patterns..

Given the continuous signal $x(t)$ we sample it every $\Delta_k$ time units to generate a sampled data sequence.

We assume that

$$\Delta_{k+N} = \Delta_k \tag{1.7.12}$$

for some given $N$. We want to investigate the spectral content of the generated sampled data sequence and its relation to the original spectrum of the continuous signal.

Define the sampling instants $\{t_k\}$ such that

$$t_k - t_{k-1} = \Delta_k \quad \text{with} \quad t_0 = 0 \tag{1.7.13}$$

By analogy with (1.6.11) we convert the sample values into a discrete pulse stream as follows:

$$x^s(t) = x(t)s(t, \{\Delta_i\}) \tag{1.7.14}$$

where $x(t)$ is a continuous-time signal and $s(t, \{\Delta_i\})$ is a $\{\Delta_i\}$ sequence pulse stream analogous to (1.5.1) i.e.

$$s(t, \{\Delta_i\}) := \sum_{k=-\infty}^{\infty} \Delta_k \delta(t - t_k) \tag{1.7.15}$$

The continuous signal given in (1.7.14) has a continuous-time Fourier transform given by

$$X^s(\omega) = \int_{-\infty}^{\infty} x^s(t)e^{-j\omega t}dt$$

$$= \sum_{k=-\infty}^{\infty} \Delta_k x[k]e^{-j\omega t_k} \tag{1.7.16}$$

We will use this to define a discrete-time Fourier transform for the case of periodic sampling. Thus we define the DTFT of a non-uniform sampled sequence by:

$$\boxed{X^d(\omega) = \sum_{k=-\infty}^{\infty} \Delta_k x[k]e^{-j\omega t_k}} \tag{1.7.17}$$

The transform given in (1.7.17) holds for arbitrary sampling periods. In the special case of a periodic sampling pattern, we can get a simpler expression. Let the period of the sampling sequence be $N$ and define

$$T = \sum_{l=1}^{N} \Delta_l \tag{1.7.18}$$

To emphasise the fact that the sampling sequence is periodic, we write (1.7.15) as $s\left(t; \Delta_1 \ \ldots, \ \Delta_N\right)$. We next note that this function is periodic with period $T$. This can be seen as follows:

$$s\left(t + T \ ; \ \Delta_1, \ \ldots, \ \Delta_N\right) = \sum_{k=-\infty}^{\infty} \Delta_k \delta(t + T - t_k) \qquad (1.7.19)$$

Using (1.7.13)

$$s\left(t + T \ ; \ \Delta_1, \ \ldots, \ \Delta_N\right) = \sum_{k=-\infty}^{\infty} \Delta_k \delta(t - t_{k-N})$$

$$= \sum_{k=-\infty}^{\infty} \Delta_{k+N} \delta(t - t_k)$$

$$= \sum_{k=-\infty}^{\infty} \Delta_k \delta(t - t_k)$$

$$= s\left(t; \Delta_1, \ \ldots, \ \Delta_N\right) \qquad (1.7.20)$$

Since $s\left(t; \Delta_1, \ \ldots, \ \Delta_N\right)$ is periodic it has a Fourier series representation; i.e.

$$s\left(t; \Delta_1, \ \ldots, \ \Delta_N\right) = \frac{1}{T} \sum_{k=-\infty}^{\infty} a_k e^{jk\frac{2\pi}{T}t} \qquad (1.7.21)$$

with

$$a_k = \int_0^T s\left(t, \Delta_1, \ \ldots, \ \Delta_N\right) e^{-jk\frac{2\pi}{T}t} dt$$

$$= \int_0^T \sum_{m=-\infty}^{\infty} \Delta_m \delta(t - t_m) e^{-jk\frac{2\pi}{T}t} dt$$

or

$$a_k = \sum_{m=1}^{N} \Delta_m e^{-jk\frac{2\pi}{T}t_m} \tag{1.7.22}$$

Using Property 1.6.2 we see that the continuous Fourier integral transform of $s(t; \Delta_1, \ldots, \Delta_N)$ is

$$S(\omega; \Delta_1, \ldots, \Delta_N) = 2\pi/T \sum_{k=-\infty}^{\infty} a_k \delta(\omega - k2\pi/T) \tag{1.7.23}$$

Hence using the modulation property of Fourier transforms in equation (1.7.14) we have

$$X^d(\omega) = X^s(\omega)$$

$$= \frac{1}{T} \int_{-\infty}^{\infty} \sum_{k=-\infty}^{\infty} a_k \delta\left[\eta - k\frac{2\pi}{T}\right] X(\omega - \eta) d\eta \tag{1.7.24}$$

where $X(\omega)$ is the continuous Fourier transform of the original continuous-time signal $x(t)$.

From (1.7.24) we see that

$$X^d(\omega) = \frac{1}{T} \sum_{k=-\infty}^{\infty} a_k X(\omega - k2\pi/T) \tag{1.7.25}$$

Comparing (1.7.25) with (1.6.12) we see that periodic sampling also produces repetition in the frequency domain, but the repeated transforms are weighted by the Fourier series coefficients $\{a_i\}$ of $s(t; \Delta_1, \ldots, \Delta_N)$ and are divided by the scaling factor $T$.

**Example 1.7.1** Suppose $x(t) = \cos(1.2\pi t)$ is sampled every 1 second.

(a) Draw the spectrum of the resulting sequence.

(b) Suppose an additional sample is introduced between every two existing samples so that $\{t_i\} = \{ 0, \ 0.2, \ 1, \ 1.2, \ 2, \ \ldots \}$.

Draw the spectrum of the new sequence.

Solution :

(a) Since $X(\omega) = \pi(\delta(\omega - 1.2\pi) + \delta(\omega + 1.2\pi))$, by (1.6.12) we have

$$X^d(\omega) = \pi \sum_{k=-\infty}^{\infty} [\delta(\omega - 1.2\pi - 2\pi k) + \delta(\omega + 1.2\pi - 2\pi k)]$$

and this is drawn in Figure 1.7.2 (a).

(b) Here we have $\Delta_1 = 0.2$sec, $\Delta_2 = 0.8$sec and $T = 1$ sec so, by (1.7.22)

$$a_0 = \Delta_1 + \Delta_2 = 1$$

$$a_1 = \Delta_1 e^{-j2\pi 0.2} + \Delta_2 e^{-j2\pi.1}$$

$$= 0.2e^{-j0.4\pi} + 0.8 = 0.8618 - 0.1902j$$

$$a_2 = 0.6382 - 0.1176j$$

$$a_3 = 0.6382 + 0.1176j$$

$$a_4 = 0.8618 + 0.1902j$$

So, by (1.7.25) we get

$$X^d(\omega) = \pi \sum_{k=-\infty}^{\infty} \Big[ \, a_0\big[\delta(\omega - 1.2\pi - 2\pi k) + \delta(\omega + 1.2\pi - 2\pi k)\big]$$

$$+ \, a_1\big[\delta(\omega - 3.2\pi - 2\pi k) + \delta(\omega - 0.8\pi - 2\pi k)\big]$$

$$+ \, a_2\big[\delta(\omega - 5.2\pi - 2\pi k) + \delta(\omega - 2.8\pi - 2\pi k)\big]$$

$$+ \, a_3\big[\delta(\omega - 7.2\pi - 2\pi k) + \delta(\omega - 4.8\pi - 2\pi k)\big]$$

$$+ \, a_4\big[\delta(\omega - 9.2\pi - 2\pi k) + \delta(\omega - 6.8\pi - 2\pi k)\big] \, \Big]$$

Clearly $X^d(\omega)$ has a periodicity of $10\pi$ and is drawn in Figure 1.7.2 (b).

ΛΛΛ

A special case of interest is when the elements of the periodic sampling sequence are all integer multiples of some 'base period' $\Delta$. In this case we can write

$$\Delta_k = M_k \Delta \tag{1.7.26}$$

where $\{M_k\}$ is a periodic sequence of integers; i.e.

$$M_{k+N} = M_k \tag{1.7.27}$$

Denote

$$\sum_{k=1}^{N} M_k = K \tag{1.7.28}$$

then

$$T = K\Delta \tag{1.7.29}$$

and

$$t_m = \Delta \sum_{l=1}^{m} M_l \tag{1.7.30}$$

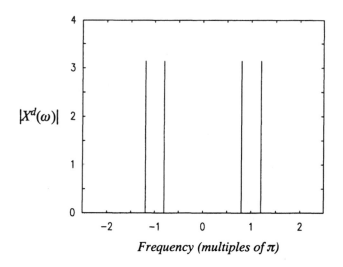

*Figure 1.7.2 (a) : Spectrum for uniform sampling.*

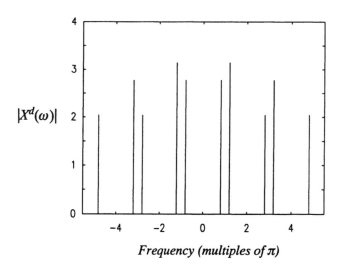

*Figure 1.7.2 (b) : Spectrum for periodic sampling.*

Then, substituting in (1.7.25) and (1.7.22) we get

$$X^d(\omega) = \frac{1}{K\Delta} \sum_{k=-\infty}^{\infty} a_k X\left[\omega - k\frac{2\pi}{K\Delta}\right] \qquad (1.7.31)$$

and

$$a_k = \Delta \sum_{m=1}^{N} M_m e^{-jk\frac{2\pi}{K}\sum_{l=1}^{m} M_l} \qquad (1.7.32)$$

We observe the following properties for $a_k$:

$$a_{k+K} = a_k \text{ for all } k \qquad (1.7.33)$$

$$a_{nK} = a_0 = K\Delta = T \text{ for all integers } n \qquad (1.7.34)$$

and

$$a_{K-r} = \bar{a}_r \text{ (complex conjugate) for all } r = 0, \ \ldots \ ,K \qquad (1.7.35)$$

Going back to (1.7.31) we can write

$$X^d(\omega) = \frac{1}{K\Delta} \sum_{m=-\infty}^{\infty} \sum_{l=1}^{K} a_{mK+l} X\left[\omega - m\frac{2\pi}{\Delta} - l\frac{2\pi}{K\Delta}\right]$$

$$= \frac{1}{K\Delta} \sum_{l=1}^{K} a_l \sum_{m=-\infty}^{\infty} X\left[\omega - l\frac{2\pi}{K\Delta} - m\frac{2\pi}{\Delta}\right] \qquad (1.7.36)$$

where we have used (1.7.33).

Denote

$$X_1^d(\omega) = \sum_{m=-\infty}^{\infty} X(\omega - m2\pi/\Delta) \qquad (1.7.37)$$

as the spectrum of the sequence resulting from sampling $x(t)$ at the uniform period $\Delta$. Then

$$X^d(\omega) = \frac{1}{K\Delta} \sum_{l=1}^{K} a_l X_1^d \left[ \omega - l\frac{2\pi}{K\Delta} \right] \qquad (1.7.38)$$

and using the fact that $X_1^d(\omega)$ is periodic with period $2\pi/\Delta$, we can write

$$\begin{bmatrix} X^d(\omega) \\ X^d(\omega - \dfrac{2\pi}{K\Delta}) \\ \cdot \\ \cdot \\ X^d(\omega - (K-1)\dfrac{2\pi}{K\Delta}) \end{bmatrix} = \frac{1}{K\Delta} \begin{bmatrix} a_0 & a_1 & \cdot & \cdot & a_{K-1} \\ a_{K-1} & a_0 & \cdot & \cdot & a_{K-2} \\ \cdot & & & & \\ \cdot & & & & \\ a_1 & a_2 & \cdot & \cdot & a_0 \end{bmatrix}$$

$$\times \begin{bmatrix} X_1^d(\omega) \\ X_1^d(\omega - \dfrac{2\pi}{K\Delta}) \\ \cdot \\ \cdot \\ \cdot \\ X_1^d(\omega - (K-1)\dfrac{2\pi}{K\Delta}) \end{bmatrix} \qquad (1.7.39)$$

We will return to this equation in the next chapter when we discuss signal reconstruction.

Finally note that $X^d(\omega)$ is also periodic with period $2\pi/\Delta$.

## 1.8   Further Reading and Discussion

There are many excellent books available on Fourier analysis techniques. For example, the basic concepts are introduced in Stuart (1961), Edwards (1980), Bracewell (1986), Dym and McKean (1972), Körner (1988), and Chui

(1992, Chapter 2). Also, books on digital signal processing, such as Oppenheim and Schafer (1989), and Oppenheim, Willsky and Young (1983) contain extensive treatments of Fourier analysis especially DFT and fast Fourier transform methods.

We have principally focused on the basic concepts in this chapter. Many of the results have a long and interesting history in the mathematics literature. For example, Property 1.5.1 and its consequences has only recently been resolved - see for example the discussion in Butzer and Nessel (1971) and Körner (1988).

The representation of $X_1^d(\omega)$ on the right hand side of equation (1.7.39) as a vector $\left[X_1^d(\omega), \ X_1^d(\omega - 2\pi/K\Delta), \ ..., \ X_1^d(\omega - (K-1)2\pi/K\Delta)\right]^T$ is a key element in the analysis of periodic discrete systems. It is sometimes called the 'alias component representation' (Smith and Barnwell (1987) or the 'modulation representation' (Verttarli (1989)).

We have also not treated some topics at all such as fast methods for evaluating Fourier transforms. Our reason for doing this is that fast implementations of Fourier methods are now available as standard options in a number of software packages; e.g. Matlab (Mathworks) and Ptolemy (1992). This enables the user to concentrate on interpretations of the results rather than needing to worry about detailed programming issues.

A word of caution is that Fourier transforms are defined in many different ways in the literature with a host of scaling factors such as $2\pi, N, \sqrt{N}, T, \sqrt{T}$ etc. Also, there are often incompatibilities between the definitions of say fast Fourier transforms (FFT's) and continuous-time Fourier integrals. In the development in this chapter, we have made *all* transforms compatible with the basic Fourier integral. Our motivations for doing this are to avoid confusion, to make the results easier to accept and to allow free mixing together in the analysis of hybrid (i.e. mixed continuous and sampled data) systems. It is suggested however that, when using standard packages, one first runs a simple example so that the appropriate scaling factors used in the implementation are identified.

## 1.9   Problems

**1.1**   Compute the Fourier transform of each of the following signals (we use $\mu(t)$ to denote the unit step function):

(a)   $\left(e^{-\alpha t} \cos \omega_0 t\right) \mu(t) \quad \alpha > 0$

(b)   $e^{-2t} \left[\mu(t+4) - \mu(t-1)\right]$

(c)   $\left(t^2 e^{-t} \sin 2t\right) \mu(t)$

(d)   $\dfrac{\sin \alpha \pi t}{\alpha \pi t} \cdot \dfrac{\sin 2\pi(t-1)}{\pi(t-1)}$

**1.2**   Given the following Fourier transforms of continuous-time signals find corresponding continuous-time signals:

(a)   $\dfrac{2 \sin(\omega - \omega_0)T_1}{\omega - \omega_0}$

(b)   $\displaystyle\sum_{k=-2}^{2} a_k \, \delta(\omega - k\omega_0) \, , \quad a_k = e^{-jk\Delta}$

(c)   $\dfrac{j\omega + 1}{6 - \omega^2 + 5j\omega}$

(d)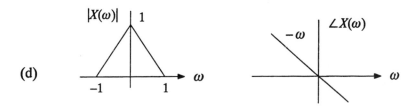

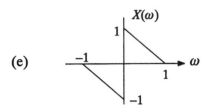

(e)

1.3  (a)  Let $x(t)$ be a real, odd signal. Show that $X(\omega) = F\{x(t)\}$ is pure imaginary and odd.

(b)  What can you say about $X(\omega) = F\{x(t)\}$ if $x(-t) = x^*(t)$ ?

(c)  Define $y(t) = \text{Re}\{x(t)\}$. Express $F\{y(t)\}$ in terms of $X(\omega) = F\{x(t)\}$.

(d)  Let $X(\omega) = F\{x(t)\}$. Define $y(t) = \dfrac{d^2x(t)}{dt^2}$.

(i)  Let $X(\omega) = \begin{cases} 1 & |\omega| < 1 \\ 0 & |\omega| > 1 \end{cases}$.  Find the value of $\displaystyle\int_{-\infty}^{\infty} |y(t)|^2 dt$.

(ii)  What is the inverse Fourier transform of $y(a\omega)$ ?

1.4  Consider the pair $x(t) \overset{F}{\leftrightarrow} X(\omega)$. Find the corresponding continuous-time signals $y_1(t), y_2(t)$ and $y_3(t)$ if

$$|Y_1(\omega)| = |Y_2(\omega)| = |Y_3(\omega)| = X(\omega)$$

and

$$\angle Y_1(\omega) = -\angle X(\omega)$$
$$\angle Y_2(\omega) = \angle X(\omega) + a\omega$$
$$\angle Y_3(\omega) = -\angle X(\omega) + a\omega$$

1.5  Find the Fourier series representations for each of the following signals:

(a)  $e^{j\alpha t}$

(b)  $\cos 2t + \sin 5t$

(c)  $x(t)$ is periodic and $x(t) = \begin{cases} \cos \pi t & 0 \leq t \leq 2 \\ 0 & 2 \leq t \leq 4 \end{cases}$

(d)

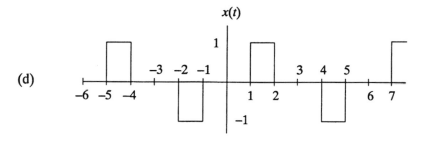

(e)  $x(t) = \displaystyle\sum_{k=-\infty}^{\infty} x_1(t-2k)$ :

1.6  Let $x(t)$ be a periodic signal, with fundamental period $T$ and Fourier series coefficients $\overline{X}_k$. Consider each of the following signals. Express the Fourier series coefficients of each signal in terms of $\overline{X}_k$.

(a)  $x(t-t_0)$

(b)  $x(-t)$

(c)  $x^*(t)$

(d)  $\displaystyle\int_{-\infty}^{t} x(\tau)d\tau$  (assume here that $\overline{X}_0 = 0$ )

(e)  $\dfrac{dx(t)}{dt}$

(f)  $x(\alpha t)$ , $\alpha > 0$

1.7  (a)   A continuous-time periodic signal $\tilde{x}(t)$ with period $T$ is said to be *odd-harmonic* if in its Fourier series representation

$$\tilde{x}(t) = \frac{1}{T} \sum_{k=-\infty}^{\infty} \overline{X}_k e^{jk\left(\frac{2\pi}{T}\right)t} \ ; \quad \overline{X}_k = 0 \text{ for every even integer } k \, .$$

Show that $\tilde{x}(t)$ is odd harmonic if and only if $\tilde{x}(t) = -\tilde{x}(t + T/2)$.

(b)   Suppose that $\tilde{x}(t)$ is an odd-harmonic periodic signal with period 2, such that $\tilde{x}(t) = t$ for $0 < t < 1$. Sketch $\tilde{x}(t)$ and calculate its Fourier series coefficients.

(c)   Similarly, we could define an even-harmonic function as one for which $\overline{X}_k = 0$ for all $k$ odd. Could $T$ be the fundamental period for such a signal? Explain your answer.

(d)   More generally, show that $T$ is the fundamental period of $\tilde{x}(t)$ if one of two things happens:

   1.   Either $\overline{X}_1$ or $\overline{X}_{-1}$ is nonzero.

   or

   2.   There are two integers $k$ and $l$ that have no common factors and are such that both $\overline{X}_k$ and $\overline{X}_l$ are nonzero.

1.8  Let $\tilde{x}(t)$ and $\tilde{y}(t)$ be periodic signals with period $T_0$ and Fourier series representations $\tilde{x}(t) = \dfrac{1}{T_0} \sum_{k=-\infty}^{\infty} \overline{X}_k e^{jk\omega_0 t} \ ; \quad \tilde{y}(t) = \dfrac{1}{T_0} \sum_{k=-\infty}^{\infty} \overline{Y}_k e^{jk\omega_0 t}$.

(a)   Show that the Fourier series coefficients of the signal $\tilde{z}(t) = \tilde{x}(t)\tilde{y}(t) = \dfrac{1}{T_0} \sum_{k=-\infty}^{\infty} \overline{Z}_k e^{jk\omega_0 t}$ are given by the discrete convolution $\overline{Z}_k = T_0 \sum_{n=-\infty}^{\infty} \overline{X}_n \overline{Y}_{k-n}$. (This is the *modulation property* of Fourier series.)

(b)   Let $x(t)$ be defined as in Problem 1.5(d). Find the Fourier coefficients for $z(t) = x(t)\cos(8\pi t)$.

(c)  Use the result of part (a) to show that $\displaystyle\int_0^{T_0} |x(t)|^2 dt = \frac{1}{T_0} \sum_{k=-\infty}^{\infty} |\overline{X}_k|^2$ .

1.9  Compute the DTFT of each of the following sequences (the associated sampling period for all the given sequences is $\Delta$ ):

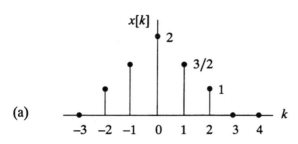

(a)

(b)  $2^k \mu[-k]$

(c)  $|a|^k \sin \omega_0 \Delta k$ ,  $|a| < 1$

(d)  $(1/3)^k \{\mu[k+4] - \mu[k-1]\}$

(e)  $\displaystyle\sum_{m=-\infty}^{\infty} (1/3)^m \delta[k-2m]$

(f)  $x[k] = \begin{cases} \cos(\pi k/3) & -4 \le k \le 4 \\ 0 & \text{elsewhere} \end{cases}$

(g)  $k(1/3)^{|k|}$

(h)  $\left[ \dfrac{\sin\left(\frac{\pi k}{2}\right)}{\pi k} \right] \left[ \dfrac{\sin\left(\frac{\pi k}{4}\right)}{\pi k} \right]$

1.10  Find the sequences having the following DTFTs.

(a)  $\begin{cases} 0 & 0 \le |\omega + 2\pi k| \le W \\ 1 & W < |\omega + 2\pi k| \le \pi \end{cases}$  with  $\Delta = 1$

(b)    Same as in (a) but $\Delta = 1/2$

(c)    $0.1\left(1 - 2e^{-j3\omega} + 4e^{j2\omega}\right), \quad \Delta = 0.1$

(d)    $\cos^2(0.2\omega), \quad \Delta = 0.2$

(e)

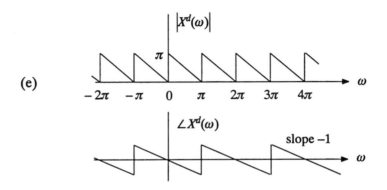

Consider the cases of $\Delta = 1$ and $\Delta = 0.5$.

1.11   Let $X^d(\omega)$ denote the DTFT of $x[n]$ in the figure below. Perform the following calculations without explicitly evaluating $X^d(\omega)$:

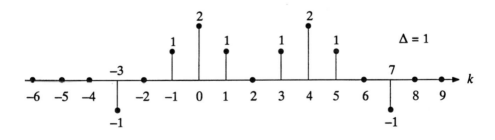

(a)    Evaluate $X^d(0)$.

(b)    Find $\angle X^d(\omega)$.

(c)    Evaluate $\displaystyle\int_{-\pi}^{\pi} X^d(\omega)\,d\omega$.

(d)   Find $X^d(\pi)$ .

(e)   Determine and sketch the signal whose DTFT is Re $\{X^d(\omega)\}$ .

(f)   Evaluate:

(i)   $$\int_{-\pi}^{\pi} |X^d(\omega)|^2 d\omega$$

(ii)   $$\int_{-\pi}^{\pi} \left| \frac{dX(\omega)}{d\omega} \right|^2 d\omega$$

1.12  Given the following list of properties:

(1)   Re $\{X^d(\omega)\} = 0$

(2)   Im $\{X^d(\omega)\} = 0$

(3)   There exists a real $\alpha$ such that $e^{j\alpha\omega\Delta} X(\omega)$ is real,

(4)   $$\int_{-\frac{\pi}{\Delta}}^{\frac{\pi}{\Delta}} X^d(\omega) d\omega = 0$$

(5)   $X^d(\omega)$ is periodic,

(6)   $X(0) = 0$

and the following list of discrete signals (let $\Delta$ be the sampling period)

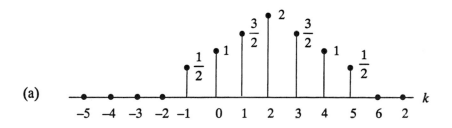

(b)  $x[k] = \begin{cases} 0 & k = 3m \\ 1 & k = 3m - 1 \\ -1 & k = 3m + 1 \end{cases}$

(c)  $x[k] = (1/2)^k \mu[k]$

(d)  $x[k] = (1/2)^{|k|}$

(e)  $x[k] = \delta[k-1] + \delta[k+2]$

(f)  $x[k] = \delta[k-1] + \delta[k+3]$

(g)  $x[k] = \begin{cases} 2 & k = 0 \\ 1 & k = 4 \\ -1 & k = -4, -3, \ 1 \\ 0 & \text{otherwise} \end{cases}$

(h)  $x[k] = \begin{cases} 1 & k = \pm \ 1 \\ -1 & k = \pm \ 3, \pm \ 4 \\ 2 & k = \pm \ 6 \\ 0 & \text{otherwise} \end{cases}$

(i)  $x[k] = \delta[k-1] - \delta[k+1]$

Complete the table overleaf by putting a '+' where a property holds for the signal, and a '−' when it does not.

1.13  Let $x[k]$ be the discrete sequence with DTFT $X^d(\omega)$ illustrated below. Sketch the DTFT of $z[k] = x[k]\varrho[k]$ for each $\varrho[k]$ as follows (assume $\Delta$ is the sampling period for both $x[k]$ and $\varrho[k]$):

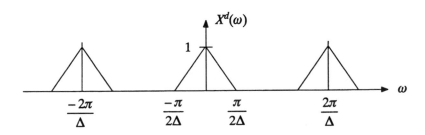

|   | a | b | c | d | e | f | g | h | i |
|---|---|---|---|---|---|---|---|---|---|
| 1 |   |   |   |   |   |   |   |   |   |
| 2 |   |   |   |   |   |   |   |   |   |
| 3 |   |   |   |   |   |   |   |   |   |
| 4 |   |   |   |   |   |   |   |   |   |
| 5 |   |   |   |   |   |   |   |   |   |
| 6 |   |   |   |   |   |   |   |   |   |

(a)  $\varrho[k] = \cos \pi k$

(b)  $\varrho[k] = \cos(\pi k/2)$

(c)  $\varrho[k] = \sin(\pi k/2)$

(d)  $\varrho[k] = \Delta \sum\limits_{m=-\infty}^{\infty} \delta[k - 2m]$

(e)  $\varrho[k] = \Delta \sum\limits_{m=-\infty}^{\infty} \delta[k - 4m]$

1.14  Let $x(t) = \cos(2\pi t) + 2\cos(3\pi t)$, and assume $\Delta = 1$ is the sampling period. Calculate and sketch all entries in the diagram in Figure 1.6.2 and confirm its commutativity.

1.15  Suppose we have the following sampling pattern: $\{t_i\} = 0, 2, 3, 5, 6, 8, 9, \dots$

Draw the resulting $|X^d(\omega)|$ for the following continuous signals:

(a)  $x(t) = \cos \pi t$

(b)    $x(t)$ which has $X(\omega)$ given by
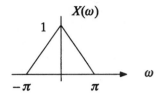

# Chapter 2

# Sampling and Reconstruction

## 2.1  Introduction

In this chapter we will explore the relationship between a continuous-time signal and its sampled values. A key question that we will examine is whether or not it is possible to reconstruct the continuous signal from the samples. This problem is very important since it tells us about the information loss incurred by the sampling process. This has significant implications in both control and signal processing based on sampled-data.

We will show that, provided the continuous-time signal has its frequency content limited to a particular range, then the continuous signal can be exactly recovered from the sample values by an infinite-dimensional non-causal filter. We also examine practical methods of reconstructing the continuous signal by use of various hold functions.

The material presented here is central to sampled-data signal processing and control and hence appears in most treatments of these subjects. One extra topic that we cover is that of reconstructing a continuous signal from sampled-data obtained with irregular, but periodic, sampling patterns. The latter result is important when irregular sampling is used e.g. in the case of multi-rate control studied later in the book.

## 2.2  Sampled Data Sequences – A Representation of Continuous Signals

A simple, yet very important, observation is that there is usually a loss of information in the sampling process. To demonstrate this loss of information let us consider a single sinewave, i.e.

$$y(t) = G\cos(\omega_0 t) \qquad (2.2.1)$$

Suppose we sample this signal at period $\Delta$ which corresponds to the frequency

$$\omega_s = 2\pi/\Delta \qquad (2.2.2)$$

We refer to this frequency as the sampling frequency.

The sampled signal is then given by

$$y[k] = y(k\Delta) = G\cos(\omega_0 k\Delta) \qquad (2.2.3)$$

Now consider another sinewave whose frequency is shifted from $\omega_0$, by an integer multiple of $\omega_s$. Namely, consider

$$y'(t) = G\cos((\omega_0 + n\omega_s)t) \qquad (2.2.4)$$

where $n$ is a fixed integer. By sampling $y'(t)$ with the same sampling period $\Delta$, we obtain

$$y'[k] = y'(k\Delta) = G\cos\left[(\omega_0 + n\omega_s)k\Delta\right] \qquad (2.2.5)$$

Since both $k$ and $n$ are integers it follows from (2.2.2), (2.2.3) and (2.2.5) that for all $n$ and $k$

$$y[k] = y'[k] \qquad (2.2.6)$$

Thus the same sample values are obtained for all signals whose frequencies differ by an integer multiple of the sampling frequency $\omega_s$. This phenomenon is known as *frequency folding* or *aliasing* and will be discussed in more detail below. This aliasing phenomenon is illustrated in Figure 2.2.1 where 10Hz and 90Hz signals are shown with a sampling period of 25 ms. Note that the sample values of the two signals are the same.

Aliasing or frequency folding, is the result of sampling at a rate that is too slow, i.e. *undersampling*. (The term 'aliasing' is apparently due to Tukey who used the term since high frequency components assume the 'alias' of lower frequency components.)

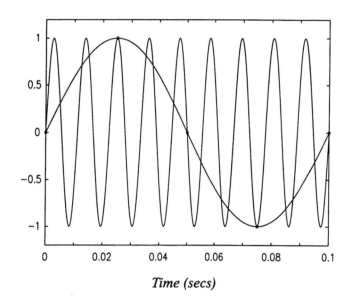

*Figure 2.2.1 : Example of aliasing with 10Hz and 90 Hz signals sampled at 40Hz.*

Sampling at a higher rate, say 200 Hz, results in the sequences marked in Figure 2.2.2 which are clearly distinct. We note that the 200 Hz sampling rate is more than twice the highest signal frequency of 90 Hz. We shall show below that this is in fact, a special case of a general result that signals can be uniquely distinguished from their sample values provided one samples at a rate which exceeds two times the highest frequency component.

The folding of high frequency components to lower frequencies by sampling is particularly troublesome in the case of noise. To avoid this difficulty one should *always* precede the sampler by a low-pass filter to reduce unwanted high frequency noise components. This filter is commonly referred to as an *anti-aliasing filter*.

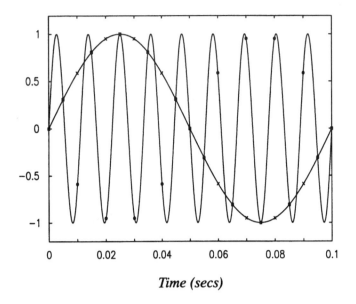

*Figure 2.2.2 : Sampling 10Hz and 90Hz signals at 200Hz,*
*eliminating aliasing.*

## 2.3    Continuous Signal Reconstruction from a Sampled Data Sequence.

The above discussion has dealt with the problem of transforming a continuous-time signal into a sequence. We next address the complementary problem of transforming a sequence into a continuous function. As we have indicated, necessary information in addition to the sequence is the associated sampling interval. With this information we can place the sequence values at their appropriate continuous-time instances. We can also form a continuous-time signal by using impulses at the sampling times of appropriate strength. Thus, as in equation (1.6.9) we can form $x^s(t)$ using $\{x[k]\}$ where

$$x^s(t) = \sum_{k=-\infty}^{\infty} \Delta x[k]\delta(t - k\Delta) \qquad (2.3.1)$$

This continuous-time function is shown in Figure 2.3.1 for a particular example where $\Delta = 1$.

To complete the process of signal recovery we must somehow fill up the intervals between the samples. Methods for achieving this are outlined below.

## 2.4    Shannon's Reconstruction Theorem

We have seen in Section 2.2 that, once aliasing occurs, then it is impossible to distinguish between different signals from their sample values.

We also saw in the example that aliasing was avoided if the sampling frequency was twice the highest frequency component in the signal. Given that this condition holds we might suspect that we may be able to perfectly reconstruct the signal from its sampled-data sequence. This is indeed the case as is made precise in the following result:

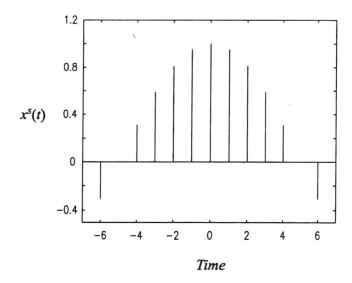

*Figure 2.3.1 : First step in reconstruction of a continuous-time function from a sequence.*

**Theorem 2.4.1** *Shannon's Reconstruction Theorem*

Any signal $y(t)$ consisting of frequency components in the range $(-\omega_s/2, \omega_s/2)$ can be exactly reconstructed from the corresponding sampled-data sequence $y[k]$, (where $\omega_s = 2\pi/\Delta$ is the sampling frequency and $\Delta$ the sampling period). The appropriate reconstruction is

$$y(t) = \sum_{k=-\infty}^{\infty} y[k] \cdot \frac{\sin\left[\left(\frac{\omega_s}{2}\right)(t-k\Delta)\right]}{\left(\frac{\omega_s}{2}\right)(t-k\Delta)} \tag{2.4.1}$$

**Proof :** If $y(t)$ has its frequency content limited to the range $(-\omega_s/2, \omega_s/2)$ then we see from (1.6.12) that

$$Y^s(\omega) = Y(\omega) \quad \text{for} \quad (-\omega_s/2 < \omega < \omega_s/2) \tag{2.4.2}$$

Hence all we need do to recover $Y(\omega)$ from $Y^s(\omega)$ is to multiply $Y^s(\omega)$ by the frequency domain function

$$H_s(\omega) = \begin{cases} 1 & \text{for } |\omega| \leq \dfrac{\omega_s}{2} \\ 0 & \text{otherwise} \end{cases} \tag{2.4.3}$$

That is we form

$$Y(\omega) = H_s(\omega)Y^s(\omega) \tag{2.4.4}$$

The corresponding reconstruction in the time domain is obtained via convolution as

$$y(t) = \int_{-\infty}^{\infty} h_s(\sigma)y^s(t-\sigma)d\sigma$$

$$= \int_{-\infty}^{\infty} h_s(\sigma) \sum_{k=-\infty}^{\infty} \Delta y[k]\delta(t-\sigma-k\Delta)$$

$$= \sum_{k=-\infty}^{\infty} y[k]\Delta h_s(t-k\Delta) \qquad (2.4.5)$$

where $h_s(\sigma) \overset{F}{\leftrightarrow} H_s(\omega)$, i.e.

$$h_s(\sigma) = \frac{1}{\Delta} \left\{ \frac{\sin\left[\left(\frac{\omega_s}{2}\right)\sigma\right]}{\left(\frac{\omega_s}{2}\right)\sigma} \right\} \qquad (2.4.6)$$

$\wedge\!\wedge\!\wedge$

We can express the result in Theorem 2.4.1 in a slightly different way as follows. We will say that a signal is bandlimited if its frequency content is limited to some upper frequency, say $\omega_{\max}$. Now, if the sampling frequency is greater than $2\omega_{\max}$, it follows from Theorem 2.4.1 that we can reconstruct the continuous signal from the sample values. We give a special name to the frequency $2\omega_{\max}$. We call this the *Nyquist rate* for the signal.

Note that the construction (2.4.1) is achieved by passing the impulse sampled signal $y^s(t)$ through a non-causal filter having impulse response

$$h_s(t) = \frac{1}{\Delta} \frac{\sin\left(\frac{\omega_s}{2}\right)t}{\left(\frac{\omega_s}{2}\right)t} \qquad (2.4.7)$$

In principle this solves the signal reconstruction problem. However, this result is really only of theoretical interest since the summation given in (2.4.1) is numerically ill-conditioned. However, it is intuitively clear that the continuous signal could be very closely recovered by practical methods of interpolation provided the signal does not change too rapidly between samples (i.e. provided the sampling frequency is high relative to the frequency content of the signal). These more practical methods of reconstruction are studied in the next section.

## 2.5    Practical Methods of Reconstruction

The non-causal low pass filter given in (2.4.7) cannot be used in practice since, among other things, it is non causal. However, there are various other low pass filters which are quite simple to realize. These do not give exact signal reconstruction but due to their simple form are easy to use.

### 2.5.1    Zero-Order-Hold (ZOH)

One possible practical reconstruction method is to use a zero-order-hold. In this reconstruction method we pass the impulse samples $x^s(t)$ through a filter having the following impulse response

$$h_0(t) = \begin{cases} \dfrac{1}{\Delta} & \text{for } 0 \le t \le \Delta \\ 0 & \text{elsewhere} \end{cases} \tag{2.5.1}$$

This leads to a staircase approximation $\hat{x}(t)$ to $x(t)$ as shown in Figure 2.5.1.

The zero-order-hold reconstruction can also be evaluated in the time domain using the convolution theorem:

$$h_0(t) \otimes x^s(t) = \int_{-0}^{\Delta} \frac{1}{\Delta} \sum_{k=-\infty}^{\infty} \Delta x[k]\delta(t - \Delta k - \sigma)d\sigma$$

$$= \sum_{k=-\infty}^{\infty} x[k][\mu(t - k\Delta) - \mu(t - k\Delta - \Delta)]$$

$$= \hat{x}(t) \tag{2.5.2}$$

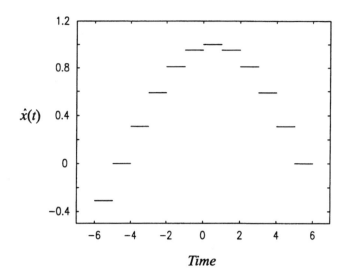

*Figure 2.5.1 : Reconstruction with zero-order-hold.*

where $\hat{x}(t)$ has the form shown in Figure 2.5.1 when $x^s(t)$ is as in Figure 2.3.1.

## 2.5.2 First-Order-Hold (FOH)

Another way of reconstruction is by connecting every two consecutive samples. The result can be seen in Figure 2.5.2 for $x^s(t)$ as in Figure 2.3.1.

This reconstruction is not causal since to fill up each sampling interval one must know already the next sample value. This reconstruction can be accomplished by passing $x^s(t)$ through the (non causal) FOH filter $h_1(t)$ defined by

$$h_1(t) = \begin{cases} \dfrac{1}{\Delta^2}\left(\Delta - |t|\right) & \text{for } |t| \le \Delta \\ 0 & \text{elsewhere} \end{cases} \tag{2.5.3}$$

We next see how this result might be made causal. Indeed, a causal version of the FOH can be achieved by passing a line between every two consecutive

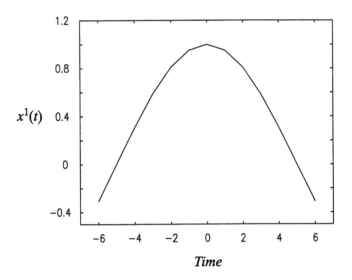

*Figure 2.5.2 : Reconstruction with non-causal first-order-hold.*

sample values and then continuing this line into the next sampling interval. Figure 2.5.3 illustrates this procedure for $x^s(t)$ as in Figure 2.3.1.

It can be seen that this result is actually obtained by passing $x^s(t)$ through the causal, FOH filter $h_2(t)$ defined by

$$h_2(t) = \begin{cases} \dfrac{1}{\Delta^2}t + \dfrac{1}{\Delta} & \text{for } 0 \le t < \Delta \\[2ex] -\dfrac{1}{\Delta^2}t + \dfrac{1}{\Delta} & \text{for } \Delta \le t < 2\Delta \\[2ex] 0 & \text{elsewhere} \end{cases} \qquad (2.5.4)$$

Both $h_1(t)$ and $h_2(t)$ are shown in Figure 2.5.4.

The corresponding frequency domain functions are obtained by finding the Fourier transform of $h_0(t), h_1(t)$ and $h_2(t)$. The following relationships are easily derived:

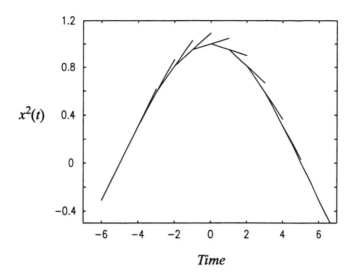

*Figure 2.5.3 : Reconstruction with Causal FOH.*

$$h_0(t) \stackrel{F}{\leftrightarrow} H_0(j\omega) = \frac{1 - e^{-j\omega\Delta}}{j\omega\Delta}$$

(2.5.5)

$$h_1(t) \stackrel{F}{\leftrightarrow} H_1(j\omega) = \frac{2}{\Delta^2\omega^2}(1 - \cos\omega\Delta)$$

(2.5.6)

and

$$h_2(t) \stackrel{F}{\leftrightarrow} H_2(j\omega) = (1 + j\omega\Delta)e^{-j\omega\Delta}\frac{\sin^2\left(\frac{\omega\Delta}{2}\right)}{\left(\frac{\omega\Delta}{2}\right)^2}$$

(2.5.7)

Figure 2.5.5 summarizes the various holds and gives their time and frequency domain characteristics for the case $\Delta = 1$.

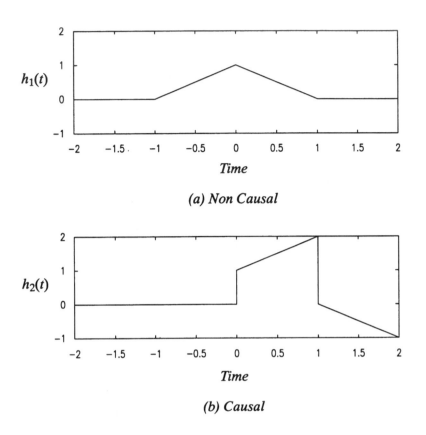

*(a) Non Causal*

*(b) Causal*

*Figure 2.5.4 : Impulse responses of FOH filters.*

## 2.6    Signal Reconstruction from Periodic Samples

All of the above discussion referred to those situations where a uniform sampling period, $\Delta$, was employed. The reader may therefore wonder if anything can be said about the problem of signal reconstruction when the sampling period is irregular. This problem has significance in many application areas where irregular sampling is necessitated by the physics of the problem under study.

The key reconstruction result for the case of regular sampling has been presented in Theorem 2.4.1. Inspection of the proof of this result shows that the

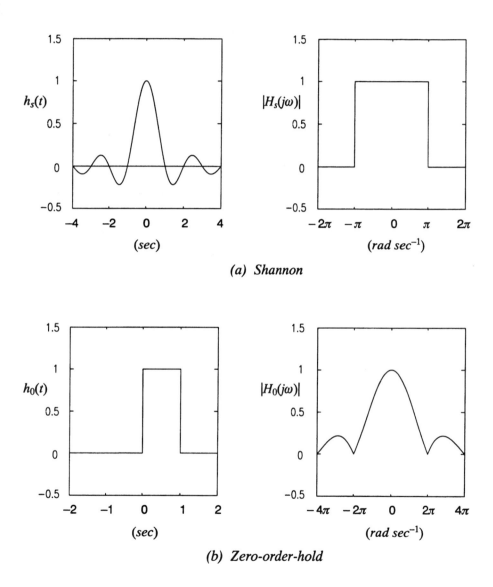

*Figure 2.5.5 : Time and frequency domain characteristics for various types of holds.*

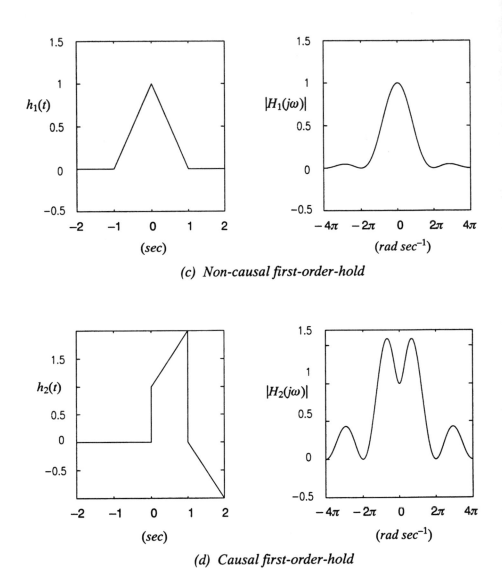

(c)  *Non-causal first-order-hold*

(d)  *Causal first-order-hold*

*Figure 2.5.5 :  (cont.) Time and frequency domain characteristics for various types of holds.*

essence of the argument hinges on the relationship between the Fourier transform, $Y(\omega)$, of the original signal and the Fourier transform $Y^s(\omega)$ of the (impulse) sampled signal. We thus might expect that the key to signal reconstruction
using irregular sampling might lie in the corresponding frequency domain relationships for irregular sampling. For the case of irregular, but periodic sampling,
these relationships have been established in Section 1.7.2. We will thus proceed
by mimicing the proof of Theorem 2.4.1 using these more intricate relationships.

Let $x(t)$ be a continuous-time signal. We denote by $\Delta$ a base sampling period. We then use $\Delta$ to construct a periodic sampling pattern $\{\Delta_k\}$, the elements
of which are multiples of $\Delta$. Thus we write

$$\Delta_k = M_k \Delta \tag{2.6.1}$$

where $\{M_k\}$ is a periodic sequence of integers, i.e.

$$M_{k+N} = M_k \tag{2.6.2}$$

We also denote

$$\sum_{k=1}^{N} M_k = K \tag{2.6.3}$$

This gives a time repetition period for the sampling strategy of $T = K\Delta$ and the
$m$th sampling instance as

$$t_m = \Delta \sum_{l=1}^{m} M_l \tag{2.6.4}$$

This set-up corresponds exactly to that used in Section 1.7.2. As before,
we use $\{x[k]\}$ to denote the given *irregularly* sampled-data and $\{x_1[k]\}$ the corresponding *regularly* sampled-data obtained with sampling period $\Delta$. Let
$X^d(\omega)$ and $X_1^d(\omega)$ denote the Fourier transforms corresponding to $\{x[k]\}$ and
$\{x_1[k]\}$ respectively.

We assume that we are given $\{x[k]\}$, or equivalently, $X^d(\omega)$. Our strategy will be to first examine how one might evaluate $X_1^d(\omega)$ from the given data $X^d(\omega)$. If we can show how to obtain $X_1^d(\omega)$ then the reconstruction problem has been converted into the problem that has already been solved in Theorem 2.4.1.

To proceed with this line of development, we recall from equation (1.7.39) that $X^d(\omega)$ is related to $X_1^d(\omega)$ by the following equation:

$$\underline{X}^d(\omega) = A\underline{X}_1^d(\omega) \tag{2.6.5}$$

where $\underline{X}^d(\omega), \underline{X}_1^d(\omega)$ are vectors formed from the transforms of the periodic sampled sequence and uniformly sampled sequence (at period $\Delta$ ); i.e.

$$\underline{X}^d(\omega) = \begin{bmatrix} X^d(\omega) \\ X^d\left[\omega - \dfrac{2\pi}{K\Delta}\right] \\ \cdot \\ \cdot \\ \cdot \\ X^d\left[\omega - \dfrac{(K-1)2\pi}{K\Delta}\right] \end{bmatrix} \tag{2.6.6}$$

$$\underline{X}_1^d(\omega) = \begin{bmatrix} X_1^d(\omega) \\ X_1^d\left[\omega - \dfrac{2\pi}{K\Delta}\right] \\ \cdot \\ \cdot \\ \cdot \\ X_1^d\left[\omega - \dfrac{(K-1)2\pi}{K\Delta}\right] \end{bmatrix} \tag{2.6.7}$$

and $A$ is the following circulant matrix

$$
A = \frac{1}{K\Delta}
\begin{bmatrix}
a_0 & a_1 & \cdots & a_{K-1} \\
a_{K-1} & a_0 & \cdots & a_{K-2} \\
\cdot & & & \\
\cdot & & & \\
\cdot & & & \\
a_1 & a_2 & \cdots & a_0
\end{bmatrix}
\tag{2.6.8}
$$

$K$ is defined in (2.6.3) and $\{a_i\}$ are the coefficients in the Fourier series expansion of $s(t; \Delta_1, \ldots, \Delta_n)$ as in (1.7.21) – see also (1.7.32) to (1.7.39).

Clearly, knowledge of the periodic sampled sequence is equivalent to knowledge of $\underline{X}^d(\omega)$. On the other hand, if we were given $X_1^d(\omega)$, then this would allow us to recover $x_1[k]$ at the sampling period $\Delta$ and hence, in view of Section 2.4, the complete continuous signal provided its bandwidth was restricted to the range $0 \leq \omega < \pi/\Delta$. If we could thus show that $A$ in equation (2.6.5) was non-singular, then it would be a simple matter to recover $X_1^d(\omega)$ from $\underline{X}^d(\omega)$. Unfortunately, as we show next, $A$ is generally singular. This means that we must have additional information about $X_1^d(\omega)$ to define a unique solution to (2.6.5).

We first investigate the structure of the matrix $A$. Define

$$
W = e^{j\frac{2\pi}{K}}
\tag{2.6.9}
$$

and recall the DFT matrix defined in Section 1.2.4, i.e.

$$
W_K = \frac{1}{\sqrt{K}}
\begin{bmatrix}
1 & 1 & 1 & 1 \\
1 & W & W^2 & W^{K-1} \\
1 & W^2 & W^4 & W^{2(K-1)} \\
\cdot & & & \\
\cdot & & & \\
1 & W^{K-1} & W^{2(K-1)} & W^{(K-1)^2}
\end{bmatrix}
\in \mathbb{C}^{K \times K}
\tag{2.6.10}
$$

We then see using the properties of Fourier series that

$$(W_K^*)A(W_K) = \text{diag}\left[\sum_{l=0}^{K-1} a_l, \sum_{l=0}^{K-1} a_l W^l, \ \ldots \ , \sum_{l=0}^{K-1} a_l W^{l(K-1)}\right] \qquad (2.6.11)$$

$$\overset{\Delta}{=} \Lambda$$

Using (1.7.32) we have for $r = 0, 1, 2, \ \ldots \ , K-1$

$$\sum_{l=0}^{K-1} a_l W^{lr} = \begin{cases} \Delta M_N & \text{for } r = 0 \\ \Delta M_m & \text{for } r = \sum_{l=1}^{m} M_l, \quad m = 1, \ \ldots \ , \ N-1 \\ 0 & \text{otherwise} \end{cases} \qquad (2.6.12)$$

Denote by $J$ the following set of integers

$$J = \left\{ 0, M_1, M_1 + M_2, \ \ldots \ , \sum_{l=1}^{N-1} M_l \right\} \qquad (2.6.13)$$

Equations (2.6.11) and (2.6.12) imply that, unless $M_l = 1, l = 1 \ \ldots \ , N$, $\Lambda_{rr}$ is only non-zero if $r \in J$. Hence, in general $A$ is singular. To study the null space of $A$ we note from (2.6.11) that

$$Av = 0 \Leftrightarrow \Lambda W_K^* v = 0 \qquad (2.6.14)$$

and hence the null space of $A$ is given by

$$N(A) = \left\{ [W_K]_{r+1}, r \notin J \right\} \qquad (2.6.15)$$

where $[W_K]_{r+1}$ is the $(r+1)$th column of $W_K$.

We can then rewrite (2.6.5) as follows:

$$W_K^* \underline{X}^d(\omega) = \Lambda W_K^* \underline{X}_1^d(\omega) \qquad (2.6.16)$$

We see that this represents only $N$ equations with $K$ unknowns, namely the elements of $\underline{X}_1^d(\omega)$.

A rather trivial possibility to obtain a unique solution is to assume that

$$\underline{X}_1^d(\omega) \perp N(A) \quad \text{for all } \omega \qquad (2.6.17)$$

As it turns out, this is equivalent to assuming that by sampling $x(t)$ with period $\Delta$ and getting the sequence $x_1[k]$ we have

$$x_1[k] = \begin{cases} x[m] & \text{if } k = mK \\ x[m+n] & \text{if } k = mK + \sum_{l=1}^{n} M_l, \quad n = 1, \ \dots \ , N-1 \\ 0 & \text{otherwise} \end{cases} \qquad (2.6.18)$$

This is equivalent to assuming that whatever is not picked up in $x[k]$ is zero – not a very realistic assumption. We thus look for more realistic assumptions that might be placed on $\{x_1[k]\}$ to achieve a unique solution to (2.6.16).

One interesting possibility is presented in the following result.

**Theorem 2.6.1** *Reconstruction Theorem for Periodic Sampling*

Let $x(t)$, $\Delta$, $\{M_l\}_1^N$, $x[k]$, $x_1[k]$, $X(\omega), X^d(\omega)$ and $X_1^d(\omega)$ be defined as above. Then:

(a) $X_1^d(\omega)$ can be uniquely recovered from $X^d(\omega)$ provided $X_1^d(\omega)$ satisfies the following constraint:

$$X_1^d(\omega) = 0 \quad \text{for} \quad \frac{N\pi}{K\Delta} \leq |\omega| \leq \frac{\pi}{\Delta} \qquad (2.6.19)$$

where $K = \sum_{l=1}^{N} M_l$.

b) Suppose

$$X(\omega) = 0 \quad \text{for} \quad |\omega| > \frac{N\pi}{K\Delta} \qquad (2.6.20)$$

then the continuous-time signal can be reconstructed from $x[k]$
by the formula

$$x(t) = \sum_{k=-\infty}^{\infty} \beta_k(t)x[k]\frac{\sin\left(\frac{\pi(t-t_k)}{K\Delta}\right)}{\frac{\pi(t-t_k)}{K\Delta}} \tag{2.6.21}$$

where $\beta_k(t)$ is a time function which depends only on the sam-

pling pattern, namely on $\{M_l\}_1^N$ and $\Delta$.

**Comment** Before formally establishing this result, it is helpful to discuss its interpretation. We define $\overline{\Delta} = K\Delta/N$ and note that $\overline{\Delta}$ is the *average* sample period. If we had regular sampling at period $\overline{\Delta}$, and the signal's frequency content was constrained to the interval $0 \le \omega < \pi/\overline{\Delta}$, then we know from Theorem 2.4.1 that the signal could be recovered from the samples. Thus, the constraint (2.6.20) implies that, provided the *average sampling period* satisfies the standard conditions of Theorem 2.4.1, then signal recovery is possible. For example, a periodic sampling sequence of the form {10, 90, 10, 90 ...} can be used to recover the same class of signals as would be recovered using a fixed sampling period of 50. This is heuristically reasonable and reduces to the standard result when the sampling is regular.

Also, we see that (2.6.19) gives exactly $(K-N)$ constraints on $X_1^d(\omega)$. We have already remarked that (2.6.16) represents $N$ equations in $K$ unknowns. Hence, the additional $(K-N)$ constraints would appear to suffice to get a unique solution. It remains to show that a solution is always possible. This will be done in the following proof where we constructively evaluate the unique solution to (2.6.16).

**Proof :**

(a)  First note that because $X_1^d(\omega)$ is periodic with period $2\pi/\Delta$, equation (2.6.19) implies that

$$X_1^d(\omega) = 0 \text{ for } \frac{N-2K}{K\Delta}\pi \le \omega \le -\frac{N\pi}{K\Delta} \qquad (2.6.22)$$

We will concentrate on the interval $\omega \in \left[-\dfrac{N\pi}{K\Delta}, \dfrac{N\pi}{K\Delta}\right)$ with the understanding that our solution for this interval is repeated every $2\pi/\Delta$ and, in view of equation (2.6.19), this will give the complete $X_1^d(\omega)$.

Let us now divide the above interval into $N$ equal subintervals and show that in each one of the subintervals the solution to (2.6.5) is unique. The subintervals are defined by

$$\left[\frac{N-2r-2}{K\Delta}\pi, \frac{N-2r}{K\Delta}\pi\right), \qquad r = 0, 1, \ \ldots, \ N-1 \qquad (2.6.23)$$

It can readily be observed that from (2.6.22) and the periodicity $(2\pi/\Delta)$ of $X_1^d(\omega)$ that

$$X_1^d\left[\omega - l\frac{2\pi}{K\Delta}\right] = 0 \quad \text{for } l = N-r, N-r+1, \ \ldots \ , K-r-1$$

$$\text{and} \quad \omega \in \left[\frac{N-2r-2}{K\Delta}\pi, \frac{N-2r}{K\Delta}\pi\right)$$

This is true since

$$\frac{N-2K}{K\Delta}\pi = \frac{N-2r-2}{K\Delta}\pi - (K-r-1)\frac{2\pi}{K\Delta}$$

$$\le \omega - l\frac{2\pi}{K\Delta} < \frac{N-2r}{K\Delta}\pi - (N-r)\frac{2\pi}{K\Delta} = -\frac{N\pi}{K\Delta} \qquad (2.6.24)$$

Corresponding to each subinterval define the matrices $A_r$ as

$$A_r = \frac{1}{K\Delta} \begin{bmatrix} a_0 & a_1 & \cdot & \cdot & a_{N-r-1} & a_{K-r} & \cdot & \cdot & & a_{K-1} \\ a_{K-1} & a_0 & \cdot & \cdot & a_{N-r-2} & a_{K-r-1} & & \cdot & \cdot & a_{K-2} \\ \cdot & & & & \cdot & \cdot & & & & \cdot \\ \cdot & & & & \cdot & \cdot & & & & \cdot \\ \cdot & & & & \cdot & \cdot & & & & \cdot \\ a_{K-N+r+1} & \cdot & \cdot & & a_0 & a_{K-N+1} & & \cdot & \cdot & a_{K-N+r} \\ a_r & a_{r+1} & \cdot & \cdot & a_{N-1} & a_0 & a_1 & \cdot & \cdot & a_{r-2} & a_{r-1} \\ \cdot & & & & & a_{K-1} & a_0 & & & a_{r-2} \\ \cdot & & & & \cdot & \cdot & & & & \cdot \\ \cdot & & & & \cdot & \cdot & & & & \cdot \\ a_1 & a_2 & \cdot & \cdot & a_{N-r} & a_{K-r+1} & & \cdot & \cdot & a_{K-1} & a_0 \end{bmatrix}$$

$$= E_r^T A E_r, \quad r = 0, 1, \ \ldots \ , N-1 \tag{2.6.25}$$

where

$$E_r = \begin{bmatrix} e_1, e_2, & \ldots, & e_{N-r}, e_{K-r+1}, & \ldots, & e_K \end{bmatrix} \in \mathbf{R}^{K \times N} \tag{2.6.26}$$

and $e_i$ is the i-th column of the $K \times K$ identity matrix.

It can be shown that the sequence of matrices $\{A_r\}$ are non singular (see Problem 2.10).

With the information in (2.6.24), for

$$\omega \in \left[ \frac{N-2r-2}{K\Delta}\pi, \ \frac{N-2r}{K\Delta}\pi \right)$$

equation (2.6.5) reduces to

$$
\begin{bmatrix}
X^d(\omega) \\
\cdot \\
\cdot \\
\cdot \\
X^d(\omega - (N-r-1)\dfrac{2\pi}{K\Delta}) \\
X^d(\omega - (K-r)\dfrac{2\pi}{K\Delta}) \\
\cdot \\
\cdot \\
X^d(\omega - (K-1)\dfrac{2\pi}{K\Delta})
\end{bmatrix}
= A_r
\begin{bmatrix}
X_1^d(\omega) \\
\cdot \\
\cdot \\
\cdot \\
X_1^d(\omega - (N-r-1)\dfrac{2\pi}{K\Delta}) \\
X_1^d(\omega - (K-r)\dfrac{2\pi}{K\Delta}) \\
\cdot \\
\cdot \\
X^d(\omega - (K-1)\dfrac{2\pi}{K\Delta})
\end{bmatrix}
\qquad (2.6.27)
$$

and because of the nonsingularity of $A_r$, this can be uniquely solved. For notational convenience, let us define

$$
b_l^r = e_1^T E_r A_r^{-1} E_r^T e_l
\qquad
\begin{cases}
l = 1, 2, \ \dots \ , \ K \\
r = 0, 1, \ \dots \ , \ N-1
\end{cases}
\qquad (2.6.28)
$$

Then clearly, from (2.6.27), we have for the above interval

$$
X_1^d(\omega) = \sum_{l=1}^{K} b_l^r X^d \left[ \omega - (l-1)\frac{2\pi}{K\Delta} \right]
\qquad (2.6.29)
$$

Combining the solutions given by (2.6.29) for each subinterval will give us the solution for the whole interval $\left[ -\dfrac{N\pi}{K\Delta}, \dfrac{N\pi}{K\Delta} \right)$, namely

$$
X_1^d(\omega) = \sum_{r=0}^{N-1} H_r(\omega) \sum_{l=1}^{K} b_l^r X^d \left[ \omega - (l-1)\frac{2\pi}{K\Delta} \right]
\qquad (2.6.30)
$$

where

$$H_r(\omega) = \begin{cases} 1 & \omega \in \left[\dfrac{N-2r-2}{K\Delta}\pi, \ \dfrac{N-2r}{K\Delta}\pi\right) \\ 0 & \text{elsewhere} \end{cases} \qquad (2.6.31)$$

Because of the periodicity of $X_1^d(\omega)$ and in view of (2.6.19) the complete $X_1^d(\omega)$ is simply a periodic repetition of (2.6.30) every $2\pi/\Delta$.

(b)  Since $X(\omega)$ is constrained to the interval $\left[-\pi/\Delta, \pi/\Delta\right)$, with (2.6.19) that is exactly what we have in equation (2.6.30). Namely

$$X(\omega) = \sum_{r=0}^{N-1} H_r(\omega) \sum_{l=1}^{K} b_l^r X^d\left[\omega - (l-1)\frac{2\pi}{K\Delta}\right] \quad \text{for all } \omega \quad (2.6.32)$$

To derive the time domain formula we employ the inverse Fourier transform (2.6.32) and use the modulation property to get

$$x(t) = \sum_{r=0}^{N-1} \sum_{l=1}^{K} b_l^r \int_{-\infty}^{\infty} x_{r,l-1}^d(\sigma) h_r(t-\sigma) d\sigma \qquad (2.6.33)$$

where

$$h_r(t) = \frac{1}{2\pi} \int_{-\infty}^{\infty} H_r(\omega) e^{j\omega t} d\omega$$

$$= \frac{1}{2\pi} \int_{\frac{N-2r-2}{K\Delta}\pi}^{\frac{N-r}{K\Delta}\pi} e^{j\omega t} d\omega = \frac{1}{2\pi j t}\left[e^{j\frac{N-r}{K\Delta}\pi t} - e^{j\frac{N-2r-2}{K\Delta}\pi t}\right]$$

$$= \frac{1}{2\pi jt} e^{j\frac{N-r-1}{K\Delta}\pi t} \left[ e^{j\frac{\pi t}{K\Delta}} - e^{-j\frac{\pi t}{K\Delta}} \right]$$

$$= e^{j\frac{N-2r-1}{K\Delta}\pi t} \frac{\sin\left(\frac{\pi t}{K\Delta}\right)}{\pi t}. \tag{2.6.34}$$

and

$$x_{r,l-1}^d(t) = \frac{1}{2\pi} \int_{-\infty}^{\infty} X^d \left[ \omega - (l-1)\frac{2\pi}{K\Delta} \right] e^{j\omega t} d\omega$$

$$= e^{j(l-1)\frac{2\pi}{K\Delta}t} x^s(t)$$

Recall (from (1.7.14)) that

$$x^s(t) = \Delta \sum_{k=-\infty}^{\infty} M_k x[k]\delta(t - t_k)$$

So

$$\int_{-\infty}^{\infty} x_{r,l-1}^d(\sigma)h_r(t-\sigma)d\sigma$$

$$= \Delta \sum_{k=-\infty}^{\infty} M_k x[k] \int_{-\infty}^{\infty} e^{j(l-1)\frac{2\pi}{K\Delta}\sigma}\delta(\sigma - t_k)e^{j\frac{N-2r-1}{K\Delta}\pi(t-\sigma)} \frac{\sin\left(\frac{\pi(t-\sigma)}{K\Delta}\right)}{\pi(t-\sigma)} d\sigma$$

$$= \Delta \sum_{k=-\infty}^{\infty} M_k x[k] e^{j(l-1)\frac{2\pi}{K\Delta}t_k} e^{j\frac{N-2r-1}{K\Delta}\pi(t-t_k)} \frac{\sin\left(\frac{\pi(t-t_k)}{K\Delta}\right)}{\pi(t-t_k)}$$

Substituting in (2.6.33) we get

$$x(t) = \Delta \sum_{r=0}^{N-1} \sum_{l=1}^{K} b_l^r \sum_{k=-\infty}^{\infty} M_k x[k] e^{j(l-1)\frac{2\pi}{K\Delta}t_k} e^{j\frac{N-2r-1}{K\Delta}\pi(t-t_k)} \frac{\sin\left(\frac{\pi(t-t_k)}{K\Delta}\right)}{\pi(t-t_k)}$$

and we get the reconstruction formula as

$$x(t) = \sum_{k=-\infty}^{\infty} \beta_k(t) x[k] \frac{\sin\left(\frac{\pi(t-t_k)}{K\Delta}\right)}{\frac{\pi(t-t_k)}{K\Delta}} \tag{2.6.35}$$

where

$$\beta_k(t) = \frac{1}{K} M_k \sum_{r=0}^{N-1} \sum_{l=1}^{K} b_l^r e^{j(l-1)\frac{2\pi}{K\Delta}t_k} e^{j\frac{N-2r-1}{K\Delta}\pi(t-t_k)} \tag{2.6.36}$$

and clearly $\beta_k(t)$ is a function only of the sampling pattern (namely, of $\{M_k\}, \Delta$ ).

$$\wedge\wedge\wedge$$

It is instructive to check the validity of the formula (2.6.35) in a special case. Thus we consider the case when $MN = K$ and the periodic samples are actually all $M\Delta$ time units apart (i.e. we actually have uniform sampling at period $M\Delta$ ). In this case from equation (1.7.32) we have

$$a_k = \Delta \sum_{m=1}^{N} M e^{-jk\frac{2\pi}{N}m} = \Delta \begin{cases} NM & \text{for } k = 0 \\ 0 & \text{otherwise} \end{cases} \tag{2.6.37}$$

but this implies equation (2.6.38) (see over).

Also, from (2.6.25) we have

$$A_r = I, \qquad r = 0, \ \ldots, \ N-1 \tag{2.6.39}$$

and from (2.6.28)

$$b_l^r = \delta[1-l]$$

Substituting in equation (2.6.36) we get

$$\beta_k(t) = \frac{1}{N} \sum_{r=0}^{N-1} e^{-j\frac{2r\pi(t-kM\Delta)}{MN\Delta}} e^{j\frac{(N-1)2\pi(t-kM\Delta)}{MN\Delta}}$$

$$
A = 
\begin{bmatrix}
\begin{array}{cccc|cccc|ccccc}
1 & 0 & \cdots & 0 & 1 & 0 & \cdots & 0 & & 1 & 0 & \cdots & 0 \\
0 & 1 & \cdots & 0 & 0 & 1 & & \vdots & \cdots & 0 & 1 & \cdots & 0 \\
\vdots & \vdots & \ddots & \vdots & \vdots & & \ddots & & & \vdots & \vdots & \ddots & \vdots \\
0 & \cdots & 0 & 1 & 0 & \cdots & & 1 & & 0 & \cdots & 0 & 1 \\
\hline
1 & 0 & \cdots & 0 & 1 & 0 & \cdots & 0 & & 1 & 0 & \cdots & 0 \\
0 & 1 & & \vdots & 0 & 1 & & \vdots & & 0 & 1 & & \vdots \\
\vdots & \vdots & \ddots & \vdots & \vdots & & \ddots & & \cdots & \vdots & \vdots & \ddots & \vdots \\
0 & \cdots & & 1 & 0 & \cdots & & 1 & & 0 & \cdots & & 1 \\
\hline
& \vdots & & & & \vdots & & \vdots & & & \vdots \\
\hline
1 & 0 & \cdots & 0 & & & & & & 1 & 0 & \cdots & 0 \\
0 & 1 & & \vdots & & \cdots & & \cdots & & 0 & 1 & & \vdots \\
\vdots & \vdots & \ddots & \vdots & & & & & & \vdots & \vdots & \ddots & \vdots \\
0 & \cdots & & 1 & & & & & & 0 & \cdots & & 1 \\
\end{array}
\end{bmatrix}
$$

$$
= 
\begin{bmatrix}
I & I & \cdots & I \\
I & I & \cdots & I \\
\vdots & & \ddots & \\
I & I & \cdots & I
\end{bmatrix}
\tag{2.6.38}
$$

and we also have

$$\sum_{r=0}^{N-1} e^{-j\frac{2r\pi(t-kM\Delta)}{MN\Delta}} = \frac{1 - e^{-j\frac{2N\pi(t-kM\Delta)}{MN\Delta}}}{1 - e^{-j\frac{2\pi(t-kM\Delta)}{MN\Delta}}}$$

$$= e^{j\frac{\pi(t-kM\Delta)}{MN\Delta}} e^{-j\frac{N\pi(t-kM\Delta)}{M\Delta}} \frac{\sin\left[\frac{\pi}{M\Delta}(t-kM\Delta)\right]}{\sin\left[\frac{\pi}{MN\Delta}(t-kM\Delta)\right]}$$

$$= e^{j\frac{(1-N)\pi(t-kM\Delta)}{MN\Delta}} \frac{\sin\left[\frac{\pi}{M\Delta}(t-kM\Delta)\right]}{\sin\left[\frac{\pi}{MN\Delta}(t-kM\Delta)\right]}$$

Substituting into (2.6.35) we get for this case

$$x(t) = \sum_{k=-\infty}^{\infty} x[k] e^{j\frac{(1-N)\pi(t-kM\Delta)}{MN\Delta}} \frac{\sin\left[\frac{\pi}{M\Delta}(t-kM\Delta)\right]}{\sin\left[\frac{\pi}{MN\Delta}(t-RM\Delta)\right]} e^{j\frac{(N-1)\pi(t-kM\Delta)}{MN\Delta}} \frac{\sin\left[\frac{\pi}{MN\Delta}(t-km\Delta)\right]}{\frac{\pi}{M\Delta}(t-RM\Delta)}$$

$$x(t) = \sum_{k=-\infty}^{\infty} x[k] e^{j\frac{(1-N)\pi(t-kM\Delta)}{MN\Delta}} \frac{\sin\left[\frac{\pi}{M\Delta}(t-kM\Delta)\right]}{\frac{\pi}{M\Delta}(t-kM\Delta)} e^{j\frac{(N-1)\pi(t-kM\Delta)}{MN\Delta}}$$

We thus end up with

$$x(t) = \sum_{k=-\infty}^{\infty} x[k] \frac{\sin\left[\frac{\pi}{M\Delta}(t-kM\Delta))\right]}{\frac{\pi}{M\Delta}(t-kM\Delta)} \tag{2.6.40}$$

Of course, this is precisely the result in Theorem 2.4.1 when $x(t)$ is uniformly sampled at period $M\Delta$. This gives a measure of confidence in the formula (2.6.21), (2.6.36) for the irregular sampling case.

The frequency domain reconstruction formula given in (2.6.32) is shown as a linear combination of filters in Figure 2.6.1. This should be compared with the very simple filter needed in the case of regular sampling.

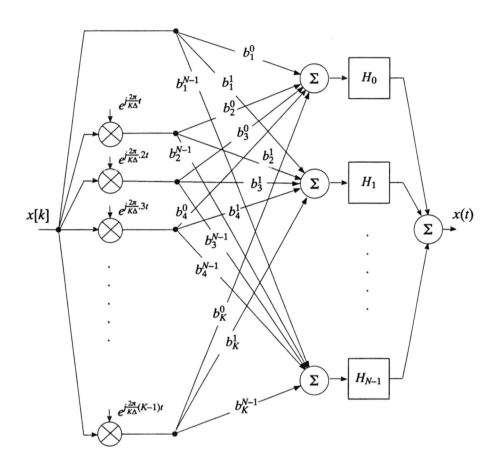

*Figure 2.6.1 : Reconstruction from periodically sampled data.*

We conclude this section by illustrating the application of Theorem 2.6.1 to a particular example.

**Example 2.6.1** Consider the signal

$$x(t) = \cos(0.1\pi t) + 2\cos(0.3\pi t)$$

This signal is sampled at the time instants $\{t_i\}$ $i = 0, 1, 2$ ...   where

$$\{t_i\} = \left\{ 0, \ 2, \ 5, \ 7, \ 10, \ 12, \ 15, \ ... \right\} \text{ seconds.}$$

Clearly the sampling pattern is periodic - the sampling intervals form a 2-periodic sequence: 2, 3, 2, 3, ... , seconds.

(a) Show that if one retains only every second sample (i.e. at 0, 5, 10, ... ) the signal is not reconstructable.

(b) Show that the signal can be reconstructed if all samples are retained and the sampling pattern known.

Solution

(a) The case when one retains only every second sample corresponds to uniform sampling with sampling period $\tilde{\Delta} = 2 + 3 = 5$ sec. Then the sampling frequency is $\omega_s = 0.4 \ \pi$ rad/sec. However, $x(t)$ contains two frequencies, $\omega_1 = 0.1\pi$ and $\omega_2 = 0.3 \ \pi$ rad/sec . Clearly, $\omega_2 > \omega_s/2$ , hence, the signal *cannot* be reconstructed. Indeed, in this case

$$x(5k) = x_1(5k)$$

where $x_1(t)$ depends only on the first frequency component; i.e.

$$x_1(t) = 3 \cos(0.1\pi t)$$

(b) Consider next the periodic sampling pattern. To show that the signal can be reconstructed from the data in this case we will follow the steps outlined in Theorem 2.6.1. We let $\Delta = 1$ sec, $M_1 = 2$, $M_2 = 3$ so $K = 5$ and $T = 5$ sec.

Using (1.7.22)

$$a_k = \sum_{n=1}^{2} \Delta_n e^{-jk\frac{2\pi}{5}t_n}$$

where $\Delta_1 = 2$ sec, $\Delta_2 = 3$ sec, $t_1 = 2$ sec, $t_2 = 5$ sec. So we get $a_0 = 5$, $a_1 = \bar{a}_4 = 2e^{-j\frac{4\pi}{5}} + 3$, $a_2 = \bar{a}_3 = 2e^{j\frac{2\pi}{5}} + 3$. Then, the spectrum of the sampled data is given by equation (1.7.38). The result is

$$X^d(\omega) = \frac{\pi}{5} \sum_{m=-\infty}^{\infty} \Bigg[ (a_0 + 2a_1)\delta(\omega - 0.1\pi - 2\pi m) + (a_0 + 2a_4)\delta(\omega + 0.1\pi - 2\pi m)$$

$$+ (a_1 + 2a_0)\delta(\omega - 0.3\pi - 2\pi m) + (a_4 + 2a_0)\delta(\omega + 0.3\pi - 2\pi m)$$

$$+ (a_1 + 2a_3)\delta(\omega - 0.5\pi - 2\pi m) + (a_4 + 2a_2)\delta(\omega + 0.5\pi - 2\pi m)$$

$$+ (a_2 + 2a_1)\delta(\omega - 0.7\pi - 2\pi m) + (a_3 + 2a_4)\delta(\omega + 0.7\pi - 2\pi m)$$

$$+ (a_2 + 2a_3)\delta(\omega - 0.9\pi - 2\pi m) + (a_3 + 2a_2)\delta(\omega + 0.9\pi - 2\pi m) \Bigg]$$

This is plotted in Figure 2.6.2.

For the reconstruction we need the matrices $A_0, A_1$ (see (2.6.25))

$$A_0 = \begin{bmatrix} a_0 & a_1 \\ a_4 & a_0 \end{bmatrix}, \quad A_1 = \begin{bmatrix} a_0 & a_4 \\ a_1 & a_0 \end{bmatrix} = \bar{A}_0$$

and from (2.6.28)

$$\begin{bmatrix} b_1^0 & b_2^0 & b_3^0 & b_4^0 & b_5^0 \end{bmatrix} = [0.2303, \ -0.0637 + 0.0542j, 0, 0, 0]$$

$$= \begin{bmatrix} \bar{b}_1^1, & \bar{b}_2^1, & b_3^1, & \bar{b}_4^1, & \bar{b}_5^1 \end{bmatrix}$$

Thus, in the frequency domain we have from equation (2.6.32) for the interval $\left[0, \ 2\pi/5\right)$

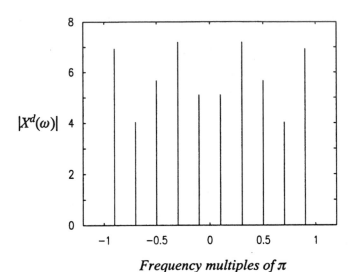

*Figure 2.6.2 : Spectrum for Example 2.6.1.*

$$X(\omega) \; = \; X_1^d(\omega) \; = \; b_1^0 X^d(\omega) + b_2^0 X^d(\omega - 0.4\pi)$$

$$= \frac{\pi}{5} \Big\{ b_1^0 \big[ (a_0 + 2a_1)\delta(\omega - 0.1\pi) + (a_1 + 2a_0)\delta(\omega - 0.3\pi) \big]$$

$$+ b_2^0 \big[ (a_4 + 2a_0)\delta(\omega - 0.1\pi) + (a_0 + 2a_4)\delta(\omega - 0.3\pi) \big]$$

$$= \pi \big[ \delta(\omega - 0.1\pi) + 2\delta(\omega - 0.3\pi) \big] \Big\}$$

Using the symmetricity of $X(\omega)$ we clearly observe that we have recovered the original signal.

$$\wedge\wedge\wedge$$

## 2.7    Further Reading and Discussion

All books on digital signal processing, e.g. Oppenheim and Schafer (1989), Oppenheim, Willsky and Young (1983) contain extensive treatments of sampling and signal reconstruction.

This chapter has gone somewhat beyond the standard engineering treatments of this topic. For example, the result in Section 2.6 on reconstruction from periodic samples does not appear in the usual engineering texts on digital signal processing. The method we have used to develop this result is believed to be novel. However, the result itself is available in the early signal processing literature. Actually, even more general results on signal reconstruction may be found in Jerri (1977) and associated references. Butzer and Stens (1992) give a historical review of sampling theory. Tutorials on the topic can be found in Butzer, Splettstösser and Stens (1988), Vaidyanathan (1990) and Higgins (1977). The recent book by Chui (1992) contains a chapter by Benedetto (1992) which develops a theory of irregular sampling. The latter chapter also contains "a 'sampling' of works of some authors who publish 'regularly' in the field".

## 2.8    Problems

2.1    Consider the two signals

$$y(t) = \cos(0.1\pi t) + 2\cos(0.3\pi t)$$

$$y_1(t) = 3\cos(0.1\pi t)$$

They are sampled at the period $\Delta = 5$ sec.

(a)    Sketch the resulting $Y^s(\omega)$ and $Y_1^s(\omega)$. What do you observe?

(b)    Compute the resulting two sampled data sequences. Can you support your observation in (a) in the time domain?

(c)    Repeat (a) with $\Delta = 1$ sec. What is your conclusion here?

2.2    Given $y_1(t)$ and $y_2(t)$ such that

$$Y_1(\omega) = 0 \text{ for } |\omega| > W_1$$

$$Y_2(\omega) = 0 \text{ for } |\omega| > W_2$$

we define

$$y(t) = y_1(t)y_2(t)$$

What is the Nyquist rate for *y(t)*?

2.3   The requirement for sampling at a rate larger than twice the highest fre-
      quency of band limited signal can be quite conservative at times. Consider
      a signal *x(t)* with $X(\omega)$ such that

$$X(\omega) = 0 \text{ for } \underline{w} < |\omega| < \bar{\omega}$$

A signal like this is called *bandpass* signal.

The Nyquist rate requirement here is $\omega_s > 2\bar{\omega}$. However, if we assume
$2\underline{\omega} > \bar{\omega}$ it is possible to reconstruct *x(t)* at a slow rate. Let *x[k]* be the sequence resulting from sampling *x(t)* at a rate $\Delta$. Passing the resulting $x^s(t)$
through a filter $H(\omega)$ defined by

$$H(\omega) = \begin{cases} K & \omega_1 < |\omega| < \omega_2 \\ 0 & \text{elsewhere} \end{cases}$$

(ideal bandpass filter) we get $\hat{x}(t)$.

(a)   Find the range of $\Delta$ and possible values of $K, \omega_1, \omega_2$ expressed in
      terms of $\underline{\omega}$ and $\bar{\omega}$, for which $\hat{x}(t) = x(t)$.

(b)   Compare the range of $\Delta$ to the Nyquist sampling period in this case.

(c)   Derive the reconstruction formula in this case.

2.4   In Problem 2.3 we described one approach to bandpass signals. Another
      approach is based on modulation of the original signal. Let *x(t)* and $X(\omega)$
      be such that $X(\omega) \neq 0$ only in the range $\underline{\omega} < |\omega| < \bar{\omega}$ where, again,
      $2\underline{\omega} > \bar{\omega}$. Multiply *x(t)* by a cosine signal to get $x_1(t)$ where

$$x_1(t) = x(t)\cos(\overline{\omega}t)$$

and pass $x_1(t)$ through the filter $H(\omega)$, where

$$H(\omega) = \begin{cases} 2/\Delta & |\omega| < \overline{\omega} - \underline{\omega} \\ 0 & \text{elsewhere} \end{cases}$$

(ideal lowpass filter), to give us the signal $x_2(t)$.

(a) How can one reconstruct $x(t)$ from the sequence $x_2[k]$ resulting from sampling $x_2(t)$ at a rate $\Delta$ ? Describe the reconstruction in the frequency domain and derive the expression in the time domain.

(b) What is the range of $\Delta$ for which the reconstruction in (a) is possible?

2.5 The aliasing effect is seen in many Western movies when the camera tracks rapidly moving stagecoach wheels.

Suppose the wheels of the coach rotate at $\omega_0$ rad/sec. The movie is projected with a frame every $\Delta$ seconds and our vision system acts as an ideal lowpass filter with cut off frequency at $W$. Assuming $W > |\omega_0 - 2\pi/\Delta|$ how do the wheels seem to rotate when:

(a) $\omega_0 > 2\pi/\Delta$

(b) $\omega_0 < 2\pi/\Delta$

(c) $\omega_0 = 2\pi/\Delta$

2.6 Given $y_1(t)$ and $y_2(t)$ as in Problem 2.2, we define

$$y(t) = y_1(t) \otimes y_2(t)$$

(a) What is the Nyquist rate for $y(t)$?

(b) Suppose $y_1(t), y_2(t)$ and $y(t)$ are all sampled at a period $\Delta$. What is the condition on $\Delta$ so that $y[k] = y_1[k] \otimes y_2[k]$ ?

2.7  Suppose we have stored $N$ data points $x[0]$, ... ,$x[N-1]$ so that
$x[k] = \cos(2\pi k/N)$. Let $\tilde{x}[k]$ be the periodic extension of $x[k]$, namely

$$\tilde{x}[k+mN] = \tilde{x}[k] = x[k] \text{ for } k = 0, 1, \ldots, N-1$$

and pick $\underline{\Delta} \le \Delta \le \overline{\Delta}$. Define

$$x^s(t) = \sum_{k=-\infty}^{\infty} \Delta\tilde{x}[k]\delta(t-k\Delta)$$

and pass $x^s(t)$ through the ideal lowpass filter $H(\omega)$ with cutoff frequency
$W$ to result in $x(t)$.

(a)  Assuming $W = \pi/\Delta$ what is $x(t)$?

(b)  Assuming $\dfrac{2\pi(N-1)}{N\Delta} < W < \dfrac{2\pi}{\Delta}$ what is $x(t)$?

2.8  Let $x(t) = e^{-at}\mu(t)$ ($\mu(t)$ is the step function) be sampled at a period $\Delta$.
The resulting $x^s(t)$ is then passed through a hold filter to give $\hat{x}(t)$. Consider three possible hold filters, ZOH, noncausal FOH and causal FOH.

(a)  For each case derive an expression for $\epsilon = \displaystyle\int_0^{\infty} [x(t) - \hat{x}(t)]^2 dt$.

(b)  Choose $a=1$ and use your results in (a) to calculate $\epsilon$ in all three cases
for $\Delta = 1, 0.1, 0.01$.

2.9  Show that (2.6.17) does indeed imply (2.6.18).

2.10  Show that $A_r$ as defined in (2.6.25) is non-singular by going through the
following steps

(a)  Define

$$E = \left[e_1, \ e_{M_1+1}, \ e_{M_1+M_2+1}, \ \ldots, \ e_{M_1+M_2+\ldots+M_{N-1}+1}\right]$$

where $e_i$ is the i-th column of the $K \times K$ identity matrix.

Show that $A = W_K E \tilde{\Lambda} E^T W_K^*$ where $A$ is defined in (2.6.8), $W_K$ in (2.6.10) and

$$\tilde{\Lambda} = \begin{bmatrix} M_N & & & & \\ & M_1 & & & 0 \\ & & M_2 & & \\ & & & \ddots & \\ 0 & & & & M_2 \end{bmatrix}$$

(b)   Show that the matrix $E_r^T W_K E, E_r$ as in (2.6.26), is nonsingular. (Hint: Permute the rows of the matrix and premultiply it by

$$\begin{bmatrix} 1 & & & & \\ & W^{M_1 r} & & 0 & \\ & & W^{(M_1+M_1)r} & & \\ & & & \ddots & \\ 0 & & & & W^{(M_1+ \ldots +M_{N-1})r} \end{bmatrix} \qquad W = e^{j\frac{2\pi}{K}}$$

to get a Vandermonde type matrix).

(b)   Combine (a) and (b) to argue that $A_r$ is nonsingular.

2.11   Show that (2.6.29) is the solution of equation (2.6.27).

2.12   Calculate $\{a_k\}, \{A_r\}$ and $\{b_l^r\}$ for the following sampling patterns:

(a)   $M_1 = 1, \; M_2 = 5, \; \Delta = 1$ sec

(b)   $M_2 = 2, \; M_2 = 4, \; \Delta = 1$ sec

(c)   $M_2 = M_2 = 3, \; \Delta = 1$ sec

(d)   $M_1 = 1, \; M_2 = 2, \; M_3 = 3, \; \Delta = 1$ sec

2.13   Consider the signal $x(t) = 2\cos(0.1\pi T) + \cos(0.3\pi t)$. The signal is sampled in the pattern of Problem 2.12 (c).

(a)   Draw $|X^d(\omega)|$.

(b)   Verify that the signal can be reconstructed from the sampled data.

(c)   Draw the reconstruction configuration as in Figure 2.6.1 as it should be used for the above data.

(d)   Draw the output of each one of the filters $H_r, r = 0, \ldots, N-1$.

2.14  Repeat Problem 2.13 for $x(t)$ with $X(\omega)$ as in Figure 2.8.1.

2.15  Establish Lemma 2.5.1.

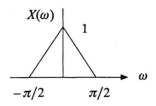

*Figure 2.8.1 : Signal for Problem 2.14.*

# Chapter 3

# Analysis of Discrete-Time Systems

## 3.1 Introduction

Our focus in this book is on digital signal processing algorithms of various types. Examples of such systems are digital filters and digital control systems. Our ultimate objective is to form an understanding of the effect that such algorithms have on the underlying *continuous-time* system. In this chapter, however, we will focus entirely on the 'at-sample' response. That is, we treat digital signal processing algorithms as simply a way of manipulating an input *sequence* $\{u[k]\}$ to produce an output *sequence* $\{y[k]\}$. For the moment therefore, we will not explicitly focus on the intersample response. This will be done in later chapters.

## 3.2 Shift Operator Models

The most commonly used tool for the analysis of discrete-time signals is the *shift* operator, $q$. This operator is defined as follows:

$$\boxed{qu[k] = u[k + 1]} \qquad (3.2.1)$$

The reader may have seen this operator previously in a course on signals and systems, digital signal processing or introductory digital control. For completeness we will present a brief review. The operator is very useful in the analysis of discrete systems since it allows one to express operations on sequences. For example, the following linear discrete-time state space model

$$x[k+1] = A_q x[k] + B_q u[k] \tag{3.2.2}$$

$$y[k] = C_q x[k] \tag{3.2.3}$$

can be expressed using the shift operator as

$$qx[k] = A_q x[k] + B_q u[k] \tag{3.2.4}$$

$$y[k] = C_q x[k] \tag{3.2.5}$$

Equation (3.2.4) can be interpreted as giving a simple formula for the state at the $(k+1)^{th}$ sample time in terms of the state and input at the $k^{th}$ sample time.

It is also possible to use $n^{th}$ order shift equations of the form:

$$\tilde{a}_n y[k+n] + \tilde{a}_{n-1} y[k+n-1] + \; \ldots \; + \tilde{a}_0 y[k]$$

$$= \tilde{b}_{n-1} u[k+n-1] + \; \ldots \; + \tilde{b}_0 u[k] \tag{3.2.6}$$

This type of equation can also be expressed compactly using the shift operator as follows:

$$\tilde{A}(q)y[k] = \tilde{B}(q)u[k] \tag{3.2.7}$$

where $\tilde{A}$ and $\tilde{B}$ are polynomials in the operator $q$

$$\tilde{A}(q) = \tilde{a}_n q^n + \tilde{a}_{n-1} q^{n-1} + \; \ldots \; + \tilde{a}_0 \tag{3.2.8}$$

$$\tilde{B}(q) = \tilde{b}_{n-1} q^{n-1} + \; \ldots \; + \tilde{b}_0 \tag{3.2.9}$$

Equations of the form (3.2.6) can be solved in several ways. For example, we can convert them to sequential state-space form; i.e. with $\tilde{a}_n$ normalized to unity we have

$$
q \begin{bmatrix} x_1[k] \\ x_2[k] \\ \cdot \\ \cdot \\ \cdot \\ x_n[k] \end{bmatrix} = \begin{bmatrix} -\tilde{a}_{n-1} & & & 1 \\ & \cdot & & \cdot \\ & \cdot & & \cdot \\ & \cdot & & \\ & & & 1 \\ -\tilde{a}_0 & & 0 & 0 \end{bmatrix} \begin{bmatrix} x_1[k] \\ x_2[k] \\ \cdot \\ \cdot \\ \cdot \\ x_n[k] \end{bmatrix} + \begin{bmatrix} \tilde{b}_{n-1} \\ \tilde{b}_{n-2} \\ \cdot \\ \cdot \\ \cdot \\ \tilde{b}_0 \end{bmatrix} u[k] \quad (3.2.10)
$$

$$
y[k] = \begin{bmatrix} 1 & 0 & \cdots & 0 \end{bmatrix} \begin{bmatrix} x_1[k] \\ x_2[k] \\ \cdot \\ \cdot \\ \cdot \\ x_n[k] \end{bmatrix} \qquad (3.2.11)
$$

To verify the equivalence of (3.2.6) and (3.2.10), (3.2.11) we note that the first line of (3.2.10) is equivalent to

$$
qy[k] = -\tilde{a}_{n-1}y[k] + x_2[k] + \tilde{b}_{n-1}u[k] \qquad (3.2.12)
$$

In an effort to eliminate $x_2[k]$ from this equation, we apply the operator $q$ to both sides and then use the second equation in (3.2.10). This gives

$$
q^2 y[k] = -\tilde{a}_{n-1}qy[k] + qx_2[k] + \tilde{b}_{n-1}qu[k]
$$

$$
= -\tilde{a}_{n-1}qy[k] - \tilde{a}_{n-2}y[k] + x_3[k]
$$

$$
+ \tilde{b}_{n-1}qu[k] + \tilde{b}_{n-2}u[k] \qquad (3.2.13)
$$

Proceeding in this way ultimately leads to (3.2.6).

Given $\{u[k]\}$ and the initial conditions $x[0]$, we can iterate the equivalent state-space model (3.2.10) step by step to generate $\{y[k]\}$.

## 3.3    z-Transforms

The reader is probably familiar with the use of Laplace transforms to solve ordinary linear differential equations. The equivalent transform for shift operator discrete-time equations is the z-transform.

The *one-sided* z-transform, $X_q(z)$, of the sequence $\{x[k]\}$ is defined as follows:

$$X_q(z) = Z\{x[k]\} \overset{\Delta}{=} \sum_{k=0}^{\infty} z^{-k} x[k] \qquad (3.3.1)$$

A key property of z-transforms is the transform of a shifted sequence. In particular, it is easy to show that

$$Z\{qx[k]\} = zX_q(z) - x[0] \qquad (3.3.2)$$

Thus, ignoring initial conditions, the shift operator $q$ is simply replaced by z when taking transforms. Equation (3.3.2) can be iterated to give further results; e.g.

$$Z\{q^2 x[k]\} = zZ\{qx[k]\} - x[1]$$

$$= z^2 X_q(z) - zx[0] - x[1] \qquad (3.3.3)$$

We complete this brief review of shift operators and z-transforms with two simple examples.

**Example 3.3.1** Determine the one-sided z-transform of the sequence $\{a^k\}$.

**Solution**

$$Z\{a^k\} = \sum_{k=0}^{\infty} z^{-k} a^k$$

$$= \frac{1}{1 - z^{-1}a} \quad \text{for} \quad |z| > |a|$$

$$= \frac{z}{z-a} \tag{3.3.4}$$

∧∧∧

**Example 3.3.2** Solve the following difference equation

$$\left(q^2 + 3q + 2\right)y[k] = u[k]$$

where $y[0] = y[1] = 0$, and $\{u[k]\}$ is a unit step applied at time 0.

**Solution**   Taking $z$-transforms of the equation gives

$$\left(z^2 + 3z + 2\right)Y_q(z) - zy[0] - y[1] - 3y[0] = U_q(z)$$

where applying the given initial conditions yields

$$Y_q(z) = \frac{U_q(z)}{z^2 + 3z + 2}$$

From Example 3.3.1, $U_q(z) = \dfrac{z}{z-1}$ and hence

$$Y_q(z) = \frac{z}{(z-1)(z^2 + 3z + 2)}$$

or

$$\frac{Y_q(z)}{z} = \frac{1}{(z-1)(z+2)(z+1)}$$

$$= \frac{1}{6}\left[\frac{1}{z-1}\right] + \frac{1}{3}\left[\frac{1}{z+2}\right] - \frac{1}{2}\left[\frac{1}{z+1}\right]$$

$$Y_q(z) = \frac{1}{6}\left[\frac{z}{z-1}\right] + \frac{1}{3}\left[\frac{z}{z+2}\right] - \frac{1}{2}\left[\frac{z}{z+1}\right]$$

Finally, again using Example 3.3.1, we have

$$y[k] = \frac{1}{6} + \frac{1}{3}[-2]^k - \frac{1}{2}[-1]^k \; ; \; k \geq 0$$

ᐯᐯᐯ

In the next section, we extend the notion of the shift operator to a closely related operator (called the delta operator).

## 3.4    The Delta Operator

The shift operator (and associated $z$-transform), as outlined above, allow us to solve many problems in discrete systems theory. However, we will find it useful to go beyond this operator when we want to connect continuous and discrete signals and study the limiting behaviour of discrete systems as the sample period, $\Delta$, approaches zero. The sampling period $\Delta$ is only implicitly included in $z$-transforms and shift-operator models. Moreover, the notion of a forward-shift as in (3.2.1) does not have a sensible continuous-time counterpart. Consequently, discrete-time representations using the $q$ operator do not converge smoothly to the underlying continuous-time result as the sampling period goes to zero. Also, the results bear little, or no, resemblance to the familiar continuous results. Finally, precisely because the limiting case, as the sampling rate goes to infinity, is ill-defined, we should expect numerical problems with shift operators at fast sampling rates relative to the dynamics of the underlying continuous-time system.

The resolution of these difficulties would seem to necessitate the use of a different operator for discrete-time systems, which has a direct counterpart in continuous systems analysis. Looking at continuous system theory we find that the basic operators are differentiation and integration. We therefore seek a discrete-time operator, which inter-alia, has the following properties:

   (i)  It leads to models which bear a one-to-one linear relationship
        to shift operator models;

(ii)   It converges to a continuous-time derivative as the sample period goes to zero; and

(iii)   It is such that the inverse operator is causal.

There is a unique operator with the above properties. This is the *delta operator* which is defined as follows:

$$\delta u[k] = \frac{u[k+1] - u[k]}{\Delta}$$

(3.4.1)

This operator has a long history in the numerical analysis field where it is known as the first divided difference operator. It also has been deployed as a computational device in the area of control with reference going back to at least 1947. (See references given at the end of the chapter.)

From (3.4.1) we see that the $\delta$ operator approximates the derivative. For example, if $u[k] = u(k\Delta)$ results from sampling a continuous-time differentiable signal $u(t)$ at period $\Delta$, then

$$\delta u[k] \simeq \frac{du(t)}{dt} \bigg|_{u = u(k\Delta)}$$

(3.4.2)

with the approximation becoming better as the sampling period tends to zero.

Because the $\delta$ operator has a continuous-time counterpart, models for systems expressed in terms of the $\delta$ operator are very similar to models expressed with the differentiation operator $d/dt$, or the Laplace transform variable $s$. Because of this, the use of the $\delta$ operator permits continuous-time intuition and insights to be used in discrete-time systems. Furthermore, it provides equivalent flexibility in the context of discrete-time modelling as does the shift operator $q$.

The inverse operator to $\delta$ is the lower Riemann sum defined as follows:

$$\delta^{-1}u = \sum_{m=0}^{k-1} \Delta u[m] \tag{3.4.3}$$

Provided $\{u[k]\}$ has certain properties, then the above sum converges to the Riemann integral as $\Delta$ approaches zero. The reader will note that we have used Riemann sums of the type given in (3.4.3) when defining discrete-time Fourier transforms in Chapter 1. This was done precisely to maintain a close connection with continuous-time definitions and analysis.

One might expect that there exists a calculus for operators $\delta$ and $\delta^{-1}$ which is very similar to that which holds for $d/dt$ and $(d/dt)^{-1}$ – differentiation and integration respectively. This is indeed the case. The following results follow immediately from the definitions.

Lemma 3.4.1

$$\delta \left\{ \sum_{m=0}^{k-1} \Delta f[m] \right\} = f[k] \tag{3.4.4}$$

$$\delta \left\{ \sum_{m=k}^{L} \Delta f[m] \right\} = -f[k] \tag{3.4.5}$$

$$\sum_{m=\alpha}^{\beta} \Delta \delta f[m] = f[\beta + 1] - f[\alpha] \tag{3.4.6}$$

$$\delta(fg) = (\delta f)g + f(\delta g) + \Delta(\delta f)(\delta g) \tag{3.4.7}$$

Proof : (3.4.4) to (3.4.6) are immediate from the definitions. Equation (3.4.7) is a rule for differencing (differentiating) a product. We argue from the definitions as follows:

$$\delta(fg) = \frac{f[k+1]g[k+1] - f[k]g[k]}{\Delta}$$

$$= \frac{(f[k+1] - f[k])g[k] + f[k](g[k+1] - g[k])}{\Delta} +$$

$$+ \frac{\Delta(f[k+1] - f[k])(g[k+1] - g[k])}{\Delta^2}$$

$$= (\delta f)g + f(\delta g) + \Delta(\delta f)(\delta g)$$

$$\wedge\wedge\wedge$$

Notice that as the sampling period tends to zero, the result for generalized differentiation of a product of two functions tends to the well known product rule for continuous-time derivatives as expected.

We will also use the following result on the 'differentiation' of a particular matrix inverse.

**Lemma 3.4.2** Consider a square invertible matrix $A$ whose generalized derivative can be written via the vector $B$ as:

$$\delta A = BB^T$$

Then the generalized derivative of $A^{-1}$ is given by:

$$\delta A^{-1} = \frac{-A^{-1}BB^TA^{-1}}{1 + \Delta B^TA^{-1}B}$$

**Proof :** Using (3.4.7) we have

$$\delta(AA^{-1}) = (\delta A)A^{-1} + A(\delta A^{-1}) + \Delta(\delta A)(\delta A^{-1}) = 0$$

from which it follows that

$$\delta A^{-1} = -(A + \Delta BB^T)^{-1}BB^TA^{-1}$$

The result then follows from the following matrix identity:

$$\left(A + \Delta BB^T\right)^{-1} = A^{-1} - \Delta A^{-1}B\left(I + \Delta B^T A^{-1}B\right)^{-1}B^T A^{-1}$$

This identity is commonly called the *Matrix Inversion Lemma*.

$$\wedge\wedge\wedge$$

In the remainder of this chapter we show how one can employ the above operator to develop discrete system theory. A distinctive feature of our analysis is that the corresponding continuous-time case will almost always turn out to be an appropriate limiting case as the sample period approaches zero.

We present below a collection of results (some of which have been established in Lemma 3.4.1) which follow from the definition of the $\delta$ operator:

(i)  *Difference of a product*

$$\delta(x[k]y[k]) = (\delta x[k])y[k] + x[k](\delta y[k]) + \Delta(\delta x[k])(\delta y[k]) \qquad (3.4.8)$$

Note that the following are also true

$$\delta(x[k]y[k]) = (\delta x[k])y[k] + x[k + 1](\delta y[k]) \qquad (3.4.9)$$

$$\delta(x[k]y[k]) = (\delta x[k])y[k + 1] + x[k](\delta y[k]) \qquad (3.4.10)$$

(ii)  *Difference of an inverse*

$$\delta\left(A^{-1}[k]\right) = -A^{-1}[k + 1](\delta A[k])A^{-1}[k]$$

$$= -A^{-1}[k](\delta A[k])A^{-1}[k + 1] \qquad (3.4.11)$$

(iii)  *Difference of a sum*

$$\delta\left[\sum_{m=0}^{k-1}\Delta f[m]\right] = f[k] \qquad (3.4.12)$$

(iv) *Sum of a difference*

$$\sum_{m=\alpha}^{\beta} \Delta \delta f[m] = f[\beta + 1] - f[\alpha] \qquad (3.4.13)$$

(v) *Difference of discrete exponential*

$$\delta\left((I + \Delta A)^k\right) = A(I + \Delta A)^k = (I + \Delta A)^k A \qquad (3.4.14)$$

$$\delta\left((I + \Delta A)^{-k}\right) = -A(I + \Delta A)^{-k-1} = -(I + \Delta A)^{-k-1} A \qquad (3.4.15)$$

(vi) *Summation by parts*

$$\sum_{k=\alpha}^{\beta-1} \Delta(\delta f[k]) g[k] = f[\beta] g[\beta] - f[\alpha] g[\alpha] - \sum_{k=\alpha}^{\beta-1} \Delta f[k + 1](\delta g[k]) \qquad (3.4.16)$$

## 3.5    Difference Equations in Delta Operator Form

Difference equations of the form (3.2.6) can equivalently be expressed in terms of the delta operator. For example, the $n^{th}$ order linear difference equation (3.2.6) taking a sequence $\{u[k]\}$ to a sequence $\{y[k]\}$ has the following form using the $\delta$ operator:

$$\bar{a}_n \delta^n y[k] + \bar{a}_{n-1} \delta^{n-1} y[k] + \ \dots \ + \bar{a}_0 y[k] = \bar{b}_{n-1} \delta^{n-1} u[k] + \ \dots \ + \bar{b}_0 u[k]$$

$$(3.5.1)$$

The connection is established by using the substitution $\delta = (q - 1)/\Delta$ or $q = \Delta\delta + 1$. Thus the coefficients in (3.5.1) and (3.2.6) are simply related as follows:

$$
\begin{bmatrix} \bar{a}_n \\ \bar{a}_{n-1} \\ \cdot \\ \cdot \\ \cdot \\ \cdot \\ \bar{a}_2 \\ \bar{a}_1 \\ \bar{a}_0 \end{bmatrix} = \begin{bmatrix} C_n^n \Delta^n & 0 & 0 & 0 & 0 \\ C_{n-1}^n \Delta^{n-1} & C_{n-1}^{n-1} \Delta^{n-1} & 0 & 0 & 0 \\ \cdot & \cdot & \cdot & \cdot & \cdot \\ \cdot & \cdot & \cdot & \cdot & \cdot \\ \cdot & \cdot & \cdot & \cdot & \cdot \\ \cdot & \cdot & \cdot & \cdot & \cdot \\ C_2^n \Delta^2 & C_2^{n-1} \Delta^2 & C_2^{n-2} \Delta^2 & 0 & 0 \\ C_1^n \Delta & C_1^{n-1} \Delta & C_1^{n-2} \Delta & C_1^1 \Delta & 0 \\ 1 & 1 & 1 & 1 & 1 \end{bmatrix} \begin{bmatrix} \tilde{a}_n \\ \tilde{a}_{n-1} \\ \cdot \\ \cdot \\ \cdot \\ \cdot \\ \tilde{a}_2 \\ \tilde{a}_1 \\ a_0 \end{bmatrix} \qquad (3.5.2)
$$

where $C_j^k = \dfrac{k!}{j!(k-j)!}$ .

The matrix on the right hand side of (3.5.2) is triangular and is always invertible.

Thus any model expressed in terms of $q$ can be readily converted to a model in $\delta$ and vice versa. The reader may therefore wonder what advantages, if any, this change of notation makes. The key point is that since $\delta$ is like a derivative, then continuous-time intuitions approximately apply to $\delta$ models. Thus, for example, we might guess that if the sequences $\{u[k]\}$ and $\{y[k]\}$ result from sampling fast relative to the bandwidth of signals of interest $y(t)$ and $u(t)$, then the solution of the difference equation (3.5.1) will be "close to" the solution of the differential equation obtained by replacing $\delta$ by $d/dt$ : this is indeed the case.

An obvious question is "How does one interpret and solve a difference equation of the form of (3.5.1)?" There are two basic methods : a) to use an interative algorithm on a computer, or b) to use discrete transform methods.

Solution by computer iteration is straightforward. We first note that with $\bar{a}_n$ normalized to one, then (3.5.1) is simply a compact way of writing the following state space model:

$$\delta\underline{x}[k] = \begin{bmatrix} -\bar{a}_{n-1} & 1 & \cdots & 0 \\ & & & \\ & & & \\ & \ddots & & \\ & & & 1 \\ -\bar{a}_0 & 0 & \cdots & 0 \end{bmatrix} \underline{x}[k] + \begin{bmatrix} \bar{b}_{n-1} \\ \\ \\ \vdots \\ \\ \bar{b}_0 \end{bmatrix} u[k]$$

$$= A_\delta \underline{x}[k] + B_\delta u[k] \tag{3.5.3}$$

$$y[k] = [1 \ 0 \ \cdots \ 0]\underline{x}[k] \tag{3.5.4}$$

where $\underline{x}[k]$ is an $n$-dimensional vector.

To verify the equivalence of the difference equation (3.5.1) and the state-space model (3.5.3), (3.5.4) we argue as in the derivation of (3.2.10). We note that the first line of (3.5.3) is equivalent to

$$\delta y[k] = -\bar{a}_{n-1}y[k] + x_2[k] + \bar{b}_{n-1}u[k] \tag{3.5.5}$$

In an effort to eliminate $x_2[k]$ from this equation, we difference it and use the second equation in (3.5.3). This gives

$$\delta^2 y[k] = -\bar{a}_{n-1}\delta y[k] + \delta x_2[k] + \bar{b}_{n-1}\delta u[k]$$

$$= -\bar{a}_{n-1}\delta y[k] - \bar{a}_{n-2}y[k] + x_3[k] + \bar{b}_{n-1}\delta u[k] + \bar{b}_{n-2}u[k] \tag{3.5.6}$$

Proceeding in this way, we finally arrive at (3.5.1). We may thus conclude

*The difference equation model (3.5.1) is a short hand way of writing the state space model (3.5.3), (3.5.4).*

The state space form shows how one can solve the difference equation (3.5.1) on a computer. We simply note that (by definition)

$$\delta\underline{x}[k] = \frac{1}{\Delta}[\underline{x}[k+1] - \underline{x}[k]] \tag{3.5.7}$$

Hence, (3.5.3) is equivalent to the following shift state space model

$$\underline{x}[k+1] = \underline{x}[k] + \Delta[A_\delta \underline{x}[k] + B_\delta u[k]] \tag{3.5.8}$$

Thus, if we are given the input sequence $u[0], u[1], \ldots$ plus the initial state $\underline{x}[0]$; then (3.5.8) can be solved recursively for $\underline{x}[1], \underline{x}[2], \ldots$ as for (3.2.10). Then (3.5.4) gives $y[0], y[1], y[2], \ldots$.

Finally we note the dependence of the solution (3.5.1) on $n$ initial conditions.

## 3.6    Discrete Delta Transform

We next turn to the solution of (3.5.1) via discrete transform techniques. Recall that in continuous-time, the appropriate transform is the (one-sided) Laplace transform defined as follows

$$Y(s) = L\{y(t)\} = \int_0^\infty e^{-st} y(t) dt \tag{3.6.1}$$

In discrete-time, it is natural to replace this transform by an equivalent transform defined using a summation to replace the integral. Thus we are lead to define the $\delta$ transform of the sequence $\{y(k\Delta)\}$ by

$$Y'(s) = \sum_{k=0}^\infty \Delta e^{-sk\Delta} y(k\Delta) \tag{3.6.2}$$

We note in the above formula that the variable $s$ always appears in the form $e^{s\Delta}$. We will see in the sequel that it is desirable to change the argument of the transform by replacing $e^{s\Delta}$ by a new variable $\gamma$ defined as follows:

$$e^{s\Delta} = 1 + \Delta\gamma \tag{3.6.3}$$

Substituting (3.6.3) into (3.6.2) gives the following final form of the (one-sided) *delta transform* $Y_\delta(\gamma)$ of the sequence $\{y[k]\}$ :

$$Y_\delta(\gamma) = D\{y[k]\} \overset{\Delta}{=} \sum_{k=0}^{\infty} \Delta(1 + \Delta\gamma)^{-k} y[k] \qquad (3.6.4)$$

We note that for $\Lambda$ small, $1 + \Delta s \simeq e^{s\Delta}$, and hence comparing (3.6.4) and (3.6.2), (3.6.1) we may guess that $s$ and $\gamma$ are closely related. For example, since (3.6.4) is a Riemann sum approximation to the Laplace transform, we might expect that the delta transform of a sampled function would converge to the Laplace transform of the original function as $\Delta \rightarrow 0$. This is indeed true under very general conditions.

We will illustrate the result above for a special case.

**Example 3.6.1** Consider a continuous-time exponential; i.e.

$$y(t) = e^{at}\mu(t) \qquad (3.6.5)$$

Find the delta transform of the sampled values of this signal sampled with period $\Delta$.

**Solution** The sampled form of this signal is

$$y[k] = e^{ak\Delta}\mu[k] \qquad (3.6.6)$$

For future reference, we note that the Laplace transform of the continuous-time function (3.6.5) is

$$Y(s) = \frac{1}{s-a} \qquad (3.6.7)$$

To derive the delta transform of the sampled sequence (3.6.6), we use the definition given in (3.6.4):

$$Y_\delta(\gamma) = \sum_{k=0}^{\infty} \Delta(1 + \gamma\Delta)^{-k} e^{ak\Delta}$$

$$= \frac{\Delta}{1 - (1 + \Delta\gamma)^{-1} e^{a\Delta}}$$

$$= \frac{\Delta(1 + \Delta\gamma)}{1 + \Delta\gamma - e^{a\Delta}} \tag{3.6.8}$$

$\wedge\wedge\wedge$

We may thus conclude :

*The following are a delta transform pair :*

$$e^{ak\Delta}\mu[k] \quad \overset{D}{\to} \quad \frac{1 + \Delta\gamma}{\gamma - \left(\frac{e^{a\Delta}-1}{\Delta}\right)} \tag{3.6.9}$$

*or equivalently :*

$$(\Delta\beta + 1)^k\mu[k] \quad \overset{D}{\to} \quad \frac{1 + \Delta\gamma}{\gamma - \beta} \tag{3.6.10}$$

The above transform pair is essentially the only one needed to solve most (simple) problems using delta transforms.

We note in passing that

$$\lim_{\Delta \to 0} \frac{1 + \Delta\gamma}{\gamma - \left[\frac{e^{a\Delta}-1}{\Delta}\right]} = \frac{1}{\gamma - a} \tag{3.6.11}$$

As expected, this is precisely the Laplace transform of (3.6.5) as given in (3.6.7).

The inverse transform used to recover a sequence from its transform is given by the following integral:

$$x[k] = \frac{1}{2\pi j} \oint X_\delta(\gamma)(1 + \Delta\gamma)^{k-1} d\gamma \qquad (3.6.12)$$

where the contour of integration $C$ is inside the region of convergence of $X_\delta(\gamma)$ and encircles all singularities of $X_\delta(\gamma)$ once in the anticlockwise sense. In practice, however, one usually uses pre-tabulated results for common transforms (see Section 3.10).

As might be expected, the delta transform defined above is closely related to the z-transform defined previously in equation (3.3.1). As before, we use $Y_q(z)$ to denote the $z$-transforms of a sequence $y[k]$, then comparing (3.3.1) with (3.6.4) we see that the $z$-transform $Y_q(z)$ and delta transform $Y_\delta(\gamma)$ are related as follows:

$$Y_q(z) = \frac{1}{\Delta} Y_\delta(\gamma) \bigg|_{\gamma = \frac{z-1}{\Delta}} \qquad (3.6.13)$$

We also have the converse to (3.6.13), namely,

$$Y_\delta(\gamma) = \Delta Y_q(z) \bigg|_{z = \Delta\gamma + 1} \qquad (3.6.14)$$

Most contemporary books on digital control and signal processing work entirely in terms of $Y_q(z)$. However, we prefer to supplement this transform with the delta transform, $Y_\delta(\gamma)$, because of the close connection with Laplace transforms. This connection is lost if only z-transforms are used.

To further illustrate the connection between delta and Laplace transforms we will consider the discrete transform of a unit step and discrete $1/\Delta$ pulse.

As in Section 1.7, we use $\delta_d[k]$ to denote a discrete $1/\Delta$ pulse; defined as follows:

$$\delta_d[k] \; = \; \begin{cases} 1/\Delta \;\; \text{for } k = 0 \\ 0 \;\; \text{otherwise} \end{cases} \qquad\qquad (3.6.15)$$

Note that if $\Delta$ is the sample period and we think of the $1/\Delta$ pulse as holding for one sample period, then we get a continuous-time function of height $1/\Delta$ and width $\Delta$ which clearly has unit area. Thus the $1/\Delta$ discrete pulse is analogous to a continuous-time impulse.

As before, we use $\delta(t)$ to denote the continuous-time impulse function. We also use $\mu(t)$ and $\mu[k]$ to denote the continuous and discrete unit step function respectively. We then have the following Laplace and delta transform pairs:

$$L\{\mu(t)\} = \frac{1}{s} \qquad\qquad (3.6.16)$$

$$D\{\mu[k]\} = \frac{1 + \Delta\gamma}{\gamma} \qquad\qquad (3.6.17)$$

$$L\{\delta(t)\} = 1 \qquad\qquad (3.6.18)$$

$$D\{\delta_d[k]\} = 1 \qquad\qquad (3.6.19)$$

The close connection between Laplace and Delta Transforms is again evident in these results.

## 3.7    Use of Discrete Delta Transforms to Solve Difference Equations

We recall that a key property used in applying Laplace transforms to solve differential equations is the result for the transform of a derivative; i.e.

$$L\left\{\frac{d}{dt}y(t)\right\} = sY(s) - y(0) \qquad (3.7.1)$$

By direct calculation using (3.6.4) we obtain the analogous result for the delta transform of a differenced sequence :

$$D\{\delta y[k]\} = \sum_{k=0}^{\infty} \Delta(1 + \Delta\gamma)^{-k}\left[\frac{y[k+1] - y[k]}{\Delta}\right]$$

$$= \frac{1}{\Delta}\left\{(1 + \Delta\gamma)\sum_{k=1}^{\infty}\Delta(1 + \Delta\gamma)^{-k}y[k] - \sum_{k=0}^{\infty}(1 + \Delta\gamma)^{-k}y[k]\right\}$$

That is,

$$D\{\delta y[k]\} = \gamma Y_\delta(\gamma) - (1 + \Delta\gamma)y[0] \qquad (3.7.2)$$

where $Y_\delta(\gamma)$ is the delta transform of $\{y[k]\}$. Note that (3.7.2) is analogous to (3.3.2) for $z$-transforms.

The close link between the result given in (3.7.2) and that given in (3.7.1) is self evident.

The transforms of high order differences are now obtained by repeated application of (3.7.2); i.e.

$$D\{\delta^2 y[k]\} = \gamma D\{\delta y[k]\} - (1 + \Delta\gamma)\delta y[0]$$

$$= \gamma^2 Y_\delta(\gamma) - \gamma(1 + \Delta\gamma)y[0] - (1 + \Delta\gamma)\delta y[0] \qquad (3.7.3)$$

and so on. Notice that the initial condition terms are always multiplied by $(1 + \Delta\gamma)$.

We observe that (ignoring initial condition terms) we simply replace $\delta$ by $\gamma$ when taking the delta transform. This was the principal reason for the change of variable introduced in (3.6.3) when defining the delta transform.

Now consider the difference equation (3.5.1). Taking delta transforms we obtain

$$\bar{A}(\gamma)Y_\delta(\gamma) = \bar{B}(\gamma)U_\delta(\gamma) + (1 + \Delta\gamma)\beta(\gamma) \tag{3.7.4}$$

where we assume, without loss of generality, that $\bar{A}(\gamma)$ is normalized so that $\bar{a}_n = 1$, i.e.

$$\bar{A}(\gamma) = \gamma^n + \bar{a}_{n-1}\gamma^{n-1} + \ldots + \bar{a}_0 \tag{3.7.5}$$

$$\bar{B}(\gamma) = \bar{b}_{n-1}\gamma^{n-1} + \ldots + \bar{b}_0 \tag{3.7.6}$$

and where $(1 + \Delta\gamma)\beta(\gamma)$ denotes the initial condition terms.

Equation (3.7.4) can be rearranged into the form:

$$Y_\delta(\gamma) = \bar{G}(\gamma)U_\delta(\gamma) + \frac{1 + \Delta\gamma}{\bar{A}(\gamma)}\beta(\gamma) \tag{3.7.7}$$

where $\bar{G}(\gamma)$ is given by

$$\bar{G}(\gamma) = \frac{\bar{B}(\gamma)}{\bar{A}(\gamma)} \tag{3.7.8}$$

If we are given $U_\delta(\gamma)$ plus initial conditions, then (3.7.7) can be used to evaluate $Y_\delta(\gamma)$. By use of partial fraction expansions and (3.6.9) we can then evaluate $\{y[k]\}$ as a sequence. Clearly, if $\bar{A}(\gamma)$ has repeated roots (3.6.9) will not be sufficient and one would need $D\{k^p e^{ak\Delta}\mu[k]\}$ for integer values for $p$.

We will illustrate the above discussion by using delta transforms to solve a difference equation.

**Example 3.7.1** Solve the following difference equation:

$$(\delta + 1)y[k] = u[k] \qquad (3.7.9)$$

where $u[k]$ is a unit step at time 0 and $y[0] = 0$.

**Solution** Taking transforms we have

$$(\gamma + 1)Y_\delta(\gamma) = U_\delta(\gamma) \qquad (3.7.10)$$

Using (3.6.17) we get

$$U_\delta(\gamma) = \frac{1 + \Delta\gamma}{\gamma} \qquad (3.7.11)$$

Hence

$$Y_\delta(\gamma) = \frac{1 + \Delta\gamma}{\gamma(\gamma + 1)}$$

or

$$\frac{Y_\delta(\gamma)}{1 + \Delta\gamma} = \frac{1}{\gamma(\gamma + 1)}$$

$$= \frac{1}{\gamma} - \frac{1}{\gamma + 1}$$

Hence

$$Y_\delta(\gamma) = \frac{1 + \Delta\gamma}{\gamma} - \frac{1 + \Delta\gamma}{\gamma + 1}$$

and using (3.6.9) and (3.6.17) we have

$$y[k] = \left[1 - (1 - \Delta)^k\right]u[k] \qquad (3.7.12)$$

$\wedge\wedge\wedge$

## 3.8    The Discrete Transfer Function

Referring again to equation (3.7.7), we note that for zero initial conditions, the delta transform of the forcing sequence $\{u[k]\}$ is related to the delta transform of the resulting sequence $\{y[k]\}$ as follows:

$$Y_\delta(\gamma) = \overline{G}(\gamma)U_\delta(\gamma) \qquad\qquad (3.8.1)$$

When the difference equation in (3.5.1) represents a model of a system, by analogy with the continuous case we call $\overline{G}(\gamma)$ the discrete transfer function of the system.

The discrete transfer function $\overline{G}(\gamma)$ can be given an alternative interpretation. We first recall from (3.6.19) that the delta transform of a $1/\Delta$ discrete pulse is unity. It then follows immediately from (3.8.1) that

---

*The discrete transfer function of a system is delta transform of the response (with zero initial conditions) to a discrete $1/\Delta$ pulse.*

---

The reader may note the similarity between this statement and the corresponding one for continuous-time systems : the continuous transfer function of a system is the Laplace transform of the response (with zero initial conditions) to a unit impulse.

## 3.9    Summary of Delta Transform Properties

For future reference, we summarize below various transform pairs such as those given in (3.6.17) and (3.6.19). We also present some useful properties of delta transforms.

| *Sequence* | *Delta transform* |
|:---:|:---:|
| $x[k]$ | $X_\delta(\gamma)$ |
| $y[k]$ | $Y_\delta(\gamma)$ |
| $ax[k] + by[k]$ | $aX_\delta(\gamma) + bY_\delta(\gamma)$ |
| $x[k - k_0]\mu[k - k_0]$ , $k_0 > 0$ | $(1 + \Delta\gamma)^{-k_0}X_\delta(\gamma)$ |
| $x[k] \otimes y[k]$ | $X_\delta(\gamma)Y_\delta(\gamma)$ |
| $\delta x[k]$ | $\gamma X_\delta(\gamma) - (1 + \Delta\gamma)x[0]$ |
| $\delta^{-1}x[k] = \sum_{m=0}^{k-1} \Delta x[m]$ | $\dfrac{1}{\gamma}X_\delta(\gamma)$ |
| $k\Delta x[k]$ | $-(1 + \Delta\gamma)\dfrac{d}{d\gamma}X_\delta(\gamma)$ |
| $\dfrac{x[k]}{k\Delta}$ | $\displaystyle\int_\gamma^\infty \dfrac{X_\delta(\eta)}{1 + \Delta\eta}\,d\eta$ |
| $(1 + \Delta a)^k x[k]$ | $X_\delta\left(\dfrac{\gamma - a}{1 + \Delta a}\right)$ |

| | |
|:---|:---|
| *Initial value theorem* | $\lim_{k\to 0} x[k] = \lim_{\gamma\to\infty}\left[\dfrac{\gamma X_\delta(\gamma)}{1 + \Delta\gamma}\right]$ |
| *Final value theorem* | if $\lim_{k\to\infty} x[k]$ exists then $\lim_{k\to\infty} x[k] = \lim_{\gamma\to 0}\gamma X_\delta(\gamma)$ |

*Table 3.9.1 : Properties of delta transform.*

| Sequence | Description | Delta transform |
|:---:|:---:|:---:|
| $p[k]$ | unit pulse | $\Delta$ |
| $\delta_d[k]$ | $1/\Delta$ pulse | $1$ |
| $\mu[k]$ | unit step | $\dfrac{1 + \Delta\gamma}{\gamma}$ |
| $k\Delta\mu[k]$ | ramp | $\dfrac{1 + \Delta\gamma}{\gamma^2}$ |
| $k^2\Delta^2\mu[k]$ | parabola | $\dfrac{(1 + \Delta\gamma)(2 + \Delta\gamma)}{\gamma^3}$ |
| $e^{ak\Delta}\mu[k]$ | exponential | $\dfrac{1 + \Delta\gamma}{\gamma - \left(\frac{e^{a\Delta}-1}{\Delta}\right)}$ |
| $(I + \Delta A)^k\mu[k]$ | | $(I + \Delta\gamma)(\gamma I - A)^{-1}$ |
| $k\Delta(I + \Delta A)^k\mu[k]$ | | $(1 + \Delta\gamma)(I + \Delta A)(\gamma I - A)^{-2}$ |
| $k^2\Delta^2(I + \Delta A)^k\mu[k]$ | | $(1 + \Delta\gamma)(2 + \Delta\gamma)(I + \Delta A)^2(\gamma I - A)^{-3}$ |
| $\sin(\omega k\Delta)$ | | $\dfrac{(1 + \Delta\gamma)\omega\ \text{sinc}(\omega\Delta)}{\gamma^2 + \Delta\phi(\omega,\Delta)\gamma + \phi(\omega,\Delta)}$ <br> where $\ \text{sinc}(\omega\Delta) = \dfrac{\sin(\omega\Delta)}{\omega\Delta}$ <br> and $\ \phi(\omega,\Delta) = \dfrac{2(1 - \cos(\omega\Delta))}{\Delta^2}$ |
| $\cos(\omega k\Delta)$ | | $\dfrac{(1 + \Delta\gamma)(\gamma + \frac{\Delta}{2}\phi(\omega,\Delta))}{\gamma^2 + \Delta\phi(\omega,\Delta)\gamma + \phi(\omega,\Delta)}$ |

*Table 3.9.2 : Delta transforms of some common sequences.*

| Delta transform | Description | Sequence |
|---|---|---|
| 1 | $1/\Delta$ – *pulse* | $\delta_d[k]$ |
| $\gamma^{-1}$ | *Delayed step* | $\mu[k-1]$ |
| $\gamma^{-2}$ | *Delayed ramp* | $(k-1)\Delta\mu[k-1]$ |
| $\gamma^{-3}$ | *Delayed parabola* | $\dfrac{1}{2}(k-1)(k-2)\Delta^2\mu[k-1]$ |
| $(\gamma I - A)^{-1}$ | *Delayed exponential* | $(I+\Delta A)^{k-1}\mu[k-1]$ |
| $(\gamma I - A)^{-2}$ | | $(k-1)\Delta(I+\Delta A)^{k-2}\mu[k-1]$ |
| $(\gamma I - A)^{-3}$ | | $\dfrac{1}{2}(k-1)(k-2)\Delta^2(I+\Delta A)^{k-3}\mu[k-1]$ |
| $\left(\gamma^2+\Delta\phi\gamma+\phi\right)^{-1}$  $\left(0<\Delta^2\phi<4\right)$ | *Delayed sine wave* | $\dfrac{1}{\omega\,\mathrm{sinc}(\omega\Delta)}\sin(\omega\Delta(k-1))\mu[k-1]$  $\omega = 1/\Delta\cos^{-1}\!\left(1-\left(\Delta^2\phi\right)/2\right)$ |
| $\dfrac{\gamma}{\gamma^2+\Delta\phi\gamma+\phi}$  $\left(0<\Delta^2\phi<4\right)$ | *Delayed cosine wave* | $\Big[\cos(\omega\Delta(k-1))-\dfrac{\Delta\phi}{2\omega\,\mathrm{sinc}(\omega\Delta)}$  $\sin(\omega\Delta(k-1))\Big]\cdot\mu[k-1]$ |

*Table 3.9.3 : Inverse delta transforms of common transforms.*

## 3.10  Stability of Discrete Systems

As for continuous-time systems, we will say that a discrete linear system is exponentially stable if the output decays (exponentially fast) to zero for all initial conditions and zero input.

Using the general formula for the solution of a discrete model, (3.7.7), the zero input initial condition response of the difference equation (3.5.1) is

$$Y_\delta(\gamma) = \frac{(1 + \Delta\gamma)\beta(\gamma)}{\overline{A}(\gamma)} \tag{3.10.1}$$

where $\beta(\gamma)$ depends on the initial conditions.

We assume that $\overline{A}(\gamma)$ takes the form:

$$\overline{A}(\gamma) = (\gamma - \overline{p}_1)(\gamma - \overline{p}_2) \cdots (\gamma - \overline{p}_n) \tag{3.10.2}$$

where $\overline{p}_1, \ldots, \overline{p}_n$ are the system poles. Then $Y_\gamma(\gamma)$ takes the form

$$\frac{Y_\delta(\gamma)}{1 + \Delta\gamma} = \frac{\beta(\gamma)}{(\gamma - \overline{p}_1)(\gamma - \overline{p}_2) \cdots (\gamma - \overline{p}_n)} \tag{3.10.3}$$

We assume for simplicity that $\overline{p}_1, \overline{p}_2, \ldots, \overline{p}_n$ are distinct. Then we can readily expand (3.10.3) via a partial fraction expansion to give

$$\frac{Y_\delta(\gamma)}{1 + \Delta\gamma} = \frac{c_1}{\gamma - \overline{p}_1} + \cdots + \frac{c_n}{\gamma - \overline{p}_n} \tag{3.10.4}$$

Hence

$$Y_\delta(\gamma) = \sum_{i=1}^{n} \frac{c_i(1 + \Delta\gamma)}{\gamma - \overline{p}_i} \tag{3.10.5}$$

Finally, using (3.6.9) we obtain

$$y[k] = \sum_{i=1}^{n} c_i [1 + \Delta\overline{p}_i]^k \mu[k] \tag{3.10.6}$$

We therefore conclude that the system modeled by equation (3.5.1) is (exponentially) stable if and only if all poles $\overline{p}_i$, satisfy $|1 + \Delta\overline{p}_i| < 1$ ; $i = 1, 2, \ldots, n$.

The location of poles for stability is therefore as in Figure 3.10.1.

We note that as $\Delta \rightarrow 0$, the discrete stability domain becomes the left half plane which we recall is the continuous-time stability domain.

## 3.11 Discrete Frequency Response

As in the continuous case, an input of particular interest is a simple sinusoid of the form:

$$u[k] = M \cos(\omega_0 k \Delta) \tag{3.11.1}$$

To analyze the response to this input it is more convenient to initially consider

$$u'[k] = M e^{j\omega_0 k \Delta} \tag{3.11.2}$$

and then to subsequently take the real part. Using (3.6.9), the corresponding discrete transform of (3.11.2) is

$$U'_\delta(\gamma) = \frac{M[1 + \Delta\gamma]}{\gamma - \left[\frac{e^{j\omega_0\Delta}-1}{\Delta}\right]} \tag{3.11.3}$$

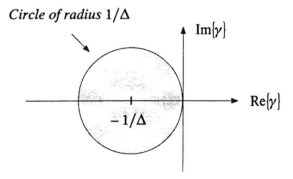

*Circle of radius* $1/\Delta$

*Figure 3.10.1 : Discrete stability domain.*

Now consider a discrete system modelled as in (3.5.1); i.e.

$$\bar{A}(\delta)y[k] = \bar{B}(\delta)u[k] \tag{3.11.4}$$

Using (3.7.7) the transform of the output response to the input given in (3.11.3) is

$$Y_{\delta}'(\gamma) = \bar{G}(\gamma) \left[ \frac{M(1 + \Delta\gamma)}{\gamma - \left[\frac{e^{j\omega_0\Delta}-1}{\Delta}\right]} \right] + \frac{(1 + \Delta\gamma)\beta(\gamma)}{\bar{A}(\gamma)} \tag{3.11.5}$$

where

$$\bar{G}(\gamma) = \frac{\bar{B}(\gamma)}{\bar{A}(\gamma)} \tag{3.11.6}$$

With $\bar{p}_1, \bar{p}_2, \ \dots \ , \bar{p}_n$ the zeros of $\bar{A}(\gamma)$ (assumed distinct for simplicity), (3.11.5) has a partial fraction expansion of the form:

$$\frac{Y_{\delta}'(\gamma)}{1 + \Delta\gamma} = \frac{c_0}{\gamma - \left[\frac{e^{j\omega_0\Delta}-1}{\Delta}\right]} + \sum_{i=1}^{n} \frac{c_i}{\gamma - \bar{p}_i} \tag{3.11.7}$$

where from the partial fraction expansion

$$c_0 = \bar{G}(\gamma)M \left. \right|_{\gamma = \frac{e^{j\omega_0\Delta}-1}{\Delta}} \tag{3.11.8}$$

Similar expressions hold for $c_1, \ \dots \ , c_n$ but the values of these constants will turn out to be unimportant in the sequel.

From (3.11.7) we immediately have

$$y'[k] = \left[ \bar{G}\left[\frac{e^{j\omega_0\Delta}-1}{\Delta}\right] M e^{j\omega_0 k\Delta} + \sum_{i=1}^{n} c_i(1 + \Delta\bar{p}_i)^k \right] \mu[k] \tag{3.11.9}$$

Assuming the system is stable, then all but the first term on the right hand side of (3.11.9) decays exponentially to zero leaving the final steady state response as

$$y'_{ss}[k] = \overline{G}\left[\frac{e^{j\omega_0\Delta} - 1}{\Delta}\right] M \, e^{j\omega_0 k\Delta} \, \mu[k] \qquad (3.11.10)$$

We call $\overline{G}(\dfrac{e^{j\omega\Delta} - 1}{\Delta})$ the *discrete frequency response*. Note that this is simply the discrete transfer function $\overline{G}(\gamma)$ evaluated on the stability boundary given in Figure 3.9.1 : $\gamma_\omega := \dfrac{e^{j\omega\Delta} - 1}{\Delta}$ for $\omega \in (0, \pi/\Delta)$.

Let $\overline{G}(\dfrac{e^{j\omega_0\Delta} - 1}{\Delta})$ be the complex number $|\overline{G}_0|e^{j\overline{\phi}_0}$ obtained when $\omega$ takes the particular value $\omega_0$. Then taking the real part in (3.11.10) gives the steady state response of the system to the input (3.11.1) as

$$y[k] = |\overline{G}_0|M \cos(\omega_0 k\Delta + \overline{\phi}_0) \qquad (3.11.11)$$

*We may then conclude that the steady state response of a stable linear discrete-time system having transfer function $\overline{G}(\gamma)$ to a sampled sinewave input of frequency $\omega$ is a sinusoidal output of the same frequency but whose magnitude has been multiplied by $|\overline{G}_0|$ and phase has been increased by $\overline{\phi}_0$ where*

$$|\overline{G}_0|e^{j\overline{\phi}_0} = \overline{G}(\gamma_\omega) \; ; \; \gamma_\omega = \frac{e^{j\omega\Delta} - 1}{\Delta}$$

We note in passing that

$$\frac{e^{j\omega\Delta} - 1}{\Delta} = j\omega \left[ 1 + \frac{j\omega\Delta}{2!} - \frac{\omega^2\Delta^2}{3!} + \cdots \right] \qquad (3.11.12)$$

It is clear from (3.11.12) that

$$\lim_{\Delta \to 0} \gamma_{\omega_0} = \lim_{\Delta \to 0} \left[ \frac{e^{j\omega_0\Delta} - 1}{\Delta} \right] = j\omega_0 \qquad (3.11.13)$$

thus recovering the well known continuous-time result.

Actually, for sampling rates greater than 20 times any frequency of interest (i.e. 20 times the system bandwidth) we have:

$$\frac{\omega\Delta}{2} < 0.15$$

and so the frequency response of a delta operator transfer function can be approximately found by substituting $\gamma = j\omega$. This is obviously intuitively appealing since it gives a direct link to the well known continuous-time result.

## 3.12  Frequency Domain Stability Criteria for Discrete-Time Systems

The reader may wonder if Nyquist stability type results can be obtained for discrete-time models. The extension is readily made by using the same argument as in continuous-time but taking account of the appropriate stability region as in Figure 3.10.1. Hence, all that is needed is to replace the usual $s$-domain curve which covers the continuous instability region (the right half-plane) by the curve given in Figure 3.12.1 which covers the discrete instability region.

As in the continuous-time case, we usually only need plot the value of

$\overline{G}(\gamma)$ for $\gamma = \dfrac{e^{j\omega\Delta} - 1}{\Delta}\, \omega \in (0, \pi/\Delta)$.

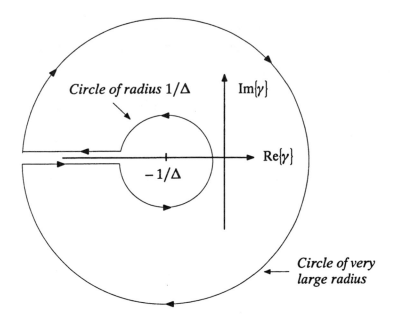

*Figure 3.12.1 : Discrete instability region.*

## 3.13   Digital Filter Implementation

In the introduction to this chapter, we suggested that it may sometimes be preferable, from a numerical point of view, to use the delta form rather than the shift form when implementing discrete transfer functions. An example of this arises when one is designing and implementing a digital filter. In this case, it is common to treat the input data as purely a sequence. However, as we have argued earlier, the sequence is typically the result of sampling a continuous signal at some rate.

Our claim is that implementing the same digital filter (irrespective of its form : low pass, high pass, notch etc.) with the sampling rate in mind and shifting the origin as is done in the delta operator formulation often results in a more robust implementation and improved performance under similar implementation constraints (e.g. finite word length for the filter coefficients). To prove this claim generally is beyond the scope of the present book. The reader is referred to the references given at the end of the chapter. However, to illustrate the claim we have chosen a simple example.

**Example 3.13.1**   Given a signal sampled at $\Delta = 0.001$ seconds, design a digital notch filter for a frequency of 10 Hz and compare the numerical accuracy in shift and delta form. (This problem arises in eccentricity compensation in gauge control in rolling mills - see references at the end of the chapter).

**Solution**  A simple way of achieving a notch filter centred at $\omega_0$ rad sec$^{-1}$ in the shift form is to place two zeros at $(\cos\omega_0\Delta \pm j\sin\omega_0\Delta)$. In our case, $\omega_0\Delta = 0.02\pi$ and hence the zeros should be at $\cos 0.02\pi \pm j\sin 0.02\pi$.

The poles are placed nearby but inside the stability domain. Thus we choose

$$\tilde{z}_1, \tilde{z}_2 = 0.99803 \pm j0.062791 \qquad (3.13.1)$$

$$\tilde{p}_1, \tilde{p}_2 = 0.99176 \pm j0.062483 \qquad (3.13.2)$$

The corresponding transfer function in *shift form* with coefficients rounded to five significant digits is

$$\tilde{F}(z) = \frac{z^2 - 1.9961z + 1}{z^2 - 1.9835z + 0.98748} \qquad (3.13.3)$$

The same filter can be converted to *delta form* by replacing $z$ by $\gamma\Delta + 1$ to obtain the following transfer function, again, rounded to five significant digits.

$$\bar{F}(\gamma) = \frac{\gamma^2 + 3.9465\gamma + 3946.5}{\gamma^2 + 16.488\gamma + 3972.1} \qquad (3.13.4)$$

Figure 3.13.1 shows the frequency response of the two implementations (3.13.3), (3.13.4).

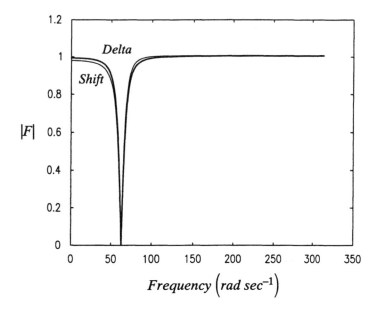

*Figure 3.13.1 : Frequency response of shift and delta filters with coefficient rounded to 5 significant digits.*

We next analyze the effect of rounding the coefficients to less digits. The significance of this is that it gives a measure of the sensitivity to the coefficients. Also, it has direct relevance to the problem of finite word-length computer implementations of the filter.

The coefficients were first rounded to four significant digits. The frequency response of the corresponding filters is shown in Figure 3.13.2. We see significant differences in performance for the shift case, whereas the performance of delta form remains essentially the same.

This trend is further highlighted when the respective coefficients are rounded to three significant digits as shown in Figure 3.13.3. Indeed, in this case, we see that the shift implementation now no longer acts as anything like a notch filter whereas the delta implementation again works well.

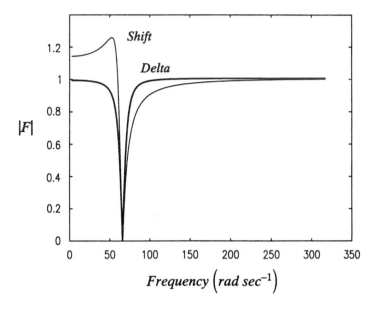

*Figure 3.13.2 : Frequency response of shift and delta filters with coefficients rounded to 4 significant digits.*

Rounding the coefficients of the delta form to two and one significant digits respectively gives the results in Figure 3.13.4, Figure 3.13.5. We see that the performance is still very close to that of the original implementation. Note however, that the shift implementation is actually *unstable* for the two significant digit case.

The principles used in the above example are quite common to many filter designs using infinite impulse response models. The important thing to note is that the differences in performance are entirely a result of the method of implementation (delta versus shift) and in no way reflect the design method.

## 3.14   Further Reading and Discussion

An interesting historical perspective on the development of discrete systems analysis is contained in the book by Åstrom and Wittenmark (1984). The

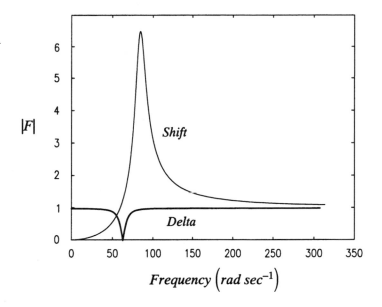

*Figure 3.13.3 : Frequency response of shift and delta filters with coefficients rounded to 3 significant digits.*

$z$-transform has a long history in the mathematics literature. For example, Lagrange (1759), (1792) and Laplace (1820) used the $z$-transform to study linear difference equations. The $z$-transform was first defined in the engineering literature by Hurewicz (1947). The transform was later defined as the $z$-transform by Ragazzini and Zadeh (1952). Early development of the theory occured in parallel in the Soviet Union, United States and United Kingdom. For example, Tsypkin (1950) called the transform the 'discrete Laplace transform' and Barker (1952) developed aspects of it. Much of the early development can be traced to Ph.D students of Ragazzini at Columbia in the 1950's.

The delta operator has its origins in the calculus of finite differences. Some brief comments on the historical development are made in Goodwin, Middleton and Poor (1992). This operator has its origins in the early 17th century where finite differences were used to obtain accurate interpolation formulae for logarithms. Later in the 18th century, Stirling (1730) proposed the discrete analog to

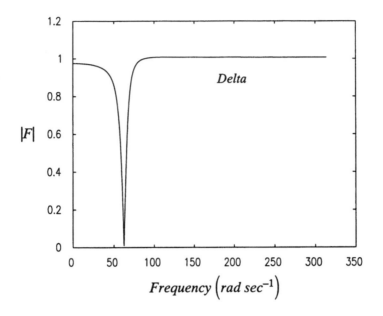

*Figure 3.13.4 : Frequency response of delta filter with*
*coefficients rounded to 2 significant digits.*
*(Note the shift filter is unstable in this case.)*

the anti-derivative or indefinite integral thus initiating divided differences (i.e. deltas) as a means of scaling finite differences of irregularly spaced data.

In the engineering literature, delta operators have been proposed at least as far back as the 1940s as a way of improving numerical properties of $z$-transforms when used with fast sampling (see for example James et. al. (1947), p.251). Also, there was early work linking $z$ and delta transforms – see for example Tschauner (1963) and Jury and Tschauner (1971). There has been a recent strong revival of interest in delta operators. This is due to several factors. Firstly, modern microprocessor technology has allowed much faster sampling rates to be employed than was previously feasible and this has led to renewed interest in numerical issues. Secondly, there is considerable current interest in relating discrete and continuous designs and the delta operator is irrefutably superior to the shift operator in this context.

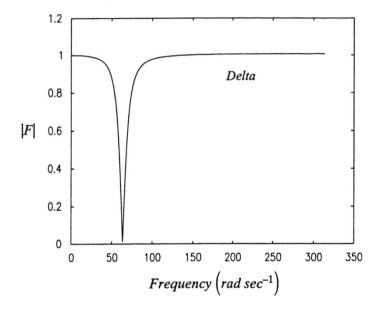

*Figure 3.13.5 : Frequency response of delta filter with*
*coefficients rounded to 1 significant digit.*
*(Note the shift filter is unstable in this case.)*

In the digital signal processing area, the delta operator has been proposed
as having low sensitivity and round-off noise performance – see Agarwal and
Burrus (1975) and Orlandi and Martinelli (1984). An interesting commentary on
the relative advantages of shift and delta operator implementation of digital filters is contained in Goodall (1990).

In the control literature, the delta operator has been used to unify continuous and discrete control theory – see Kitamori (1983). It has been used as a way
of improving numerical accuracy in microprocessor implementation of controllers – see Goodall and Brown (1985), Middleton and Goodwin (1986). Further
results on numerical and implementation issues such as finite word-length problems are discussed in Williamson (1991) and Gevers and Li (1993). The book
by Middleton and Goodwin (1990) shows how systems theory can be developed

in a unified fashion for both continuous-time and discrete-time systems using the delta operator.

Other results on the delta operator may be found in Mori et. al. (1987), Williamson (1988), Hori, Nikiforuk and Kanai (1988), Peterka (1986), Li and Gevers (1990), Kawake, Yamamura and Kanai (1990), Kanai, Kishimoto, Hori and Nikiforuk (1990), Hori, Kanai and Nikiforuk (1990), Neuman (1988), Soh (1991) and Premaratne and Jury (1992a), (1992b).

The rolling mill example discussed in Section 3.13 is described in more detail in Middleton and Goodwin (1990).

## 3.15  Problems

3.1  Derive the following properties of the $\delta$ operator :

(a)  $\delta(x[k]y[k]) = (\delta x[k])y[k] + x[k+1](\delta y[k])$    (equation (3.4.9))

$= (\delta x[k])y[k+1] + x[k](\delta y[k])$    (equation (3.4.10))

(b)  $A^{-1}[k] = -A^{-1}[k+1](\delta A[k])A^{-1}[k]$

$= -A^{-1}[k](\delta A[k])A^{-1}[k+1]$    (equation (3.4.11))

(c)  $\delta((I + \Delta A)^{-1}) = -A(I + \Delta A)^{-k-1} = -(I + \Delta A)^{-k-1}A$

(equation (3.4.15))

(d)  $\displaystyle\sum_{k=a}^{\beta-1} \Delta(\delta f[k])g[k] = f[\beta]g[\beta] - f[\alpha]g[\alpha] - \sum_{k=\alpha}^{\beta-1} \Delta f[k+1](\delta g[k])$

(equation (3.4.16))

3.2  Verify the transforms given in Table 3.9.2.

3.3  Verify the inverse transforms given in Table 3.9.3.

3.4  Show that the inverse transform of $(1 + \Delta\gamma)^{-n}$ is $\delta_d[k-n]$ where $\delta_d[k]$ is the $1/\Delta$ -pulse sequence.

3.5 Let $\underline{x}[k]$ satisfy the following equation :

$$\underline{x}[k+1] = \begin{bmatrix} \cos(\omega_0\Delta) & \dfrac{1}{\omega_0}\sin(\omega_0\Delta) \\ -\omega_0\sin(\omega_0\Delta) & \cos(\omega_0\Delta) \end{bmatrix} \underline{x}[k]$$

(a) Show that this equation in the delta operator form is :

$$\delta\underline{x}[k] = \begin{bmatrix} \dfrac{\Delta}{2}\phi(\omega_0, \Delta) & \mathrm{sinc}(\omega_0\Delta) \\ -\omega_0^2\mathrm{sinc}(\omega_0\Delta) & \dfrac{\Delta}{2}\phi(\omega_0, \Delta) \end{bmatrix} \underline{x}[k]$$

where $\phi(\omega, \Delta) = \dfrac{2(1-\cos(\omega_0\Delta))}{\Delta^2}$ , $\mathrm{sinc}(\omega_0\Delta) = \dfrac{\sin(\omega_0\Delta)}{\omega_0\Delta}$ .

(b) What happens to the two equations when $\Delta = 0$ ?

(c) What can you then conclude about the two equations for very fast sampling (very small $\Delta$ )?

3.6 Find the delta transform for the sequence

$$u[k] = 1, \ 1, \ -2, \ -2, \ 1, \ 1, \ -2, \ -2, \ ....$$

by two methods :

(a) directly, using the definition in equation (3.6.4); and

(b) by writing $u[k]$ in a discrete Fourier series form and using Table 3.9.2.

3.7 Find the inverse delta transforms for the following functions (assume $\Delta = 0.2$ ) :

(a) $\dfrac{1}{\gamma + 0.5}$

(b) $\dfrac{1}{(\gamma + 0.5)^2}$

(c) $\dfrac{1}{\gamma^2 + 0.1\gamma + 1}$

(d)  $\dfrac{\gamma}{(\gamma + 5)^2}$

3.8   Let  $X_\delta(\gamma) = \dfrac{\gamma^2 + \gamma - 2}{\gamma^3 + 2\gamma^2 + \gamma}$.  Find $x[k]$ for $\Delta = 0.5,\ 0.1,\ 0$.

3.9   The delta transform defined by (3.6.4) is the one-sided transform. The two-sided transform will then be defined by

$$D\{x[k]\} = \sum_{k=-\infty}^{\infty} \Delta(1 + \Delta\gamma)^{-1}x[k]$$

One also has to define the Region of Convergence (ROC) as part of the transform. Find the two-sided delta transforms and the corresponding ROCs for the following sequences :

(a)   $e^{a\Delta k}$

(b)   $a^k$

3.10  Consider a system modelled by the following difference equation (assume $\Delta = 0.1$ sec) :

$$\bar{A}(\delta)y[k] = \bar{B}(\delta)u[k]$$

where

$$\bar{A}(\delta) = \delta^2 + 10\delta + 79$$

$$\bar{B}(\delta = \delta + 8$$

(a)   Write the transfer function $\bar{G}(\gamma)$.

(b)   Calculate the frequency response $\bar{G}(\omega)$ and draw the Bode plots (namely, plot $20\log|\bar{G}(\omega)|$ versus $\omega$ and $\angle\bar{G}(\omega)$ versus $\omega$ ).

(c)   What would have changed in the plots in (b) if you were told that $\Delta = 0.05$ sec?

3.11  Consider the shift domain transfer function :

$$\bar{F}(z) = \frac{1}{z - 0.9995}$$

(a)  Calculate the step and frequency response.

(b)  Round the filter coefficients to 3 significant digits and repeat part (a). Comment on the result.

(c)  Truncate the filter coefficients to 3 significant digits and repeat part (a). Comment on the result.

(d)  Change the implementation to delta form with $\Delta = 0.01$ and repeat (a), (b) and (c). What are your conclusions?

3.12  Consider any shift domain stable transfer function having a pair of zeros at $\cos a \pm j \sin a$. Show that the steady-state frequency response to a sinewave of frequency, $f = \dfrac{a}{2\pi\Delta}$, is exactly zero.

# Chapter 4

# Discrete-Time Models of Continuous Deterministic Systems

## 4.1    Introduction

In Chapter 3 we considered only systems that connected an input *sequence* to an output *sequence*. However, it is often the case that we are interested in the sampled response of a *continuous-time* system. This will be the focus of the current chapter. We will show how the discrete-time transfer function relates to the continuous-time transfer function of the underlying system.

Depending on the nature of the continuous-time signal and the sampling rate, an anti-aliasing filter is usually applied to the signal before it is sampled. The resulting sampled sequence can then be viewed as the response of a purely discrete-time system to the input sequence. However, before we can evaluate the corresponding discrete-time transfer function we must specify how the input sequence is connected to the continuous system (i.e. the nature of the input hold) and the nature of the sampling process at the output.

In this chapter we consider the special case illustrated in Figure 4.1.1 where $G_p(s)$ is the continuous-time plant transfer function. For simplicity, we assume a zero-order-hold (ZOH) input and an anti-aliasing filter, $F(s)$ , prior to sampling. More general cases are treated in later chapters of the book.

It is often mathematically convenient to view both the hold part and the sample part in Figure 4.1.1 as each consisting of two parts. In the hold, the Dirac-

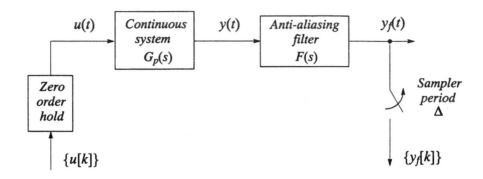

*Figure 4.1.1 : Typical connection of continuous-time
system with sampled input and output.*

pulse signal (see (1.6.9) and (2.3.1)) is first constructed from $\{u[k]\}$ using the

sampling period $\Delta$ , then this is passed through a continuous-time hold filter hav-

ing impulse response as defined in (2.5.1). Similarly, for the sampler on the other

end, the output is sampled by a $\Delta$ -impulse train (see (1.6.11)) to give a Dirac-

pulse signal out of which the corresponding sequence is extracted.

We recall that the operation of the ZOH on the input sequence is as shown

in Figure 4.1.2.

This chapter develops expressions for the discrete transfer function relat-

ing $\{y[k]\}$ to $\{u[k]\}$ as functions of the sampling period $\Delta$ , the continuous-

time system and anti-alaising filter transfer function.

We develop several alternative forms for the resulting discrete-time mod-

el. These all lead to the same final result but give different insights which will

be useful in our subsequent analysis.

## 4.2    State-Space Development

Given a state-space model for the continuous-time system and anti-alias-

ing filter, the conversion to discrete-time equivalent form proceeds as follows.

Let the state-space model for the combination of $G_p$ and $F$ be

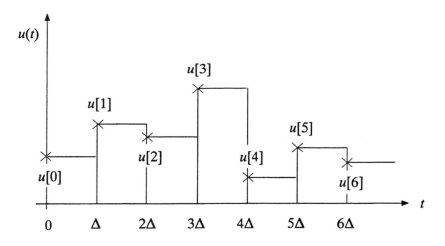

*Figure 4.1.2 : Operation of ZOH on sampled sequence*
*(×denotes sequence values).*

$$\frac{dx(t)}{dt} = Ax(t) + Bu(t) \qquad (4.2.1)$$

$$y_f(t) = Cx(t) \qquad (4.2.2)$$

Then, the sampled response to (4.2.1) with a zero-order-hold input is obtained by solving the equation over one sample period with constant input $u[k]$. This leads to

$$x[k+1] = e^{A\Delta}x[k] + A^{-1}\left[e^{A\Delta} - I\right]Bu[k] \qquad (4.2.3)$$

where $e^{A\Delta}$ is the matrix exponential $e^{A\Delta} = I + A\Delta + \frac{1}{2}A^2\Delta^2 + \cdots$

Equation (4.2.3) can be rearranged into the following form:

$$\boxed{\delta x[k] = A_\delta x[k] + B_\delta u[k]} \qquad (4.2.4)$$

$$y_f[k] = C_\delta x[k]$$

(4.2.5)

where

$$A_\delta = \frac{1}{\Delta}\left[e^{A\Delta} - I\right] = \left[I + \frac{A\Delta}{2!} + \frac{A^2\Delta^2}{3!} + \cdots\right] A$$

(4.2.6)

$$B_\delta = \frac{1}{\Delta}A^{-1}\left[e^{A\Delta} - I\right] B = \left[I + \frac{A\Delta}{2!} + \frac{A^2\Delta^2}{3!} + \cdots\right] B$$

(4.2.7)

$$C_\delta = C$$

(4.2.8)

In equation (4.2.3) we assumed that $A^{-1}$ existed. However, there is an equivalent expression which holds for general $A$. In particular, (4.2.4), (4.2.5) still hold with $A_\delta, B_\delta$ being given by the expressions on the right of (4.2.6) and (4.2.7) respectively.

There exists substantial literature on numerically robust procedures for evaluating the matrices $A_\delta, B_\delta$. It turns out that it is best not to use a power series as in (4.2.6), (4.2.7). However, since there now exists many packages which do this computation automatically (Matlab©, Matrix-x© etc.) we will not delve into these issues here.

An important observation from (4.2.6) to (4.2.8) is that

$$\lim_{\Delta \to 0} A_\delta = A$$

(4.2.9)

$$\lim_{\Delta \to 0} B_\delta = B$$

(4.2.10)

$$C_\delta = C$$

(4.2.11)

Hence, the discrete state-space model (in delta form) converges to the underlying continuous state-space model as the sampling period approaches zero.

Taking delta transforms of (4.2.4), (4.2.5) with zero initial conditions gives

$$(\gamma I - A_\delta)X_\delta(\gamma) = B_\delta U_\delta(\gamma)$$

$$Y_{f\delta}(\gamma) = C_\delta X_\delta(\gamma)$$

(4.2.12)

Hence, we immediately obtain:

---

*The discrete transfer function of a combined system and antialiasing filter having model* $dx(t)/dt = Ax(t) + Bu(t)$, $y(t) = Cx(t)$, *and ZOH input is*

$$\overline{G}_p(\gamma) = C_\delta(\gamma I - A_\delta)^{-1}B_\delta \qquad (4.2.13)$$

---

**Example 4.2.1** Consider a continuous-time plant having transfer function $G_p(s) = \dfrac{1}{\tau s + 1}$ with $F(s) = 1$ and ZOH input. Compute the corresponding discrete transfer function.

**Solution** A state-space model corresponding to $G_p(s)$ is

$$\frac{d}{dt}x(t) = -\frac{1}{\tau}x(t) + \frac{1}{\tau}u(t)$$

$$y(t) = x(t)$$

(4.2.14)

Hence, applying (4.2.6) and (4.2.7) we obtain

$$A_\delta = \frac{e^{-\frac{\Delta}{\tau}} - 1}{\Delta} \;,\quad B_\delta = \frac{1 - e^{-\frac{\Delta}{\tau}}}{\Delta} \;,\quad C_\delta = 1 \qquad (4.2.15)$$

and thus, from (4.2.13) the discrete transfer function is

$$\overline{G}_p(\gamma) = \frac{\frac{1-e^{\frac{\Delta}{\tau}}}{\Delta}}{\gamma + \frac{1-e^{\frac{\Delta}{\tau}}}{\Delta}} \tag{4.2.16}$$

As expected, we see that

$$G_p(s)F(s) = \lim_{\Delta \to 0} \overline{G}_p(s) \tag{4.2.17}$$

△△△

## 4.3    Transform Development

An alternative way of developing the discrete-time model linking $u[k]$ to $y_f[k]$ in Figure 4.1.1 is to use Laplace transform techniques. Before we do this, we replace the sequences $\{y_f[k]\}$ and $\{u[k]\}$ by Dirac-pulse signals. These, as we recall, are defined by

$$
\begin{aligned}
u^s(t) &= \sum_{k=-\infty}^{\infty} \Delta u[k]\delta(t-k\Delta) \\[2mm]
y_f^s(t) &= \sum_{k=-\infty}^{\infty} \Delta y_f[k]\delta(t-k\Delta)
\end{aligned}
\tag{4.3.1}
$$

In defining these signals we have introduced a scaling by $\Delta$ (see also the discussion leading to (1.6.9)). We note that $u[k]$ is only relevant to the input over the interval $\left(k\Delta \leq t < (k+1)\Delta\right)$. Thus, the scaling by $\Delta$ keeps the 'energy' in $u^s(t)$ roughly the same as in the zero-order-hold form of $u[k]$. This allows us to maintain a close connection between continuous and discrete signals.

By direct calculation, the Laplace transforms of $u^s(t)$ and $y_f^s(t)$ are as follows:

$$U^s(s) = L\{u^s(t)\}$$

$$= \int_0^\infty e^{-st} \left[ \sum_{k=-\infty}^\infty \Delta u[k]\delta(t - k\Delta) \right] ds$$

$$= \sum_{k=0}^\infty \Delta e^{-sk\Delta} u[k] \qquad\qquad (4.3.2)$$

and similarly,

$$Y_f^s(s) = \sum_{k=0}^\infty \Delta e^{-sk\Delta} y_f[k] \qquad\qquad (4.3.3)$$

Expressions (4.3.2), (4.3.3) are interesting since they reveal the connection between the delta transforms of $\{u[k]\}$, $\{y_f[k]\}$ and the Laplace transforms of the corresponding Dirac-pulse signals $u^s(t)$ and $y_f^s(t)$. In particular, from (4.3.2), (4.3.3) and (3.6.4) we have

$$\left. U^s(s) = U_\delta(\gamma) \right|_{\gamma = \dfrac{e^{s\Delta} - 1}{\Delta}} \qquad\qquad (4.3.4)$$

and

$$\left. Y_f^s(s) = Y_{f\delta}(\gamma) \right|_{\gamma = \dfrac{e^{s\Delta} - 1}{\Delta}} \qquad\qquad (4.3.5)$$

where $U_\delta(\gamma), Y_{f\delta}(\gamma)$ are the delta transforms of the sequences $\{u[k]\}$ and $\{y_f[k]\}$ respectively.

We observe that if $\{y_f[k]\}$ and $\{u[k]\}$ are related by a discrete transfer function $C$, i.e.

$$U_\delta(\gamma) = C(\gamma)Y_{f\delta}(\gamma)$$

(4.3.6)

then it is clear from (4.3.4), (4.3.5) that the Laplace transforms of the corresponding Dirac-pulse signals $u^s(t)$ and $y_f^s(t)$ must be related as follows :

$$U^s(s) = C(\gamma)Y_f^s(s) \; ; \; \gamma = \frac{e^{s\Delta} - 1}{\Delta}$$

(4.3.7)

This property will be exploited in Chapter 9 when we consider the action of a discrete controller on a continuous-time plant.

Returning to Figure 4.1.1 we next describe the action of the hold circuit. We recall from Section 2.5.1 that the ZOH has impulse response given by

$$h_0(t) = \begin{cases} 1/\Delta & \text{for } 0 \le t \le \Delta \\ 0 & \text{elsewhere} \end{cases}$$

and the corresponding transfer function is

$$H_0(s) = \frac{1 - e^{-s\Delta}}{s\Delta}$$

(4.3.8)

The operation of the above filter on a single scaled Dirac pulse is described in Figure 4.3.1.

Hence, if the input is the Dirac-pulse signal given in (4.3.1) then the output is precisely the continuous-time, piecewise constant signal illustrated in Figure 4.1.2.

In summary, the system shown in Figure 4.1.1 is equivalent to the representation, shown in Figure 4.3.2 which is described in terms of the Laplace transform variable $s$.

We next proceed to evaluate the equivalent discrete transfer function linking $u^s(t)$ and $y^s(t)$.

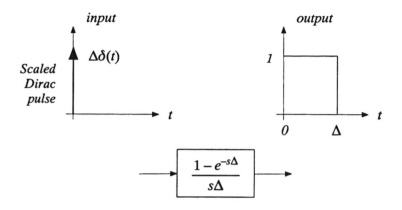

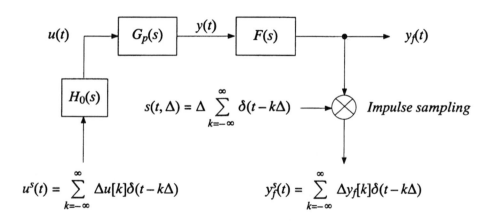

*Figure 4.3.1 : ZOH operation on a single Dirac pulse.*

*Figure 4.3.2 : The equivalent continuous-time configuration of the system shown in Figure 4.1.1.*

In Chapter 3 we have seen that the discrete transfer function of a system is the delta transform of the response of the system to a discrete $1/\Delta$ - pulse, i.e. $\delta_d[k]$. However, the special input sequence, $\delta_d[k]$, is equivalent to a Dirac-pulse signal of the form

$$u^s(t) = \sum_{k=-\infty}^{\infty} \Delta \delta_d[k]\delta(t - k\Delta)$$

$$= \Delta \left[1/\Delta\right] \delta(t)$$

$$= \delta(t) \tag{4.3.9}$$

Thus, when $u[k] = \delta_d[k]$, the corresponding output of the filter, $y_f(t)$, in Figure 4.3.2 is simply the continuous-time impulse response of $G_p(s)F(s)H_0(s)$, i.e. the response to $\delta_d[k]$ is

$$y_f(t) = L^{-1}\{F(s)G_p(s)H_0(s)\} \tag{4.3.10}$$

Let $S_\Delta$ denote the sampling operation with period $\Delta$ and $D$ the delta transform operation. Then the corresponding transform of the sampled output response for $u[k] = \delta_d[k]$ is

$$Y_{f\delta}(\gamma) = D\left\{S_\Delta\left\{L^{-1}\{F(s)G_p(s)H_0(s)\}\right\}\right\} \tag{4.3.11}$$

By this notation we mean : take the inverse Laplace transform, sample the resulting continuous-time signal and apply the delta transform on the resulting sequence (see Example 4.3.1).

We thus conclude

---

*The discrete transfer function of a continuous plant with transfer function $G_p(s)$ having hold circuit $H_0(s)$, and anti-aliasing filter $F(s)$ is*

$$\overline{G}(\gamma) = D\left\{S_\Delta\left\{L^{-1}\{F(s)G_p(s)H_0(s)\}\right\}\right\}$$

---

The above procedure is illustrated by two simple examples.

**Example 4.3.1** Consider again Example 4.2.1 where

$$G(s) = F(s)G_p(s)H_0(s) = \frac{1 - e^{-s\Delta}}{(\tau s + 1)\Delta s} \tag{4.3.12}$$

Derive the corresponding discrete transfer function.

**Solution** The impulse response of $G(s)$ is

$$g(t) = L^{-1}\{G(s)\} = \frac{1}{\Delta}\left[\left(1 - e^{-\frac{t}{\tau}}\right)\mu(t) - \left(1 - e^{-\frac{(t-\Delta)}{\tau}}\right)\mu(t - \Delta)\right] \tag{4.3.13}$$

Sampling $g(t)$ with sampling period $\Delta$ we get

$$S_\Delta\{g(t)\} = g[k] = \frac{1}{\Delta}\left[\left(1 - e^{-\frac{k\Delta}{\tau}}\right)\mu[k] - \left(1 - e^{-\frac{(k-1)\Delta}{\tau}}\right)\mu[k-1]\right]$$

Evaluating the delta transform of this sequence leads to

$$\bar{G}(\gamma) = D\{g[k]\} = \frac{1}{\Delta}\left[1 - \frac{1}{1 + \Delta\gamma}\right]\left[\frac{1 + \Delta\gamma}{\gamma} - \frac{1 + \Delta\gamma}{\gamma - \frac{e^{\frac{-\Delta}{\tau}} - 1}{\Delta}}\right]$$

$$= \frac{\frac{1 - e^{\frac{-\Delta}{\tau}}}{\Delta}}{\gamma + \frac{1 - e^{\frac{-\Delta}{\tau}}}{\Delta}} \tag{4.3.14}$$

exactly as in (4.2.16).

∿∿∿

**Example 4.3.2**   Consider a continuous-time system having the following model:

$$F(s)G_p(s) = \frac{-s+5}{s^4 + 23s^3 + 185s^2 + 800s + 2500} \qquad (4.3.15)$$

Let $\Delta = 0.01$ sec (sampling frequency 100 Hz). Derive the corresponding discrete transfer function using a ZOH input.

**Solution**   Repeating the procedure of calculating $G(s) = F(s)G_p(s)H_0(s)$, $g(t) = L^{-1}\{G(s)\}$, $g[k] = S_\Delta\{g(t)\}$ and finally $\overline{G}(\gamma) = D\{g[k]\}$ we obtain the following result (accurate to two significant digits) :

$$\overline{G}(\gamma) \simeq \frac{-0.80\gamma + 4.5}{\gamma^4 + 22\gamma^3 + 180\gamma^2 + 760\gamma + 2200} \qquad (4.3.16)$$

We note that (4.3.16) is quite similar to its continuous-time counterpart in (4.3.15). To emphasize this point we may write (4.3.16) using the shift operator and $z$-transform and get (see (3.6.14)) :

$$\tilde{G}(z) = \frac{1}{\Delta}\overline{G}(\gamma)\Big|_{\gamma = \frac{z-1}{\Delta}}$$

$$= \frac{-10^{-6}(0.15537z^3 + 0.41605z^2 - 0.47402z - 0.142)}{z^4 - 3.7777z^3 + 5.3506z^2 - 3.3674z + 0.79453} \qquad (4.3.17)$$

which bears no apparent resemblance to (4.3.15).

We will return to this example in Section 4.5 when we discuss numerical issues.

$$\triangle\triangle\triangle$$

To facilitate the evaluation of ZOH discrete equivalent of systems, we list below in tabular form several commonly encountered continuous-time transfer functions and their ZOH discrete-time equivalents.

| $F(s)G_p(s)$ | $\overline{G}(\gamma)$ |
|---|---|
| $\dfrac{1}{s}$ | $\dfrac{1}{\gamma}$ |
| $\dfrac{1}{s^2}$ | $\dfrac{\frac{\Delta}{2}\gamma + 1}{\gamma^2}$ |
| $\dfrac{1}{s^3}$ | $\dfrac{\frac{1}{6}(\Delta\gamma) + \Delta\gamma + 1}{\gamma^3}$ |
| $\dfrac{1}{s+a}$ | $\dfrac{\alpha}{a(\gamma + \alpha)} \; ; \; \alpha = \dfrac{1 - e^{-a\Delta}}{\Delta}$ |
| $\dfrac{1}{(s+a)^2}$ | $\dfrac{(\alpha - ae^{-a\Delta})\gamma + \alpha^2}{a^2(\gamma + \alpha)^2} \; ; \; \alpha = \dfrac{1 - e^{-a\Delta}}{\Delta}$ |
| $\dfrac{\omega}{(s+a)^2 + \omega^2}$ | $\dfrac{\left[\frac{\omega(1-e^{-a\Delta}\cos\omega\Delta)-ae^{-a\Delta}\sin\omega\Delta}{\Delta}\right]\gamma + \omega\left[\frac{1-2e^{-a\Delta}\cos\omega\Delta+e^{-2a\Delta}}{\Delta^2}\right]}{(a^2 + \omega^2)\left[\gamma^2 + \left[\frac{2-2e^{-a\Delta}\cos\omega\Delta}{\Delta}\right]\gamma + \left[\frac{1-2e^{-a\Delta}\cos\omega\Delta+e^{-2a\Delta}}{\Delta^2}\right]\right]}$ |
| $\dfrac{s}{(s+a)^2 + \omega^2}$ | $\dfrac{\left[e^{-a\Delta}\left(\frac{\sin\omega\Delta}{\omega\Delta}\right)\right]\gamma}{(a^2 + \omega^2)\left[\gamma^2 + \left[\frac{2-2e^{-a\Delta}\cos\omega\Delta}{\Delta}\right]\gamma + \left[\frac{1-2e^{-a\Delta}\cos\omega\Delta+e^{-2a\Delta}}{\Delta^2}\right]\right]}$ |

*Table 4.3.1 : ZOH discrete-time equivalent transfer functions.*

## 4.4     Continuous-Time and Discrete-Time Poles and Zeros

We have seen so far two methods, namely state-space and transform, for deriving the discrete-time transfer function corresponding to a continuous-time system with a ZOH at the input and a sampler at the output (Figure 4.1.1). It is, however, well known that systems with rational transfer functions are characterized by their poles and zeros. Thus in this section we will explain the relationships between the poles and zeros of the continuous-time model and those of its discrete-time counterpart.

### 4.4.1   Poles

Let us consider the system described by equations (4.2.1) – (4.2.8). The corresponding continuous-time transfer function is

$$F(s)G_p(s) = C(sI - A)^{-1}B \tag{4.4.1}$$

and the discrete-time transfer function, as shown previously in (4.2.13), is

$$\overline{G}_p(\gamma) = C_\delta(\gamma I - A_\delta)^{-1}B_\delta .$$

The poles of $F(s)G_p(s)$ and $\overline{G}_p(\gamma)$ are thus the eigenvalues of the matrices $A$ and $A_\delta$ respectively. Hence, using (4.2.6), if $p_1, \ldots, p_n$ are the poles of $F(s)G_p(s)$ (i.e. the eigenvalues of $A$) and $\overline{p}_1, \ldots, \overline{p}_n$ are the poles of $\overline{G}_p(\gamma)$ (i.e. the eigenvalues of $A_\delta$ ) we have the following relationship between discrete poles $(\overline{p}_i)$ and continuous poles $(p_i)$ :

$$\boxed{\overline{p}_i = \frac{1}{\Delta}\left(e^{p_i\Delta} - 1\right)} \tag{4.4.2}$$

Thus, for every continuous-time pole there is a corresponding discrete-time pole which can be calculated via (4.4.2). It is quite illuminating to study the mapping defined by (4.4.2).

For convenience, let us denote the continuous plane variable $s$ by

$$s = \sigma + j\omega$$

(4.4.3)

and the discrete plane variable $\gamma$ by

$$\gamma = \frac{1}{\Delta}\left(Me^{j\phi} - 1\right)$$

(4.4.4)

Then, via the mapping (4.4.2), we must have

$$M = e^{\sigma\Delta}$$

(4.4.5)

$$\phi = \omega\Delta$$

(4.4.6)

This mapping is clearly not one-to-one since every infinite strip $(2k-1)\pi/\Delta < \omega \leq (2k+1)\pi/\Delta$ maps onto the entire $\gamma$ plane. Typically, however, the sampling period $\Delta$ is chosen so that all the poles of the continuous-time transfer function are inside the strip $-\pi/\Delta < \omega \leq \pi/\Delta$ to assure a one-to-one relationship with their discrete-time counterparts.

Some typical results of this mapping are shown pictorially in Table 4.4.1 for various $s$-plane curves. These results were obtained from (4.4.4) to (4.4.6). For example, the top right hand plot shows $\gamma = (1/\Delta)\, Me^{j\overline{\phi}} - 1/\Delta$ with $\overline{\phi} = \overline{\omega}\Delta$. Notice that this is a straight line of slope $\overline{\omega}\Delta$ translated left by an offset of $-1/\Delta$. The other curves are obtained by similar reasoning.

One can see in Table 4.4.1 that as $\Delta \rightarrow 0$ each of the $\gamma$-plane curves converges to the corresponding $s$-plane curve. This is as expected since

$$\lim_{\Delta \to 0} \frac{1}{\Delta}(e^{s\Delta} - 1) = s$$

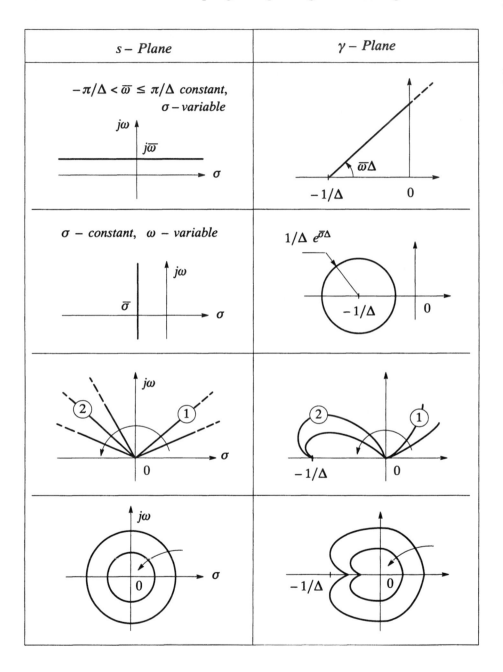

*Table 4.4.1 :   Corresponding s-plane and γ-plane curves mapped*
*via the transformation  $\gamma = 1/\Delta(e^{s\Delta} - 1)$.*

**Example 4.4.1** : Consider a continuous-time system with the following transfer function

$$F(s)G_p(s) = \frac{\omega_0^2}{s^2 + 2\zeta\omega_0 s + \omega_0^2} \tag{4.4.7}$$

Derive the corresponding discrete-time transfer function with sampling period $\Delta$. Relate the continuous and discrete pole locations.

**Solution** Sampling with ZOH at a period $\Delta$ results in

$$\bar{G}(\gamma) = \frac{\omega_0^2\left\{\gamma\left[p_1 \frac{1-e^{p_2\Delta}}{\Delta} - p_2 \frac{1-e^{p_1\Delta}}{\Delta}\right] + \frac{1}{\Delta^2}(1-e^{p_1\Delta})(1-e^{p_2\Delta})(p_1-p_2)\right\}}{p_1 p_2(p_1-p_2)\left[\gamma^2 + \frac{1}{\Delta}(2-e^{p_1\Delta}-e^{p_2\Delta})\gamma + \frac{1}{\Delta^2}(1-e^{p_1\Delta})(1-e^{p_2\Delta})\right]} \tag{4.4.8}$$

where

$$p_{1,2} = \omega_0\left[-\zeta \pm j\sqrt{1-\zeta^2}\right] \tag{4.4.9}$$

are the continuous-time poles and

$$\bar{p}_i = \frac{e^{p_i\Delta} - 1}{\Delta} \qquad i = 1, 2$$

are the corresponding discrete-time poles.

Using (4.4.2) to (4.4.6) one can readily construct the constant damping $\zeta$ and constant natural frequency $\omega_0$ loci as shown in Figure 4.4.1. In the figure, we have plotted the loci corresponding to $\zeta = \sin\theta, \theta = \{0, \pi/20, \ldots, 9\pi/20\}$, $\omega_0 = \{0, \pi/10, \ldots, 9\pi/10\}$ and $\Delta = 1$.

We see from the diagram that, in the vicinity of the origin, the discrete and continuous poles are in close agreement whereas they depart from each other away from the origin.

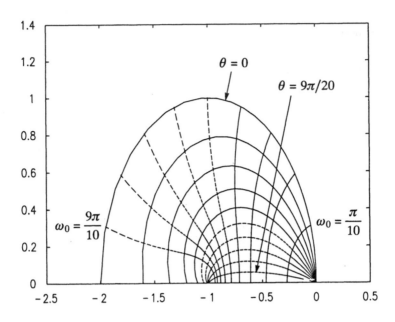

*Figure 4.4.1 : Loci of constant $\zeta$ and $\omega_0$ in the $\gamma$ plane.*

$\wedge\wedge\wedge$

### 4.4.2   Zeros

Returning to Example 4.4.1 we notice that while the continuous-time transfer function has no finite zeros its discrete counterpart does have a finite zero. This means that the sampling process may create zeros. In fact, we argue below that, generically, discrete-time models of a sampled $n$-th order continuous-time system will have $n-1$ finite zeros. Writing the continuous-time transfer function in its partial fraction form we have (assuming, for simplicity, distinct poles)

$$F(s)G_p(s) = \sum_{i=1}^{n} \frac{\alpha_i}{s - p_i} \tag{4.4.10}$$

Then, from Table 4.3.1 we have

$$\overline{G}(\gamma) = \sum_{i=1}^{n} \frac{\alpha_i \overline{p}_i}{p_i(\gamma - \overline{p}_i)}, \quad \overline{p}_i = \frac{1 - e^{-p\Delta}}{\Delta} \tag{4.4.11}$$

Since the $\overline{p}_i$ depend on the sampling period $\Delta$, one can see that the numerator resulting from the summation in (4.4.11) will, in general, be an $(n-1)$th order polynomial the coefficients of which are functions of $\Delta$. Some of these coefficients may be zero for some values of $\Delta$ but, generically, for an arbitrary choice of $\Delta$, the numerator is of order $(n-1)$. So

*The discrete-time equivalent of an nth order continuous-time system with ZOH has, generically, $(n-1)$ zeros.*

While there exists a general and simple formula (4.4.11) which maps the continuous-time poles into their discrete-time equivalents, nothing of this kind exists for the zeros. However, (4.2.9)–(4.2.11) imply that if

$$F(s)G_p(s) = \frac{B(s)}{A(s)} \tag{4.4.12}$$

and

$$\overline{G}(\gamma) = \frac{\overline{B}(\gamma)}{\overline{A}(\gamma)} \tag{4.4.13}$$

then

$$\lim_{\Delta \to 0} \overline{B}(s) = B(s) \tag{4.4.14}$$

and

$$\lim_{\Delta \to 0} \overline{A}(s) = A(s)$$

This means that if $B(s)$ is of degree $m < n$, $m$ of the $(n-1)$ zeros of $\overline{G}(\gamma)$ converge to the $m$ zeros of $F(s)G_p(s)$ as $\Delta \to 0$. This also means that the first

$(n-1-m)$ coefficients of $\bar{B}(\gamma)$ go to zero as $\Delta \to 0$ which in turn implies that the remaining $(n-1-m)$ zeros of $\bar{G}(\gamma)$ go to infinity as $\Delta \to 0$. Åstrom, Hagander and Sternby (1984) investigated the behaviour of these additional zeros, commonly referred to as *sampling zeros*, as $\Delta \to 0$. Their key result is restated in Middleton and Goodwin (1990) as follows:

**Lemma 4.4.1** Consider a continuous-time transfer function

$$F(s)G_p(s) = \frac{K\prod\limits_{i=1}^{m}(s-z_i)}{\prod\limits_{i=1}^{n}(s-p_i)}$$

where $n>m$. Then, the corresponding discrete-time transfer function has generically $(n-1)$ zeros, $\bar{z}_i$, that satisfy

(i)     $$\lim_{\Delta \to 0} \bar{z}_i = \lim_{\Delta \to 0} \frac{e^{z_i\Delta}-1}{\Delta} = z_i \quad i = 1, \ldots, m \qquad (4.4.15)$$

(ii)    $$\lim_{\Delta \to 0}\left\{(\gamma-\bar{z}_{m+1})(\gamma-\bar{z}_{m+2}) \cdots (\gamma-\bar{z}_{n-1}) - \Delta^{-(r-1)}C_r(\Delta\gamma)\right\} = 0 \qquad (4.4.16)$$

where $r = n-m$ and $C_r(x)$ denotes a polynomial of degree $(r-1)$ given by

$$C_r(x) = C_1^r x^{r-1} + C_2^r x^{r-2} + \ldots + C_r^r \qquad (4.4.17)$$

and where the coefficients satisfy:

$$C_k^r = \sum_{i=1}^{k}\sum_{l=1}^{i}(-1)^{i-l}\begin{bmatrix}r-i\\k-i\end{bmatrix}\begin{bmatrix}r+1\\i-l\end{bmatrix}l^r \qquad (4.4.18)$$

The first few polynomials $C_r(\Delta\gamma)$ are:

$$C_1(\Delta\gamma) = 1$$

$$C_2(\Delta\gamma) = \Delta\gamma + 2$$

$$C_3(\Delta\gamma) = (\Delta\gamma)^2 + 6(\Delta\gamma) + 6$$

$$C_4(\Delta\gamma) = (\Delta\gamma)^3 + 14(\Delta\gamma)^2 + 36(\Delta\gamma) + 24$$

$$C_5(\Delta\gamma) = (\Delta\gamma)^4 + 30(\Delta\gamma)^3 + 150(\Delta\gamma)^2 + 240(\Delta\gamma) + 120$$

and

(iii) $$\lim_{\Delta\to 0} \bar{z}_i = -\infty \quad \text{for} \quad i = m+1, \ \ldots, \ n-1$$

$\wedge\wedge\wedge$

(For the proof of this lemma see Åström, Hagander and Sternby (1984)).

**Example 4.4.2** : Investigate the zeros of the system in Example 4.4.1.

**Solution**    In Example 4.4.1 we saw that the discrete-time poles are $1/\Delta(e^{p_i\Delta} - 1)$. $i = 1,2$ as expected, while there appears one finite sampling zero. Using the approximation $e^x \simeq 1 + x + x^2/2$ we get for the numerator of (4.4.8) to be

$$\gamma p_1 \frac{1 - e^{p_2\Delta}}{\Delta} - \gamma p_2 \frac{1 - e^{p_1\Delta}}{\Delta} + \frac{1}{\Delta^2}\left(1 - e^{p_1\Delta} - e^{p_2\Delta} + e^{(p_1+p_2)\Delta}\right)(p_1 - p_2)$$

$$\simeq \frac{p_1 p_2 (p_1 - p_2)}{2}\gamma\Delta + p_1 p_2 (p_1 - p_2)$$

$$\simeq \frac{p_1 p_2 (p_1 - p_2)}{2}[\gamma\Delta + 2] \tag{4.4.19}$$

as expected from Lemma 4.4.1. So the sampling zero in this example, for small $\Delta$, will be approximately $-2/\Delta$ and as $\Delta \to 0$ it will go to $-\infty$, again, as stated in Lemma 4.4.1.

$\wedge\wedge\wedge$

In the following example we show how the discrete-time poles and zeros vary with the sampling period.

**Example 4.4.3** : Investigate the discrete zeros corresponding to the following continuous-time system

$$F(s)G_p(s) = \frac{s^2 + s + 1}{s(s + 1)(s^2 + 4)}$$

**Solution** In Table 4.4.2 we show how the discrete-time poles and zeros vary with $\Delta$. $\Delta$ takes on the values $0, 0.1, 0.2, 0.5, 1.0, 2.0, 5.0, 10.0$. Note that for $\Delta = 0.0$ we get the continuous-time poles and zeros.

⌇⌇⌇

## 4.5 Numerical Issues

The discussion in Section 4.4 touched upon a very important advantage that the delta operator has over the commonly used shift operator : the discrete transfer function in delta form is close to the continuous transfer function for small $\Delta$. Many algorithms are thus better conditioned using delta operator implementation than shift operator implementation. This is due to the fact that as sampling rates increase, the poles and zeros of a model represented using shift operators tend to cluster about the point $z = 1$. Thus, the shift operator discrete-time state transition matrix $A_q$ tends to the identity matrix. This may be seen by noting that if $A$ is the continuous-time state transition matrix then $A_q$ is given by:

$$A_q = e^{A\Delta} = I + \frac{A\Delta}{1!} + \frac{(A\Delta)^2}{2!} + \frac{(A\Delta)^3}{3!} + \dots$$

The dynamics of the system will be captured by the fractional part of the entries of $A_q$, but in floating point computer implementation much of the avail-

| | $\bar{z}_1$ | $\bar{z}_2, \bar{z}_3$ | $\bar{p}_1$ | $\bar{p}_2$ | $\bar{p}_3, \bar{p}_4$ |
|---|---|---|---|---|---|
| $\Delta = 0$ | $-\infty$ | $-0.5 \pm j0.87$ | $0$ | $-1$ | $\pm j2$ |
| $\Delta = 0.1$ | $-20$ | $-0.52 \pm j0.82$ | $0$ | $-0.95$ | $-0.2 \pm j1.99$ |
| $\Delta = 0.2$ | $-10$ | $-0.54 \pm j0.78$ | $0$ | $-0.91$ | $-0.4 \pm j1.95$ |
| $\Delta = 0.5$ | $-4$ | $-0.59 \pm j0.65$ | $0$ | $-0.79$ | $-0.92 \pm j1.68$ |
| $\Delta = 1.0$ | $-2.02$ | $-0.6 \pm j0.46$ | $0$ | $-0.63$ | $-1.42 \pm j0.91$ |
| $\Delta = 2.0$ | $-0.87$ | $-0.48 \pm j0.27$ | $0$ | $-0.43$ | $-0.83 \pm j0.38$ |
| $\Delta = 5.0$ | $-0.28$ | $-0.29 \pm j0.066$ | $0$ | $-0.2$ | $-0.37 \pm j0.11$ |
| $\Delta = 10.0$ | $-0.11$ | $-0.06 \pm j0.087$ | $0$ | $-0.1$ | $-0.059 \pm j0.091$ |

*Table 4.4.2 :  Discrete pole and zero locations*
*for different sample periods.*

able wordlength will be used in storing the unity part of $A_q$. Consequently, at sufficiently high sampling rate the dynamics of the filter will be lost since the remaining part will be too small to be represented with the wordlength available.

As we have seen, using the delta operator we avoid this problem. The poles of delta operator models tend to their continuous-time values as the sampling rate increases. Using delta operators, the states of a filter are updated according to (3.5.8). Evidently the dynamics of the filter are contained in the matrix $A_\delta$,

where $A_\delta = A \left[ I + \dfrac{A\Delta}{2!} + \dfrac{(A\Delta)^3}{3!} + \ \dots \ \right]$ is the state transition matrix in the

delta model. The full wordlength of the computer may be used to store the important fractional entries in $A_\delta$. Eventually, as the sampling rate increases, finite wordlength effects will cause $\Delta A_\delta$ in (3.5.8) to appear as the zero matrix and the dynamics of the filter will no longer be achieved. However, this will occur at sampling rates which are typically five to ten times faster than with shift operator.

We illustrate by a simple example.

**Example 4.5.1** : Investigate the numerical properties of the system described previously in Example 4.3.2.

**Solution** The step response of the continuous-time system given by (4.3.15) was simulated over 10 seconds as shown in Figure 4.5.1. Also shown superimposed on this is the step response for the shift and delta operator discrete-time models (4.3.16) and (4.3.17) using $\Delta = 0.01$ sec. As can be seen, the shift operator implementation is a very poor approximation to the continuous-time system,

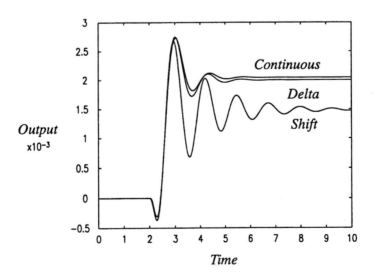

*Figure 4.5.1* : *Comparison of step response of continuous-time system to the step responses of discrete-time approximating systems using shift and operator implementations.*

while the delta operator implementation is so close to the true continuous-time response that it is difficult to tell the two apart.

Notice too that this happens despite the fact that the shift coefficient representation in (4.3.17) involves three more significant digits than does the delta operator implementation in (4.3.16). This result should be compared with that in Section 3.13 where digital filter implementation issues were discussed.

$$\wedge\!\wedge\!\wedge$$

## 4.6 Frequency Domain Development

For future use we denote the transfer function of the plant and the anti-aliasing filter (including the ZOH) as follows (see equation (4.3.11)):

$$\overline{G}_p^s(s) = \overline{G}_p(\gamma) \Big|_{\gamma = \frac{e^{s\Delta} - 1}{\Delta}}$$

$$= D\Big\{S_\Delta\big\{L^{-1}\{F(s)G_p(s)H_0(s)\}\big\}\Big\} \Big|_{\gamma = \frac{e^{s\Delta} - 1}{\Delta}} \qquad (4.6.1)$$

We have seen in Section 3.10 that the frequency response of a discrete-time system is obtained by evaluating $\overline{G}_p(\gamma)$ at the values $\gamma_\omega = \left(e^{j\omega\Delta} - 1\right)/\Delta$. Thus, using (4.6.1), the discrete frequency transfer function is precisely $\overline{G}_p^s(j\omega)$.

Hence from (4.6.1) we have

$$\overline{G}_p^s(j\omega) = D\Big\{S_\Delta\big\{L^{-1}\{F(s)G_p(s)H_0(s)\}\big\}\Big\} \Big|_{\gamma_\omega} \qquad (4.6.2)$$

For simplicity, we use the symbol $K(s)$ to denote $F(s)G_p(s)H_0(s)$ in the sequel. We also recall the following summation result (see (1.5.2)):

$$\sum_{k=-\infty}^{\infty} e^{-j\omega k\Delta} = \frac{2\pi}{\Delta} \sum_{l=-\infty}^{\infty} \delta(\omega - l2\pi/\Delta) \qquad (4.6.3)$$

We can now use the formula for the inverse Laplace transform to obtain an expression for $\overline{G}_p^s(j\omega)$. From (4.6.2) we have

$$\overline{G}_p^s(j\omega) = \sum_{k=-\infty}^{\infty} \Delta e^{-j\omega k\Delta} \left[ \frac{1}{2\pi} \int_{-\infty}^{\infty} K(j\omega_c) e^{j\omega_c k\Delta} d\omega_c \right]$$

Note that we have extended the summation to negative $k$. This does not change the result since the sequence in brackets is zero for $k < 0$. Thus

$$\overline{G}_p^s(j\omega) = \frac{1}{2\pi} \int_{-\infty}^{\infty} K(j\omega_c) \sum_{k=-\infty}^{\infty} \Delta e^{j(\omega_c-\omega)k\Delta} d\omega_c$$

$$= \int_{-\infty}^{\infty} K(j\omega_c) \sum_{l=-\infty}^{\infty} \delta(\omega_c - \omega - l2\pi/\Delta) d\omega_c$$

$$\overline{G}_p^s(j\omega) = \sum_{l=-\infty}^{\infty} K(j(\omega - l2\pi/\Delta)) \qquad (4.6.4)$$

This expression tells us that to get the discrete-time frequency response we need to repeat the continuous-time frequency response shifted by multiples of $2\pi/\Delta$ and then sum them up. We will see this result again in later chapters from a different perspective.

**Example 4.6.1** : Consider again Example 4.2.1 where

$$K(s) = \frac{1 - e^{-s\Delta}}{(\tau s + 1)\Delta s}$$

Evaluate the discrete frequency response using (4.6.4).

**Solution** From (4.2.16) and (4.6.4) we have

$$\overline{G}_p^s(j\omega) = \sum_{l=-\infty}^{\infty} \frac{1 - e^{-j\left(\omega - l\frac{2\pi}{\Delta}\right)\Delta}}{\left(\tau j(\omega - l\frac{2\pi}{\Delta}) + 1\right)\Delta\left(j(\omega - l\frac{2\pi}{\Delta})\right)}$$

We will use the following formula (Hansen (1975)):

$$\sum_{k=-\infty}^{\infty} \frac{1}{(ka + b)(kc + d)} = \frac{\pi}{ad - bc}\left(\cot\frac{\pi b}{a} - \cot\frac{\pi d}{c}\right)$$

$$= \frac{\pi}{ad - bc}\sin\left(\frac{\pi}{ac}(ad - bc)\right)\cosec\frac{\pi d}{c}\cosec\frac{\pi b}{a}$$

We want to find the sum

$$\sum_{k=-\infty}^{\infty} \frac{1}{[j\tau(\omega - k\omega_s) + 1][j(\omega - k\omega_s)]}$$

$$= \sum_{k=-\infty}^{\infty} \frac{1}{[k(-j\tau\omega_s) + (1 + j\tau\omega)][k(-j\omega_s) + j\omega]}$$

Using the above formula with $a = -j\tau\omega_s, b = 1 + j\tau\omega, c = -j\omega_s, d = j\omega$
we have

$$ad - bc = (-j\tau\omega_s)(j\omega) - (1 + j\tau\omega)(-j\omega_s)$$

$$ac = (-j\tau\omega_s)(-j\omega_s) = -\tau\omega_s^2$$

$$\sin\left(\frac{\pi}{ac}(ad - bc)\right) = \sin\left(\frac{\pi}{+\tau\omega_s^2}(+j\omega_s)\right) = \sin\left(j\frac{\pi\Delta}{\tau 2\pi}\right) = \sin\left(j\frac{\Delta}{2\tau}\right)$$

$$= \frac{e^{\frac{-\Delta}{2\tau}} - e^{\frac{\Delta}{2\tau}}}{2j}$$

$$\csc(\pi d/c) = \frac{1}{\sin\left(\pi\frac{j\omega}{-j\omega_s}\right)} = \frac{-1}{\sin\left(\pi\frac{\omega}{\omega_s}\right)} = \frac{-1}{\sin\left(\frac{\omega\Delta}{2}\right)} = \frac{2j}{-e^{j\frac{\omega\Delta}{2}} + e^{-j\frac{\omega\Delta}{2}}}$$

$$\csc(\pi b/a) = \frac{1}{\sin\left(\pi\frac{1+j\pi\omega}{-j\pi\omega_s}\right)} = \frac{+1}{\sin\left(j\frac{\Delta}{2\tau} - \frac{\omega\Delta}{2}\right)} = \frac{2j}{e^{-\left(\frac{\Delta}{2\tau}+j\frac{\omega\Delta}{2}\right)} - e^{\left(\frac{\Delta}{2\tau}+j\frac{\omega\Delta}{2}\right)}}$$

So, the infinite sum will be

$$\frac{1-e^{-j\omega\Delta}}{\Delta}\left(\frac{-\pi\Delta}{2j\pi}\right)\left(\frac{e^{\frac{-\Delta}{2\tau}}-e^{\frac{\Delta}{2\tau}}}{2j}\right)\left(\frac{2j}{e^{-j\frac{\omega\Delta}{2}}-e^{j\frac{\omega\Delta}{2}}}\right)\left(\frac{2j}{e^{-\left(\frac{\Delta}{2\tau}+j\frac{\omega\Delta}{2}\right)}-e^{\left(\Delta 2\tau+j\frac{\omega\Delta}{2}\right)}}\right)$$

$$= \frac{(1-e^{-j\omega\Delta})e^{+\frac{\Delta}{2\tau}}(e^{-\frac{\Delta}{\tau}}-1)}{e^{j\frac{\omega\Delta}{2}}(1-e^{-j\omega\Delta})\left(e^{-\frac{\Delta}{2\tau}}e^{-j\frac{\omega\Delta}{2}}-e^{\frac{\Delta}{2\tau}}-e^{j\frac{\omega\Delta}{2}}\right)}$$

$$= \frac{e^{-\frac{\Delta}{\tau}}-1}{\left(e^{-\frac{\Delta}{\tau}}-e^{j\omega\Delta}\right)}$$

Thus

$$\overline{G}_p^s(j\omega) = \frac{\Delta\left(1-e^{\frac{-\Delta}{\tau}}\right)}{e^{j\omega\Delta}+e^{\frac{-\Delta}{\tau}}}$$

Note that this is exactly the same result as would be obtained from (4.2.16) or (4.3.14) using (4.6.1).

$\wedge\wedge\wedge$

It is instructive to evaluate (4.6.4) by an alternative route. To do this we consider the continuous-time system described, again, in state-space form by

$$\frac{d}{dt}x(t) = Ax(t) + Bu(t) \tag{4.6.5}$$

$$y(t) = Cx(t) \tag{4.6.6}$$

The corresponding continuous-time transfer function is

$$F(s)G_p(s) = C(sI - A)^{-1}B \tag{4.6.7}$$

and the continuous-time frequency response is

$$F(j\omega)G_p(j\omega) = C(j\omega I - A)^{-1}B \tag{4.6.8}$$

Now in Section 4.2 we showed that the corresponding discrete-time transfer function with ZOH input is given by

$$\overline{G}_p(\gamma) = C_\delta(\gamma I - A_\delta)^{-1}B_\delta \tag{4.6.9}$$

where

$$C_\delta = C; \quad A_\delta = \frac{1}{\Delta}\left[e^{A\Delta} - I\right]; \quad B_\delta = \frac{1}{\Delta}A^{-1}\left[e^{A\Delta} - I\right]B \tag{4.6.10}$$

(See (4.2.7) for the case where $A$ is singular).

The corresponding discrete frequency response is obtained by evaluating $\overline{G}_p(\gamma)$ at $\gamma_\omega = \left(e^{j\omega\Delta} - 1\right)/\Delta$.

Substituting this into (4.6.9) gives the discrete frequency response as

$$\overline{G}_p(\gamma_\omega) = C_\delta \left(\frac{e^{j\omega\Delta} - 1}{\Delta}I - A_\delta\right)^{-1} B_\delta$$

$$= C\left[e^{j\omega\Delta}I - e^{A\Delta}\right]^{-1} B_0 \tag{4.6.11}$$

where

$$B_0 = \Delta\left(I + \frac{\Delta A}{2!} + \frac{\Delta^2 A^2}{3!} + \frac{\Delta^3 A^3}{4!} + \dots\right)B$$

$$= A^{-1}\left[e^{A\Delta} - I\right]B \quad \text{(for nonsingular } A\text{)} \tag{4.6.12}$$

We next wish to compare the result (4.6.11) with the earlier result of (4.6.4):

$$\overline{G}_p^s(j\omega) = \sum_{l=-\infty}^{\infty} K(j(\omega - l2\pi/\Delta)) \tag{4.6.13}$$

where

$$K(s) = F(s)G_p(s)H_0(s) \tag{4.6.14}$$

and

$$H_0(s) = \frac{1 - e^{-s\Delta}}{s\Delta} \tag{4.6.15}$$

To show directly that (4.6.11) and (4.6.13) are indeed equal we establish the following result.

Lemma 4.6.1

$$\sum_{k=-\infty}^{\infty} \left( j(\omega - k\frac{2\pi}{\Delta})I - A \right)^{-1} BH_0 \left( j(\omega - k\frac{2\pi}{\Delta}) \right) = \left[ e^{j\omega\Delta}I - e^{A\Delta} \right]^{-1} B_0 \tag{4.6.16}$$

Proof : Since on both sides of (4.6.16) we have periodic function of $\omega$ (with period $2\pi/\Delta$) we can represent them in a series (dual to the Fourier series) of the form $\dfrac{\Delta}{2\pi} \displaystyle\sum_{p=-\infty}^{\infty} B_p e^{j\omega p\Delta}$. To prove the lemma all we need is to show that the $B_p$'s on both side are equal for every $p$.

On the right side we get

$$(B_p)_{RHS} = \int_{-\frac{\pi}{\Delta}}^{\frac{\pi}{\Delta}} \left[ e^{j\omega\Delta}I - e^{A\Delta} \right]^{-1} e^{-j\omega p\Delta} d\omega B_0$$

$$= \frac{2\pi}{\Delta} e^{-A\Delta(p+1)} B_0 \quad \text{for } p \le -1$$

$$= 0 \quad \text{for } p \le -1 \tag{4.6.17}$$

On the left hand side we get

$$(B_p)_{LHS} = \int_{-\frac{\pi}{\Delta}}^{\frac{\pi}{\Delta}} \sum_{k=-\infty}^{\infty} \left( j(\omega - k\frac{2\pi}{\Delta})I - A \right)^{-1} e^{-j\omega p\Delta} H_0 \left( j(\omega - k\frac{2\pi}{\Delta}) \right) d\omega B$$

$$= \int_{-\infty}^{\infty} [j\omega I - A]^{-1} e^{-j\omega p\Delta} H_0(j\omega) d\omega B$$

$$= \int_{-\infty}^{\infty} (j\omega I - A)^{-1} B H_o(j\omega) e^{j\omega t} d\omega \bigg|_{t=-p\Delta}$$

$$= 2\pi F^{-1} \left[ (j\omega I - A)^{-1} B H_0(j\omega) \right] \bigg|_{t=-p\Delta}$$

$$= 2\pi \int_{-\infty}^{t} e^{A(t-\sigma)} B h_0(\sigma) d\sigma \bigg|_{t=-p\Delta}$$

$$= \begin{cases} \frac{2\pi}{\Delta} e^{-A\Delta(p+1)} \int_0^{\Delta} e^{A(\Delta-\sigma)} B d\sigma & \text{for } p \le -1 \\ 0 & \text{for } p > -1 \end{cases}$$

$$= \begin{cases} \frac{2\pi}{\Delta} e^{-A\Delta(p+1)} B_0 & \text{for } p \le -1 \\ 0 & \text{for } p > -1 \end{cases}$$

which is the same as (4.6.17). This completes the proof.

$$\wedge\wedge\wedge$$

## 4.7    Further Reading and Discussion

This chapter has developed various ways of deriving the discrete-time model corresponding to a given continuous-time system with ZOH input. Although the different approaches lead to the same end result, we consider it insightful to show different aspects of the result starting from alternative problem descriptions and using alternative methods.

The extended set of references provided in Section 3.14 of Chapter 3 also relate to the results in the present chapter. The key additional reference is that of Åstrom, Hagander and Sternby (1984) where Lemma 4.4.1 is established. This result is also discussed from a different perspective in Middleton and Goodwin (1990).

## 4.8    Problems

4.1    Consider the following differential equation:

$$\frac{d^2y(t)}{dt^2} + \omega_0^2 y(t) = 0, \quad y(0) = a, \quad \frac{dy(0)}{dt} = b$$

(a)    Find the solution for this equation.

(b)    Write a state-space form of this differential equation.

(c)    The state vector in (b) is sampled every $\Delta$ seconds to give the sequence $\{\underline{x}[k]\}$ .

Show that $\underline{x}[k]$ satisfies $\underline{x}[k+1] = A_q \underline{x}[k]$ and the sampled output $y[k]$ satisfies $y[k] = C\underline{x}[k]$ . Find $A_q$ and $C$.

(d)    Rewrite the equations in (c) in the form

$$\delta \underline{x}[k] = A_\delta \underline{x}[k]$$

$$y[k] = C\underline{x}[k]$$

and hence find $A_\delta$ .

(e)    Verify that (4.2.9) holds.

4.2    Consider a system with anti-aliasing filter given by the transfer function

$$F(s)G_p(s) = \frac{s+6}{(s+2)(s+3)(s+1)}$$

(a)    Write a minimal state-space realization of the given transfer function.

(b)    The output is sampled every $\Delta = 0.1$ seconds and a ZOH is used at the input. Find the resulting $C_\delta$, $A_\delta$, $B_\delta$, $D_\delta$ ..

(c)    Calculate the discrete transfer function from your results in (b).

4.3    Use Table 3.3.2 and the formula

$$\bar{G}(\gamma) = D\left\{ S_\Delta \left[ L^{-1}\{F(s)G_p(s)H_0(s)\} \right] \right\}$$

to verify the entries in Table 4.3.1.

4.4    Consider again the transfer function given in Problem 4.2:

$$F(s)G_p(s) = \frac{s+6}{(s+2)(s+3)(s+1)}$$

The output was sampled with period 0.1 sec and a ZOH input was used. Calculate the corresponding discrete transfer function.

4.5    In Table 4.3.1 consider the discrete transfer functions which correspond to

$$\frac{\omega}{(s+a)^2 + \omega^2}$$

$$\frac{s}{(s+a)^2 + \omega^2}$$

(a)   For each resulting discrete transfer function calculate the resulting poles and zeros for the values $\omega\Delta = 1.0, 0.8, 0.6, 0.4, 0.3, 0.2, 0.1, 0.0$. Compare the results to the corresponding continuous-time poles and zeros. What are your conclusions?

(b)   Calculate the corresponding poles and zeros in the $z$-domain.

Compare the results obtained by delta transform and $z$-transform approaches. What do you observe from these results?

4.6   Suppose $y(t)$ satisfies $\delta^i y[k]\big|_{k=0} = 0$ for $i = 0, \ldots, l-1$, show that

$y[i] = 0$ for $i = 0, 1, \ldots, l-1$.

4.7   Show that if a continuous-time system has a zero at $s = 0$ then the discrete transfer function has also a zero at $\gamma = 0$.

Hint: A zero at the origin for $C(sI - A)^{-1}B$ occurs only if $CA^{-1}B = 0$.

4.8   Consider Example 4.4.1. Rewrite the numerator in (4.4.8) as

$$f(\Delta)[(\Delta\gamma) + g(\Delta)]$$

Show that $\lim_{\Delta\to 0} g(\Delta) = 2$.

# Chapter 5

# Optimal Linear Estimation with Finite Impulse Response Filters

## 5.1  Introduction

In this and the next chapter we turn to estimation problems. These problems have a major impact on many problems in signal processing and control.

In this chapter we consider a simplified estimation problem : estimating a signal, $s'(t)$, which is measured in the presence of white measurement noise $v'(t)$. We call this the 'SV problem'. Our filter will use only a *finite time* window of data, say from time $t - T$ to $t$. The resulting filter will be in a special form known as a lattice filter.

An important feature of lattice filters is that there is a simple way of increasing the window length. Hence one can easily examine filters of different complexity to judge which is best for a given application. Also, the design of a lattice filter depends only on the spectral properties of the signal $s'(t)$ and in this sense it is non-parametric.

We will develop the discrete lattice filter and show how this reveals further interesting interconnections between discrete and continuous-time signal processing. In particular we will show that for fixed *window length* (in real time), as the output is sampled more rapidly, the discrete lattice filter converges to an underlying continuous-time lattice filter. Our limiting procedure (in passing from discrete-time to continuous-time) involves considering a fixed time interval and allowing the number of sample points in that interval to increase.

## 5.2    Problem Description

We consider the problem of estimating the value of a continuous-time signal process $s'(\cdot)$ at time $t$ based on measurements of a related process $y'(\cdot)$ over a finite observation window of length $T$.

We assume that the processes $y'(\cdot)$ and $s'(\cdot)$ are related by

$$\boxed{y'(\tau) = s'(\tau) + v'(\tau), \quad \forall \tau \in R} \tag{5.2.1}$$

We also assume that the process $s'(\cdot)$ is zero mean and wide-sense stationary with known covariance function $W'(\cdot)$, $v'(\cdot)$ is a zero mean stationary "white noise" process with incremental covariance $\Gamma' d\tau$, and $s'(\cdot)$, $v'(\cdot)$ are uncorrelated, i.e.

$$\boxed{E\{s'(t)s'(t-\tau)\} = W'(\tau), \quad \forall t, \tau \in R} \tag{5.2.2}$$

$$\boxed{E\{v'(t)v'(t-\tau)\} = \Gamma'\delta(\tau), \quad \forall t, \tau \in R} \tag{5.2.3}$$

$$\boxed{E\{s'(t)v'(\tau)\} = 0, \quad \forall t, \tau \in R} \tag{5.2.4}$$

The reader will recall that the Fourier transform of the covariance is called the *spectral density* of the process. Thus from (5.2.2) the spectral density of the process $s'$ is given by

$$S_s(\omega) = \int_{-\infty}^{\infty} W'(\tau)e^{-j\omega\tau} \, d\tau \tag{5.2.5}$$

Similarly, the spectral density of the process $v'$ is given by

$$S_{y'}(\omega) = \int_{-\infty}^{\infty} \Gamma' \delta(\tau) e^{-j\omega\tau} \, d\tau$$

$$= \Gamma' \tag{5.2.6}$$

In the sequel we will develop the optimal filter in terms of the covariance functions. However, it follows from (5.2.5), (5.2.6) that we could equally use the spectral densities.

## 5.3　Sampled Model

We first wish to consider the situation where sampled signals are used. However, it makes no sense to directly sample $y'(\cdot)$ since this would lead to a discrete-time process of unbounded variance. This problem is resolved by replacing the impractical ideal sampler by an anti-aliasing filter before the sampling process.

Before deriving a model for the corresponding sampled system it is useful to consider what kind of anti-aliasing filter might be useful in practice. Ideally we do not want the anti-aliasing filter to significantly interact with the system dynamics but we do want it to smooth the noise between samples. This suggests that we use an anti-aliasing filter having unity d.c. gain and settling time of the order of $\Delta$. A very simple filter that we have seen before having this characteristic is one having the following impulse response.

$$h_0(\tau) = \begin{cases} 1/\Delta & \text{for } 0 \le \tau < \Delta \\ 0 & \text{otherwise} \end{cases}$$

In the sequel we assume that the above anti-aliasing filter is employed.

The sampled output is then given by

$$\bar{y}[k+1] = \frac{1}{\Delta} \int_{k\Delta}^{(k+1)\Delta} y'(t)\ dt \tag{5.3.1}$$

The following Lemma then establishes statistical properties of the sampled signals.

**Lemma 5.3.1** The resulting sampled process satisfies

$$\bar{y}[k] = s[k] + v[k] \tag{5.3.2}$$

where the process $\{s[k]\}$ is zero mean and wide sense stationary with known covariance function $W[m]$ satisfying

$$W[j-k] = \frac{1}{\Delta^2} \int_{(j-1)\Delta}^{j\Delta} \int_{(k-1)\Delta}^{k\Delta} W'(t_2-t_1)\ dt_2 dt_1 \tag{5.3.3}$$

The process $\{v[k]\}$ is a zero mean white noise sequence with variance $\bar{\Gamma}$ where

$$\bar{\Gamma} = \frac{\Gamma'}{\Delta} \tag{5.3.4}$$

Proof :  We note from (5.3.1) that

$$E\{s[j]s[k]\} = E\left\{ \frac{1}{\Delta} \int_{(j-1)\Delta}^{j\Delta} s'(t)\ dt \ \frac{1}{\Delta} \int_{(k-1)\Delta}^{k\Delta} s'(t)\ dt \right\}$$

$$= \frac{1}{\Delta^2} \int_{(j-1)\Delta}^{j\Delta} \int_{(k-1)\Delta}^{k\Delta} E\{s'(t_1)s'(t_2)\}\ dt_2 dt_1 \tag{5.3.5}$$

Equation (5.3.3) follows immediately.

$$E\{v[j]v[k]\} = \frac{1}{\Delta^2} \int\limits_{(j-1)\Delta}^{j\Delta} \int\limits_{(k-1)\Delta}^{k\Delta} E\{v'(t)v'(\tau)\} \; dtd\tau$$

$$= \frac{\Gamma'}{\Delta^2} \int\limits_{(j-1)\Delta}^{j\Delta} \int\limits_{(k-1)\Delta}^{k\Delta} \delta(t-\tau) \; dtd\tau$$

$$= \Gamma'\delta_d[j-k] = \begin{cases} \Gamma'/\Delta & \text{for } j = k \\ 0 & \text{otherwise} \end{cases} \qquad (5.3.6)$$

⋀⋀⋀

In the sequel we take $\Gamma' = 1$. This is without loss of generality since we can always scale the data appropriately by changing units.

## 5.4    The Discrete Lattice Filter

We want to generate the optimal estimate of $s[k]$ using the measurements $\{\bar{y}[k-1], \; \dots, \; \bar{y}[k-N]\}$. The linear estimate is thus obtained from a finite impulse response filter as follows:

$$\hat{s}[k,N] = \Delta \sum_{i=1}^{N} h[i,N] \; \bar{y}[k-i] \qquad (5.4.1)$$

where we show in our notation for the filter coefficients $h$ and the estimate, $\hat{s}$, their dependence on the number of samples ($N$) considered.

Our aim is to minimize the expected (mean square) error between $\hat{s}[k,N]$ and the true value $s[k]$. Thus we define

$$J = E\{(s[k] - \hat{s}[k,N])^2\} \qquad (5.4.2)$$

Differentiating (5.4.2) with respect to the coefficients $h[i, N]$ and setting the result to zero leads to the following set of equations:

$$E\{\bar{y}[k-i](s[k]-\hat{s}[k,N])\} = 0 \quad i = 1, 2, \ldots, N \qquad (5.4.3)$$

These equations are often referred to as the 'orthgonality principle' – i.e. the optimal error must be orthogonal to all the data points.

Substituting (5.4.1) and using (5.3.2) – (5.3.4) yields

$$W[i] - \Delta \sum_{j=1}^{N} h[j, N](W[j-i] + \delta_d[j-i]) = 0 ; \quad i = 1, 2, \ldots, N \quad (5.4.4)$$

Equation (5.4.4) represents a set of $N$ equations in the $N$ unknowns $h(1,N)$, ..., $h(N,N)$. In principle, (5.4.4) can thus be solved for these unknowns.

We next define a complementary problem: Given the data $y[k]$, $y[k-1]$, ..., $y[k-N+1]$ generate the optimal estimate of $s[k-N]$.
Denoting this estimate by

$$\hat{s}[k-N,N] = \Delta \sum_{i=1}^{N} b[i, N] \; \bar{y}[k-i+1] \qquad (5.4.5)$$

and repeating the derivation leading to equation (5.4.4) we get

$$W[i] - \Delta \sum_{j=1}^{N} b[N-j+1, N] \; (W[j-1] + \delta_d[j-i]) = 0$$
$$i = 1, 2, \ldots, N \qquad (5.4.6)$$

Comparing (5.4.4) and (5.4.6) we readily observe that

$$b[N-j+1, N] = h[j, N] \qquad j = 1, 2, \ldots, N \qquad (5.4.7)$$

Thus, the reverse time problem has effectively recovered the same solution as the forward time problem.

We next study what happens when we change either the present time, $k$, or the number of samples used to form the estimate, $N$. Since we want to "differentiate" with respect to both these indices we generalize the idea of the delta operator to a partial type difference operator. We will write a subscript after the operator to indicate which variable it applies to. Thus, we write

$$\partial_j h[j,N] = \frac{1}{\Delta}\left(h[j+1,N] - h[j,N]\right) \tag{5.4.8}$$

$$\partial_N h[j,N] = \frac{1}{\Delta}\left(h[j,N+1] - h[j,N]\right) \tag{5.4.9}$$

We then have the following result

**Lemma 5.4.1** Every sequence $h[j,N]$ which satisfies (5.4.4) also satisfies

$$\partial_N h[j,N] = -h[N+1,N+1]\, h[N-j+1,N] \quad j = 1,2,\ \ldots,\ N \tag{5.4.10}$$

**Proof :** Applying $\partial_N$ on both sides of (5.4.4) we get (see Problem 5.1)

$$0 = \partial_N \sum_{j=1}^{N} h[j,N]\,\left(W[j-i] + \delta_d[i-j]\right)$$

$$= \frac{1}{\Delta} h[N+1,N+1]\,\left(W[N+1-i] + \delta_d[N+1-i]\right)$$
$$+ \sum_{j=1}^{N} \partial_N h[j,N]\,\left(W[j-i] + \delta_d[j-i]\right) \quad i = 1,2,\ \ldots,\ N$$

Clearly $\delta_d[N+1-i] = 0$ for all $i = 1,2,\ \ldots,\ N$ so we get

$$h[N+1,N+1]W[N+1-i] + \Delta \sum_{j=1}^{N} \partial_N h[j,n](W[j-i] + \delta_d[j-i]) = 0$$

Substituting equation (5.4.4) and some algebra leads to

$$\sum_{j=0}^{N}\left(h[N+1,N+1]h[N+1-j,N] + \partial_N h[j,N]\right)\left(W[j-i] + \delta_d[j-i]\right) = 0$$

$$i = 1, 2 \ \dots, \ N$$

and (5.4.10) follows.

$\triangle\triangle\triangle$

We next define the forward and backward *innovation* sequences by

$$e_f[k, 0] = e_b[k, 0] = y[k] \qquad (5.4.11)$$

$$e_f[k, N] = \bar{y}[k] - \hat{s}[k, N] \qquad (5.4.12)$$

$$e_b[k, N] = \bar{y}[k-N] - \hat{s}[k-N, N] \qquad (5.4.13)$$

To summarize, we then have the following result describing the discrete lattice filter:

**Lemma 5.4.2**

1.   $W[i] = \Delta \sum_{k=1}^{N} h[j,N] \left(W[j-i] + \delta_d[j-i]\right) = 0 \quad i = 1, 2, \ \dots, \ N$

2.   $b[N-j+1,N] = h[j,N] \quad j = 1, 2 \ \dots, \ N$

3.   $\partial_N h[j,N] = -h[N+1,N+1] \ h[N-j+1,N] \quad j = 1, \ 2, \ \dots, \ N$

4.   The innovation sequences $e_f[k,N]$ and $e_b[k,N]$ satisfy the following relationships:

$$\boxed{\partial_N e_f[k, N] = -h[N+1, N+1]e_b[k-1, N]}$$ (5.4.14)

$$\boxed{\partial_N e_b[k, N] = -h[N+1, N+1]e_f[k, N] - \partial_k e_b[k-1, N]}$$ (5.4.15)

5. The optimal estimates can be expressed entirely in terms of $e_f[\cdot, \cdot]$ and $e_b[\cdot, \cdot]$ as follows:

$$\Delta \sum_{i=1}^{N} h[i, i]e_b[k-1, i-1] = \hat{s}[k, N]$$ (5.4.16)

$$\Delta \sum_{i=1}^{N} h[i, i]e_f[k-N+i-1, i-1] = \hat{s}[k-N, N]$$ (5.4.17)

6. The covariances of the innovations sequences satisfy the following for $m \geq 0$:

$$E\{e_f[k, N]e_f[k-m, N-m]\} = \delta_d[m]\left[1 + \Delta W[0] - \Delta^2 \sum_{i=1}^{N} h[i, N]W[i]\right]$$ (5.4.18)

$$E\{e_b[k, N]e_b[k, N-m]\} = \delta_d[m]\left[1 + \Delta W[0] - \Delta^2 \sum_{i=1}^{N} h[i, N]W[i]\right]$$ (5.4.19)

7. The cross-covariance between $e_f[\cdot, \cdot]$ and $e_b[\cdot, \cdot]$ satisfies

$$E\{e_f[k, N]e_b[k-1, N]\} = W[N+1] - \Delta \sum_{i=1}^{N} h[i, N]W[N+1-i]$$ (5.4.20)

**Proof :** Parts 1, 2, 3 are a restatement of equations (5.4.6), (5.4.7) and (5.4.10).

4. From (5.4.12), (5.4.1) and Lemma 5.4.1 we have

$$\partial_N e_f[k, N] = -\partial_N \hat{s}[k, N]$$

$$= -\partial_N \Delta \sum_{i=1}^{N} h[i, N] \bar{y}[k - i]$$

$$= -h[N + 1, N + 1] \bar{y}[k - N - 1] - \Delta \sum_{i=1}^{N} \partial_N h[i, N] \bar{y}[k - i]$$

$$= -h[N + 1, N + 1] \bar{y}[k - N - 1]$$
$$+ h[N + 1, N + 1] \sum_{i=1}^{N} \Delta h[N - i + 1, N] \bar{y}[k - i]$$

$$= -h[N + 1, N + 1] \left[ \bar{y}[k - N - 1] - \sum_{i=1}^{N} \Delta h[N - i + 1, N] \bar{y}[k - i] \right]$$

and by (5.4.7), (5.4.5) and (5.4.13)

$$\partial_N e_f[k, N] = -h[N + 1, N + 1] \left[ \bar{y}[k - N - 1] - \Delta \sum_{i=1}^{N} b[i, N] \bar{y}[k - i] \right]$$

$$= -h[N + 1, N + 1] (\bar{y}[k - N - 1] - \hat{s}[k - 1 - N, N])$$

$$= -h[N + 1, N + 1] e_b[k - 1, N]$$

which establishes (5.4.14).

Similarly, from (5.4.13), (5.4.5) and (5.4.7) we get

$$\partial_N e_b[k, N] = \partial_N \bar{y}[k - N] - \partial_N \hat{s}[k - N, N]$$

$$= \partial_N \bar{y}[k - N] - \partial_N \sum_{i=1}^{N} \Delta h[i, N] \bar{y}[k - N + i]$$

$$= \partial_N \bar{y}[k - N] - h[N + 1, N + 1] \bar{y}[k - N - 1 + N + 1]$$

$$- \Delta \sum_{i=1}^{N} \partial_N (h[i, N] \bar{y}[k - N + i])$$

Also, using the rule for differencing a product (see equation (3.4.9)) we have

$$\partial_N\big(h[i,N]\bar{y}[k-N+i]\big) = \partial_N h[i,N]\bar{y}[k-N-1+i] + h[i,N]\partial_N\bar{y}[k-N+i]$$

and since

$$\partial_N\bar{y}[k-N+i] = -\partial_k\bar{y}[k-1-N+i]$$

we get from Lemma 5.4.1, (5.4.7), (5.4.12) and (5.4.13)

$$
\begin{aligned}
\partial_N e_b[k,N] &= -\partial_k\bar{y}[k-1-N] - h[N+1,N+1]\bar{y}[k] \\
&\quad + h[N+1,N+1]\Delta\sum_{i=1}^{N} h[N-i+1,N]\bar{y}[k-N-1+i] \\
&\quad + \Delta\sum_{i=1}^{N} h[i,N]\partial_k\bar{y}[k-1-N+i] \\
&= -h[N+1,N+1]\left[\bar{y}[k] - \Delta\sum_{i=1}^{N} h[i,N]\bar{y}[k-i]\right] \\
&\quad - \partial_k\left[\bar{y}[k-1-N] - \Delta\sum_{i=1}^{N} b[i,N]\bar{y}[k-i]\right] \\
&= -h[N+1,N+1]e_f[k,N] - \partial_k e_b[k-1,N]
\end{aligned}
$$

and (5.4.15) is established.

5. From (5.4.5) and (5.4.13)

$$\Delta\sum_{i=1}^{N} h[i,i]e_b[k-1,i-1] = \Delta\sum_{i=1}^{N} h[i,i]\big[\bar{y}[k-1-i+1] - \hat{s}[k-1-i+1,i-1]\big]$$

$$= \Delta\sum_{i=1}^{N} h[i,i]\bar{y}[k-i] - \Delta\sum_{i=1}^{N} h[i,i]\hat{s}[k-i,i-1]$$

$$= \Delta \sum_{i=1}^{N} h[i,i]\bar{y}[k-i]$$

$$- \Delta \sum_{i=1}^{N} h[i,i]\Delta \sum_{j=1}^{i=1} b[j,i-1]\bar{y}[k-j]$$

Hence using (5.4.10)

$$\Delta \sum_{i=1}^{N} h[i,i]e_b[k-1,i-1] = \Delta \sum_{i=1}^{N} h[i,i]\bar{y}[k-i] + \Delta^2 \sum_{i=1}^{N}\sum_{j=1}^{i-1} \partial_i h[j,i-1]\bar{y}[k-j]$$

$$= \Delta \sum_{i=1}^{N} \Delta\partial_i \sum_{j=1}^{i-1} h[j,i-1]\bar{y}[k-j]$$

$$= \Delta \sum_{j=1}^{N} h[j,n]\bar{y}[k-j] = \hat{s}[k,N]$$

where we have used (5.4.1). This establishes (5.4.16).

Establishing (5.4.17) is very similar (see Problem 5.2).

6.    $E\{e_f[k,N]e_f[k-m,N-m]\}$

$$= E\{(\bar{y}[k] - \hat{s}[k,N])(\bar{y}[k-m] - \hat{s}[k-m,N-m])\}$$

$$= E\{\bar{y}[k]\bar{y}[k-m]\} - \Delta \sum_{i=1}^{N} h[i,N]E\{\bar{y}[k-m]y[k-i]\}$$

$$- \Delta \sum_{i=1}^{N-m} h[i,N-m]E\{\bar{y}[k]\bar{y}[k-m-i]\}$$

$$- \Delta^2 \sum_{i=1}^{N}\sum_{j=1}^{N-m} h[i,N]h[j,N-m]E\{\bar{y}[k-i]\bar{y}[k-m-j]\}$$

$$= W[m] + \delta_d[m] - \Delta \sum_{i=1}^{N} h[i,N](W[m-i] + \delta_d[m-i])$$

$$- \Delta \sum_{i=1}^{N-m} h[i,N-m](W[m+i] + \delta_d[m+i])$$

$$+ \Delta^2 \sum_{i=1}^{N} \sum_{j=1}^{N-m} h[i,N]h[j,N-m](W[m+j-i] + \delta_d[m+j-i])$$

$$= \delta_d[m] + W[m] - \Delta \sum_{i=1}^{N} h[i,N](W[m-i] + \delta_d[m-i])$$

$$- \Delta \sum_{j=1}^{N-m} h[j,N-m]\Bigg[ W[m+j]$$

$$- \Delta \sum_{i=1}^{N} h[i,N](W[m+j-i] + \delta_d[m+j-1])\Bigg]$$

$$= \delta_d[m]\Bigg[ 1 + \Delta W[0] - \Delta^2 \sum_{i=1}^{N} h[i,N]W[i]\Bigg]$$

where we have used (5.4.4).

Equation (5.4.19) is obtained by a similar route :

$$E\{e_f[k,N]e_b[k-1,N]\}$$

$$= E\{(\bar{y}[k] - \hat{s}[k,N])(\bar{y}[k-N-1] - \hat{s}[k-N-1,N])\}$$

$$= E\{\bar{y}[k]\bar{y}[k-N-1]\} - \Delta \sum_{i=1}^{N} h[i,N]E\{\bar{y}[k-N-1]\bar{y}[k-i]\}$$

$$- \Delta \sum_{i=1}^{N} h[i,N]E\{\bar{y}[k]\bar{y}[k-N+i-1]\}$$

$$+ \Delta^2 \sum_{i-1}^{N} \sum_{j=1}^{N} h[i,N]h[j,N]E\{\bar{y}[k-i]\bar{y}[k-N+j-1]\}$$

$$= (W[N+1] + \delta_d[N+1]) - \Delta \sum_{i=1}^{N} h[i,N](W[N+1-i] + \delta_d[N+1-i])$$

$$- \Delta \sum_{i=1}^{N} h[i,N](W[N-i+1] + \delta_d[N-i+1])$$

$$+ \Delta^2 \sum_{i=1}^{N} h[i,N] \sum_{j=1}^{N} h[j,N](W[N-j+1-i] + \delta_d[N-j+1-i])$$

$$= W[N+1] - \Delta \sum_{i=1}^{N} h[i,N] W[N+1-i]$$

$$- \sum_{i=1}^{N} h[i,N] \left[ W[N-i+1] \right.$$

$$\left. - \sum_{j=1}^{N} h[j,N] (W[N-i+1-j] + \delta_d[N-i+1-j]) \right]$$

$$= W[N+1] - \Delta \sum_{i=1}^{N} h[i,N] W[N+1-i]$$

where we have used (5.4.4).

$\triangle\triangle\triangle$

The following result relates the innovation sequences to the coefficients $h[N+1,N+1]$.

**Lemma 5.4.3**

$$h[N+1,N+1] = \frac{1}{\Delta} \frac{E\{e_f[k,N] e_b[k-1,N]\}}{E\{e_b^2[k-1,N]\}} \tag{5.4.21}$$

Proof :  Writing (5.4.4) for $N+1$ we get

$$W[i] - \Delta \sum_{j=1}^{N+1} h[j,N+1] [W[j-1] + \delta_d[j-i]] = 0 \quad i = 1, 2, \ldots, N+1$$

For $i = N+1$ we then get

$$W[N+1] - \Delta \sum_{j=1}^{N+1} h[j,N+1] (W[N+1-j] + \delta_d[N+1-j]) = 0$$

or

$$W[N+1] - h[N+1, N+1] - \Delta \sum_{j=1}^{N+1} h[j, N+1] W[N+1-j] = 0$$

$$W[N+1] - h[N+1, N+1] - \Delta \sum_{j=1}^{N} h[j, N+1] W[N+1-j]$$
$$- \Delta h[N+1, N+1] W[0] = 0$$

$$W[N+1] - \Delta \sum_{j=1}^{N} h[j, N] W[N+1-j]$$
$$- \Delta \sum_{j=1}^{N} \big( h(j, N+1) - h[j, N] \big) W[N+1-j]$$
$$- h[N+1, N+1] \big( 1 + \Delta W[0] \big) = 0$$

Using Lemma 5.4.1, Lemma 5.4.2, and equations (5.4.10), (5.4.19), (5.4.20) we get

$$E\{e_f[k, N] e_b[k-1, N]\} - h[N+1, N+1](1 + \Delta W[0])$$
$$+ h[N+1, N+1] \Delta^2 \sum_{j=1}^{N} h[N-j+1, N] W[N+1-j] = 0$$

or

$$E\{e_f[k, N] e_b[k-1, N]\}$$
$$- h[N+1, N+1] \left[ 1 + \Delta W[0] - \Delta^2 \sum_{j=1}^{N} h[j, N] W[j] \right] = 0$$

Using (5.4.19), this yields (5.4.21).

⋀⋀⋀

The basic equations describing the discrete-time lattice filter are as given in equations (5.4.14) and (5.4.15). We have expressed these equations as partial *difference* equations so that the connection with the continuous case to be presented next is clear. Indeed, we will express the continuous lattice as a set of partial *differential* equations.

However, equations (5.4.14), (5.4.15) can also be expressed in many other alternative ways. For example, substituting (5.4.8), (5.4.9) into these equations leads to the following difference form of the discrete lattice filter:

$$e_f[k, N+1] = e_f[k, N] - K[N+1]e_b[k-1, N] \qquad (5.4.22)$$

$$e_b[k, N+1] = e_b[k-1, N] - K[N+1]e_f[k, N] \qquad (5.4.23)$$

where we have substituted

$$K[N+1] = \Delta h[N+1, N+1] \qquad (5.4.24)$$

These equations are illustrated in Figure 5.4.1.

The lattice form shown in Figure 5.4.1 appears in many texts on signal processing. However, we prefer the form shown in equations (5.4.14) and (5.4.15) since this will reveal the link to continuous-time estimation. This link is not evident from the shift form.

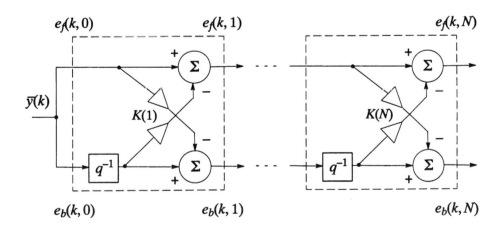

*Figure 5.4.1 : Lattice structure in discrete-time (shift operator form).*

## 5.5    Continuous-Time Lattice Structure

We next consider the corresponding *continuous-time* lattice problem. Here we assume we have measurements of the continuous signal $y'(t)$ in the time interval $[t-T, t]$ and would like to generate the estimate of the continuous signal $s'(t)$. Denote this estimate

$$\hat{s}'(t, T) = \int_0^T h'(\sigma, T) y'(t-\sigma) d\sigma \tag{5.5.1}$$

Also, as in the discrete-time case, we define the backward estimate

$$\hat{s}'(t-T, T) = \int_0^T b'(\sigma, T) y'(t-\sigma) d\sigma \tag{5.5.2}$$

and the innovation processes

$$e_f'(t, 0) = e_b'(t, 0) = y'(t) \tag{5.5.3}$$

$$e_f'(t, T) = y'(t) - \hat{s}'(t, T) \tag{5.5.4}$$

$$e_b'(t, T) = y'(t-T) - \hat{s}'(t-T, T). \tag{5.5.5}$$

Then the following result describes the continuous-time lattice.

Lemma 5.5.1

1.    $$W'(\tau) - \int_0^T h'(\sigma, T)\Big(W'(\tau-\sigma) + \delta(\tau-\sigma)\Big) d\sigma = 0 \tag{5.5.6}$$

$$0 \le \tau \le T$$

2. $\quad b'(\sigma, T) = h'(T - \sigma, T) ; \qquad 0 \le \sigma \le T$ $\hspace{2cm}$ (5.5.7)

3. $\quad \dfrac{\partial h(\sigma, T)}{\partial T} = -h(T, T)h(T - \sigma, T) ; \qquad 0 \le \sigma \le T$ $\hspace{1cm}$ (5.5.8)

(This relationship is known as the *Bellman-Krein-Siegert relation*.)

4.
$$\boxed{\dfrac{\partial e_f(t, T)}{\partial T} = -h'(T, T)e_b'(t, T)}$$
$\hspace{4cm}$ (5.5.9)

$$\boxed{\dfrac{\partial e_b'(t, T)}{\partial T} = -h'(T, T)e_f'(t, T) - \dfrac{\partial e_b'(t, T)}{\partial t}}$$
$\hspace{2cm}$ (5.5.10)

(These are the continuous-time lattice equations.)

5.
$$\hat{s}'(t, T) = \int_0^T h'(\sigma, \sigma)e_b(t, \sigma)d\sigma$$
$\hspace{3cm}$ (5.5.11)

$$\hat{s}'(t - T, T) = \int_0^T h'(\sigma, \sigma)e_f(t - T + \sigma, \sigma)d\sigma$$
$\hspace{2cm}$ (5.5.12)

6. $\quad E\{e_f(t, T)e_f(t - \tau, T - \tau)\} = 0 ; \qquad \tau > 0$ $\hspace{1.5cm}$ (5.5.13)

$\quad E\{e_b(t, T)e_b(t, T - \tau)\} = 0 ; \qquad \tau > 0$ $\hspace{2cm}$ (5.5.14)

Proof : The proof of this lemma can be directly carried out by repeating the derivations in Section 5.4 for the discrete case.

## 5.6 Relationships between the Discrete and Continuous Lattice Filters

Comparing Lemma 5.4.2 with Lemma 5.5.1 we see that the form of the discrete-time filter converges to the form of the continous-time filter as $\Delta \to 0$. In particular, all we need do is make the following associations:

$$\left. \begin{array}{c} N\Delta \to T \\[2ex] k\Delta \to t \\[2ex] \partial_k \to \dfrac{\partial}{\partial t} \\[2ex] \partial_N \to \dfrac{\partial}{\partial T} \\[2ex] \Delta\Sigma \to \displaystyle\int d\sigma \end{array} \right\} \qquad (5.6.1)$$

Beyond the relationships above between the results in discrete and continuous-time it is also interesting to ask how the coefficients $h[i, N]$ relate to the function $h'(\tau, T)$. Recall that $h[i, N]$ and $h'(\tau, T)$ are the solutions of a set of linear equations (5.4.4), and an integral equation (5.5.6), respectively.

To establish how the two are related we note first that equation (5.5.6) is a Fredholm integral equation of the second kind. A well known condition for the existence of a unique solution $h(\tau, T)$ which is continuous on $[0, T]$ is that

$$W(0)T < 1 \qquad (5.6.2)$$

In the sequel we are going to assume that (5.6.2) is satisfied. We assume that, for a given $T$, the following holds:

$$T = N\Delta \qquad (5.6.3)$$

and we define the following piecewise constant form of the discrete sequences $h[i, j]$ and $W[i]$ :

$$\tilde{h}(\tau, \sigma) = \left\{ \begin{array}{ll} h[i,j] & \text{for} \quad (i-1)\Delta \le \tau < i\Delta \\ h[j,j] & \text{for} \quad \tau = \sigma \end{array} \right\} \quad (j-1)\Delta < \sigma \le j\Delta \quad (5.6.4)$$

$$\tilde{W}(\tau) = W[i] \quad \text{for} \quad (i-1)\Delta < \tau \le i\Delta \quad\quad (5.6.5)$$

Then we have the following result which shows that the above continuous-time functions satisfy an integral equation similar to (5.5.6).

**Lemma 5.6.1** The piecewise constant function $\tilde{h}(\tau, T)$ satisfies

$$\tilde{h}(\tau, T) + \int_0^T \tilde{h}(\sigma, T)\tilde{W}(\tau - \sigma)d\sigma = \tilde{W}(\tau + \Delta) \quad\quad (5.6.6)$$

for $\tau = 0, \Delta, 2\Delta, \ldots, (N-1)\Delta$.

**Proof :** Direct substitution of (5.6.4), (5.4.6) and the use of (5.4.4) will establish (5.6.6). (See Problem 5.3).

$\triangle\triangle\triangle$

We denote for the function $f : [0, T] \rightarrow R$ the sup norm

$$\| f \|_\infty = \sup_{\sigma \in [0,T]} |f(\sigma)| \quad\quad (5.6.7)$$

We then have the following result which connects the coefficients in the optimal discrete lattice filter with those in the optimal continuous lattice filter:

**Lemma 5.6.2** With $h'(\tau, T)$ as defined in Section 5.5 and $\tilde{h}(\tau, t)$ as above, assuming $W'(\tau)$ is continuous on $[0, T]$ and (5.6.2) holds, then

$$\lim_{\Delta \to 0} \| h'(\tau, T) - \tilde{h}(\tau, T) \|_\infty = 0 \qu\quad (5.6.8)$$

Proof : Recall that given $T$ and $\tau \in [0, T)$ $N, i$ and $\Delta$ are related through

$$T = N\Delta \qquad (5.6.9)$$

$$\tau = i\Delta < N\Delta \qquad (5.6.10)$$

so as $\Delta \to 0$, $N$ and $i$ grow appropriately.

From (5.6.10), (5.5.6) and Lemma 5.6.1 we have

$$|h'(\tau, T) - \tilde{h}(\tau, T)| \leq W'(\tau) - \tilde{W}(\tau + \Delta)| +$$

$$\left| \int_0^T \left( h'(\sigma, T) W'(\tau - \sigma) - \tilde{h}(\sigma, T) \tilde{W}(\tau - \sigma) \right) d\sigma \right|$$

$$\leq \left| W'(\tau) - \tilde{W}(\tau + \Delta) \right| + \int_0^T \left| h'(\tau, T) \right| \cdot \left| W'(\tau - \sigma) - \tilde{W}(\tau - \sigma) \right| d\sigma$$

$$+ \int_0^T \left| h'(\tau, T) - \tilde{h}(\tau, t) \right| \cdot \left| \tilde{W}(\tau - \sigma) \right| d\sigma \qquad (5.6.11)$$

Since $W'(\tau)$ is continuous on $[0, T]$ there exist a $B > 0$ such that

$$\left| W'(\tau_1) - W'(\tau_2) \right| \leq B \left| \tau_1 - \tau_2 \right| \qquad \tau_1, \tau_2 \in [0, T] \qquad (5.6.12)$$

Then, clearly, using (5.6.5) and (5.6.12) it follows that

$$\left| h'(\tau, T) - \tilde{h}(\tau, T) \right| \leq 2B\Delta + B\Delta \int_0^T \left| h'(\sigma, T) \right| d\sigma + W'(0) \int_0^T \left| h'(\sigma, T) - \tilde{h}(\sigma, T) \right| d\sigma$$

or

$$\| h'(\tau, T) - \tilde{h}(\tau, T) \|_\infty$$

$$\leq B\Delta \left( 2 + T \| h'(\sigma, T) \|_\infty \right) + W'(0) T \| h'(\tau, T) - \tilde{h}(\tau, T) \|_\infty$$

As $\Delta \to 0$ we get

$$\lim_{\Delta \to 0} \| h'(\tau, T) - \bar{h}(\tau, T) \|_\infty \leq \left( W'(0)T \right) \lim_{\Delta \to 0} \| h'(\tau, T) - \bar{h}(\tau, T) \|_\infty \quad (5.6.13)$$

and since by (5.6.2) $W'(0)T < 1$, the only way (5.6.13) can hold is if (5.6.8) holds.

ΛΛΛ

The significance of Lemma 5.6.2 is in fact that it can be used to relate all the sequences generated in the discrete lattice structure to their continuous-time counterparts. In fact, passing all those sequences through ZOH the resulting signals can be shown to approach the continuous-time signals in the sense of (5.6.8) as $\Delta \to 0$. (See Problem 5.4).

## 5.7    Further Reading and Discussion

The discrete lattice filter is described in Itakura and Saito (1971). An excellent overview of linear filtering theory appears in Kailath (1974).

Other treatments of lattice filtering may be found in Haykin (1986), Widrow and Stearns (1985), Honig and Messershmitt (1984), Levy et. al (1979), Goodwin and Sin (1984).

The relationship between the continuous and discrete lattice filter is based on Weller et al. (1993).

## 5.8    Problems

5.1    (a)    Prove that

$$\partial_N \sum_{i=1}^{N} f[i, N] = \frac{1}{\Delta} f[N + 1, N + 1] + \sum_{i=1}^{N} \partial_N f[i, N]$$

Note the similarity with the differentation $\dfrac{d}{dx}\left[ \displaystyle\int_0^x f(\sigma, x)d\sigma \right]$.

5.2    Establish (5.4.17) by following the proof of (5.4.16).

5.3    Prove Lemma 5.6.1.

5.4    Using the sequences $y[k]$, $\hat{s}[k, N]$, $\hat{s}[k-N, N]$, $e_f[k, N]$, $e_b[k, N]$, define the continuous-time signals:

$$
\left.
\begin{aligned}
\tilde{y}(\sigma) &= y[i]\\[4pt]
\tilde{s}(\sigma, \tau) &= \hat{s}[i, j]\\[4pt]
\tilde{s}(\sigma - \tau, \tau) &= \hat{s}[i - j, j]\\[4pt]
\tilde{e}_f(\sigma, \tau) &= e_f[i, j]\\[4pt]
\tilde{e}_b(\sigma, \tau) &= e_b[i, j]
\end{aligned}
\right\}
\qquad
\begin{aligned}
&i\Delta \le \sigma < (i + 1)\Delta\\[6pt]
&j\Delta \le \tau < (j + 1)\Delta
\end{aligned}
$$

(a)    Show that for $t = k\Delta$, $T = N\Delta$ and $\tilde{h}(t, T)$ as defined in (5.6.4),

$$
\tilde{s}(t, T) \;=\; \int_0^T \tilde{h}(\sigma, T)\tilde{y}(t - \sigma)\, d\sigma
$$

$$
\tilde{s}(t - T, T) \;=\; \int_0^T \tilde{h}(T - \sigma, T)\,\tilde{y}(t - \sigma)\, d\sigma
$$

$$
\int_0^T \tilde{h}(\sigma, \sigma)\,\tilde{e}_b(t, \sigma)\, d\sigma \;=\; \tilde{s}(t + \Delta, T)
$$

$$
\int_0^T \tilde{h}(\sigma, \sigma)\,\tilde{e}_f(t - T + \sigma, \sigma)\, d\sigma \;=\; \tilde{s}(t - T, T)
$$

(b)   Assuming $y'(t)$ is uniformly continuous so that there exists a $K > 0$
      s.t. $\left| y'(t_1) - y'(t_2) \right| \leq K \left| t_1 - t_2 \right|$, show that $\lim_{\Delta \to 0} \bar{y}(t) = y'(t)$.

(b)   Use (a), (b) and Lemma 5.6.2 to show that for $T = N\Delta$ and
      $\tau = i\Delta < N\Delta$, $\lim_{\Delta \to 0} \left\| \hat{s}'(\tau, T) - \tilde{s}(\tau, T) \right\|_\infty = 0$.

# Chapter 6

# Optimal Linear Estimation with State-Space Filters

## 6.1  Introduction

In this chapter we will again consider estimation problems. However, whereas in Chapter 5 only covariance or spectral density descriptions were used for the signals, we will here adopt an alternative description based on state-space models. This will allow us to obtain a rather simple solution to the problem. In particular, it will no longer be necessary to restrict attention to stationary signals. Also, elegant sequential solutions to the problem will be possible.

The main result of this chapter will be to derive the optimal linear filter in state-space form. This is the celebrated Kalman filter. We will develop both its discrete and continuous form and discuss their interconnection. We will also discuss the problem of continuous state estimation with sampled measurements.

## 6.2  Signal Model

Rather than use the covariance description as in (5.2.1) – (5.2.6), we will here describe the observed continuous-time signal in state-space form:

$$\boxed{dx(t) = Ax(t)dt + dv(t)} \tag{6.2.1}$$

$$\boxed{dy(t) = Cx(t)dt + d\omega(t)} \tag{6.2.2}$$

where $v(t)$ and $w(t)$ are vector Wiener processes with zero means and incremental covariances $\Omega dt$ and $\Gamma dt$ respectively. We assume $v(t)$ and $w(t)$ are independent, and that $\Omega$ and $\Gamma$ are symmetric and positive semidefinite matrices.

For those readers not familiar with stochastic differential equations of the form (6.2.1) and (6.2.2) it will suffice, for our purposes, to think of them as ordinary differential equations driven by white noise. Thus, we may write the state equation (6.2.1) formally as

$$\frac{d}{dt}x(t) = Ax(t) + \overset{\bullet}{v}(t) \tag{6.2.3}$$

with output $y'(t)$ given by $Cx(t) + \overset{\bullet}{w}(t)$ . We use the notation $y'(t) = dy/dt$, thus

$$y'(t) = \frac{d}{dt}y(t) = Cx(t) + \overset{\bullet}{w}(t) \tag{6.2.4}$$

In (6.2.3), (6.2.4) $\overset{\bullet}{v}(t)$, $\overset{\bullet}{w}(t)$ denote independent continuous-time "white noise" processes. The covariance functions for $\overset{\bullet}{v}(t)$ and $\overset{\bullet}{w}(t)$ are Dirac pulse functions, i.e.

$$E\{\overset{\bullet}{v}(t)\,\overset{\bullet}{v}(s)^{T}\} = \Omega\delta(t-s) \tag{6.2.5}$$

$$E\{\overset{\bullet}{w}(t)\,\overset{\bullet}{w}(s)^{T}\} = \Gamma\delta(t-s) \tag{6.2.6}$$

The above description applies equally well to the non-stationary case where $A$, $C$, $\Omega$ and $\Gamma$ may all depend explicitly on time. In the simpler stationary problem, we can also evaluate signal spectral densities. Indeed, in (6.2.5), (6.2.6), it is interesting to observe that $\Omega$ and $\Gamma$ are in fact *spectral density* matrices. To see this, we recall that the Fourier transform of the covariance is the spectral density. Thus, the spectral density for $\overset{\bullet}{v}(t)$ is

$$S_{\overset{\cdot}{y}}(\omega) = \int\limits_{-\infty}^{\infty} \Omega\delta(\tau)e^{-j\omega\tau}d\tau$$

$$= \Omega \qquad\qquad (6.2.7)$$

It can also be readily shown (see Problem 6.1) that for $t \geq s$ the covariance of $y'$ satisfies:

$$E\{y'(t)y'(s)^T\} = Ce^{A(t-s)}\int\limits_{0}^{\infty} e^{A\sigma}\Omega e^{A^T\sigma}d\sigma C^T + \Gamma\delta(t-s) \qquad (6.2.8)$$

## 6.3   The Sampling Process

We wish to consider the situation where sampled signals are used. However, as in Chapter 5, it makes no sense to sample the output $y'(t)$ of the system (6.2.3), (6.2.4) directly. The reason is that the output has infinite variance, as follows from (6.2.8). Thus, direct sampling will simply lead to a signal of infinite variance which is clearly impractical.

The resolution of this problem is, as in Chapter 5, very simple : one *always* needs to use an anti-aliasing filter prior to sampling. The dynamics of this anti-aliasing filter then have to be built into the signal description. We assume the anti-aliasing filter to be a linear time invariant and causal SISO system given by its impulse response $h(t)$ and we denote by $\bar{y}(t)$ the filtered output signal - see Figure 6.3.1. In case $y'(t)$ is a vector we apply an anti-aliasing filter to each of its components.

We will use the same anti-aliasing filter as in Chapter 5. Thus, we assume a filter having the following impulse response:

$$h_0(\tau) = \begin{cases} 1/\Delta & \text{for } 0 \leq \tau < \Delta \\ 0 & \text{otherwise} \end{cases} \qquad (6.3.1)$$

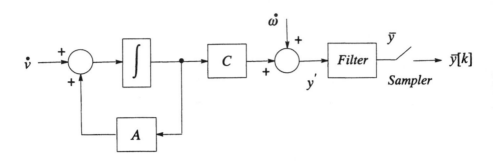

*Figure 6.3.1 : Sampling a Stochastic System.*

It is interesting to relate the sampled output $\bar{y}[k+1]$ to the continuous-time signal $y(t)$ whose derivative appears as the continuous-time output in the model (6.2.4). In particular, we have

$$\bar{y}[k+1] = \frac{1}{\Delta} \int\limits_{k\Delta}^{(k+1)\Delta} y'(\tau)d\tau$$

$$= \frac{1}{\Delta} \left[ \int\limits_{t_0}^{(k+1)\Delta} y'(\tau)\,d\tau - \int\limits_{t_0}^{k\Delta} y'(\tau)\,d\tau \right]$$

$$= \frac{1}{\Delta} \left[ \int\limits_{t_0}^{(k+1)\Delta} dy - \int\limits_{t_0}^{k\Delta} dy \right]$$

$$= \delta y[k] \tag{6.3.2}$$

where we have used (6.2.4) to link $y'$ and $dy/dt$.

Thus the sampled output $\bar{y}[k+1]$ is exactly $\delta y[k]$. This gives a nice connection to the continuous-time model (6.2.4) in which the output is $\dfrac{d}{dt}y(t)$.

## 6.4 Discrete Stochastic Model

Our next objective is to develop a corresponding discrete-time model describing the sampled response in Figure 6.3.1. We will use the notation $\delta y[k]$ for the output since this is precisely $\bar{y}[k+1]$ as shown in (6.4.1).

We have the following result:

Lemma 6.4.1 Let the system output of equations (6.2.3), (6.2.4) be sampled at an interval $\Delta$ using the anti-aliasing filter given in (6.3.1). Then the sampled output response can be described by the following discrete-time system (expressed in delta operator form):

$$\boxed{\delta x[k] = A_\delta x[k] + v_\delta[k]} \tag{6.4.1}$$

$$\boxed{\delta y[k] = C_\delta x[k] + \omega_\delta[k]} \tag{6.4.2}$$

where

$$x[k] = x(k\Delta) \tag{6.4.3}$$

$$\boxed{A_\delta = \frac{e^{A\Delta} - I}{\Delta}} \tag{6.4.4}$$

$$\boxed{C_\delta = \frac{1}{\Delta} C \int_0^\Delta e^{As} ds} \tag{6.4.5}$$

The sequences $v_\delta[k]$ and $\omega_\delta[k]$ are i.i.d. sequences having the following covariance structure

$$E\left\{\begin{bmatrix} v_\delta[k+m] \\ \omega_\delta[k+m] \end{bmatrix} \begin{bmatrix} v_\delta[k]^T \omega_\delta[k]^T \end{bmatrix}\right\} = \begin{bmatrix} \Omega_\delta & S_\delta \\ S_\delta^T & \Gamma_\delta \end{bmatrix} \delta_d[m] \tag{6.4.6}$$

where $\delta_d[m]$ is the discrete $1/\Delta$ pulse and

$$\begin{bmatrix} \Omega_\delta & S_\delta \\ S_\delta^T & \Gamma_\delta \end{bmatrix} = \frac{1}{\Delta} \int_0^\Delta \left( e^{\bar{A}_f \tau} \right) \begin{bmatrix} \Omega & 0 \\ 0 & \Gamma \end{bmatrix} \left( e^{\bar{A}_f \tau} \right)^T d\tau \qquad (6.4.7)$$

$$\bar{A}_f = \begin{bmatrix} A & 0 \\ C & 0 \end{bmatrix} \qquad (6.4.8)$$

**Proof :** From (6.2.3), (6.2.4), (6.4.1), the sampled output can be expressed as

$$\frac{d}{dt} \begin{bmatrix} x \\ y \end{bmatrix} = \begin{bmatrix} A & 0 \\ C & 0 \end{bmatrix} \begin{bmatrix} x \\ y \end{bmatrix} + \begin{bmatrix} \dot{v} \\ \dot{\omega} \end{bmatrix} \qquad (6.4.9)$$

$$\delta y[k] = \frac{y((k+1)\Delta) - y(k\Delta)}{\Delta} \qquad (6.4.10)$$

Solving equation (6.4.9) on the interval $k\Delta \leq t < (k+1)\Delta$ gives

$$\begin{bmatrix} x[k+1] \\ y[k+1] \end{bmatrix} = e^{\bar{A}_f \Delta} \begin{bmatrix} x[k] \\ y[k] \end{bmatrix} + \int_0^\Delta e^{\bar{A}_f(\Delta - \tau)} \begin{bmatrix} \dot{v}(\tau + k\Delta) \\ \dot{\omega}(\tau + k\Delta) \end{bmatrix} d\tau \qquad (6.4.11)$$

where

$$\bar{A}_f = \begin{bmatrix} A & 0 \\ C & 0 \end{bmatrix} \qquad (6.4.12)$$

$$e^{\bar{A}_f \sigma} = \begin{bmatrix} e^{A\sigma} & 0 \\ C \int_0^\sigma e^{A\tau} d\tau & I \end{bmatrix} \qquad (6.4.13)$$

The result then follows on using (6.2.5), (6.2.6)

$$\wedge\wedge\wedge$$

It is interesting to note that the above discrete-time model is compatible with the underlying continuous-time model, since it is clear from the various expressions that

$$\lim_{\Delta\to 0} A_\delta = A \qquad (6.4.14)$$

$$\lim_{\Delta\to 0} \begin{bmatrix} \Omega_\delta & S_\delta \\ S_\sigma^T & \Gamma_\delta \end{bmatrix} = \begin{bmatrix} \Omega & 0 \\ 0 & \Gamma \end{bmatrix} \qquad (6.4.15)$$

It is also interesting to note that the quantities $\Omega_\delta$, $\Gamma_\delta$, $S_\delta$ used in the discrete model are also spectral densities *not* variances. To see why this is so, we note that the Fourier transform of the covariance function is the spectral density. Hence, the (discrete) spectral density for $v_\delta[k]$ is

$$S_v^d(\omega) = \sum_{m=-\infty}^{\infty} \Delta E\{v_\delta[k+m]v_\delta[k]^T\}e^{-j\omega m\Delta}$$

$$= \sum_{m=-\infty}^{\infty} \Delta\Omega_d\delta_d[m]e^{-j\omega m\Delta}$$

$$= \Omega_d \qquad (6.4.16)$$

Some readers who have studied discrete-time stochastic models previously will know that it is usual to express these models in terms of signal variances. However, discrete variances are not well scaled for use in estimation. We will illustrate this claim by several examples.

**Example 6.4.1** : Say we have a low frequency signal that we want to estimate. We are told it is measured in wide band noise (i.e. noise that is almost white). Say the noise bandwidth is ten times that of the signal, and the signal to noise ratio of *variances* is –20 dB. Say we now keep the spectral density of the noise constant but increase its bandwidth to one hundred times that of the signal, then clearly the noise variance (the integral of spectral density) must go up by a factor of 10:1 and hence the new signal to noise ratio will fall to –40 dB. Does this mean that the achievable estimation accuracy in the second case is significantly worse than in the first case?

Solution The answer to this question is no because the 'amount' of noise that overlaps the signal bandwidth has not been changed. Indeed, the extra high frequency noise could easily be eliminated by a simple low pass filter. (Of course, the optimal filter will do this automatically for us.) Thus we see that spectral densities give a better view of the achievable accuracy in estimation problems.

△△△

Example 6.4.2 : Say we want to estimate the same low bandwidth signal as in Example 6.4.1 using sampled data. We use an initial sampling rate $1/\Delta$ which is already well above the Nyquist rate for the signal. However, somebody suggests we should sample twice as fast to double the amount of data. Is this a sensible suggestion?

Solution There are two equivalent ways of showing this suggestion is not helpful. Firstly, if we leave the anti-aliasing filter as it was for the original sampling rate, then we will find that the noise variance doubles when we sample at the faster rate due to folding. Hence we have twice as much data but with half the signal to noise ratio. Alternatively, we might consider changing the anti-aliasing filter so that it corresponds to the fast sampling rate. Now we still find that the noise variance doubles due to the extra bandwidth. (Variance is the integral of spectral density over the bandwidth).

△△△

## 6.5    The Discrete Kalman Filter

We consider the discrete-time model of (6.4.1) – (6.4.7). Our objective is to find the best linear estimate of the state $x[k]$ given the sampled data $\{\delta y[l]\ ;\ 0 \le l \le k-1\}$.

### 6.5.1   Model Simplification

Before we embark on the development of this optimal filter it is useful to eliminate the coupling term $S_\delta$ in (6.4.6). This will make subsequent expressions simpler and clearer. Since our filter will depend on the data $\delta y[l]$ we can transform the model so as to eliminate the coupling term. This is clarified in the following result.

Lemma 6.5.1   Consider the model (6.4.1), (6.4.7) with noise covariance

$$E\left\{\begin{bmatrix} v_\delta[k+m] \\ \omega_\delta[k+m] \end{bmatrix} \begin{bmatrix} v_\delta[k]^T \omega_\delta[k]^T \end{bmatrix}\right\} = \begin{bmatrix} \Omega_\delta & S_\delta \\ S_\delta^T & \Gamma_\delta \end{bmatrix} \delta_d[m] \tag{6.5.1}$$

This model can be expressed in the following equivalent form:

$$\delta x[k] = A_\delta' x[k] + v_\delta'[k] + S_\delta \Gamma_\delta^{-1} \delta y[k] \tag{6.5.2}$$

$$\delta y[k] = C_\delta x[k] + \omega_\delta[k] \tag{6.5.3}$$

where

$$A_\delta' = A_\delta - S_\delta \Gamma_\delta^{-1} C_\delta \tag{6.5.4}$$

$$E\left\{\begin{bmatrix} v_\delta'[k+m] \\ \omega_\delta[k+m] \end{bmatrix} \begin{bmatrix} v_\delta'[k]^T \omega_\delta[k] \end{bmatrix}\right\} = \begin{bmatrix} \Omega_\delta - S_\delta \Gamma_\delta^{-1} S_\delta^T & 0 \\ 0 & \Gamma_\delta \end{bmatrix} \tag{6.5.5}$$

Proof :  Equation (6.4.1) can be written as

$$\delta x[k] = A_\delta x[k] + v_\delta[k] - S_\delta \Gamma_\delta^{-1}(\delta y[k] - \delta y[k])$$

$$= A_\delta x[k] + v_\delta[k] - S_\delta \Gamma_\delta^{-1}(C_\delta x[k] + \omega_\delta[k] - \delta y[k])$$

$$= A_\delta' x[k] + S_\delta \Gamma_\delta^{-1} \delta y[k] + v_\delta'[k] \qquad (6.5.6)$$

where

$$v_\delta'[k] = v_\delta - S_\delta \Gamma_\delta^{-1} \omega_\delta[k] \qquad (6.5.7)$$

The result follows immediately.

$$\wedge\!\wedge\!\wedge$$

In the sequel we will be concerned with the optimal estimation of $x[k]$ given measurements of $\{\delta y[l]; 0 \le l \le k\}$. Hence, the term depending on $\delta y[k]$ in (6.5.2) can simply be treated as an additional known input, and, without loss of generality, we can thus simply assume that $v_\delta$ and $\omega_\delta$ are uncorrelated i.e. $S_\delta = 0$. If this is not the case, then we can always apply the preliminary transformation described in Lemma 6.5.1 to ensure that this is so.

### 6.5.2   The Optimal Filter

We assume that we are given an estimate $\hat{x}_0$ of the initial state $x[0]$ and that we know the initial mean square estimation error

$$P_0 = E\{(x[0] - \hat{x}_0)(x[0] - \hat{x}_0)^T\}$$

We also assume knowledge of $A_\delta$, $C_\delta$, $\Omega_\delta$ and $\Gamma_\delta$ and we want to estimate the whole state vector $x[k]$ from the measurement of $\bar{y}[k+1] = \delta y[k]$.

We will use a linear state-space observer of the following form:

$$\delta \hat{x}[k] = A_\delta \hat{x}[k] + H_\delta(\delta y[k] - C_\delta \hat{x}[k]); \qquad \hat{x}[0] = \hat{x}_0 \qquad (6.5.8)$$

and we want to find $H_\delta$ so that the mean square error

$$E\left\{(x[k] - \hat{x}[k])(x[k] - \hat{x}[k])^T\right\}$$

is minimized. Thus we want to minimize the matrix

$$P[k] \stackrel{\Delta}{=} E\left\{e[k]e[k]^T\right\} \qquad (6.5.9)$$

where

$$e[k] \stackrel{\Delta}{=} x[k] - \hat{x}[k] \qquad (6.5.10)$$

We then have the following result:

**Theorem 6.5.1** *The Kalman Filter*

(a) The optimal (in the mean square sense) linear estimator of the state of the system (6.4.1), (6.4.2) is given by the observer

$$\delta\hat{x}[k] = A_\delta\hat{x}[k] + H_\delta[k](\delta y[k] - C_\delta\hat{x}[k]); \qquad \hat{x}[k_0] = \hat{x}_0 \qquad (6.5.11)$$

$$H_\delta[k] = (I + \Delta A_\delta)\, P[k]C_\delta^T \left(\Gamma_\delta + \Delta C_\delta P[k]C_\delta^T\right)^{-1} \qquad (6.5.12)$$

The covariance of the optimal state estimation error satisfies the following Riccati difference equation.

$$\begin{aligned}
\delta P[k] &= A_\delta P[k] + P[k]A_\delta^T + \Omega_\delta \\
&\quad - (I + \Delta A_\delta)P[k]C_\delta^T R[k]^{-1} C_\delta P[k](I + \Delta A_\delta^T) \\
&\quad + \Delta A_\delta P[k]A_\delta^T \; ; \quad P[k_0] = P_0
\end{aligned} \qquad (6.5.13)$$

where

$$R[k] = \Gamma_\delta + \Delta C_\delta P[k]C_\delta^T \qquad (6.5.14)$$

(b)  For the special case of stationary signals, then subject to mild stabilizability and detectability assumptions, $P[k]$ converges to a constant positive definite matrix $P_\infty$ which results in an exponentially stable filter and $P_\infty$ satisfies the following algrebraic Riccati equation:

$$
\Delta A_\delta P_\infty A_\delta^T + A_\delta P_\infty + P_\infty A_\delta^T + \Omega_\delta \\
- (I + \Delta A_\delta) P_\infty C_\delta^T R^{-1} C_\delta P_\infty (I + \Delta A_\delta^T) = 0 \qquad (6.5.15)
$$

Proof :  (We present an outline only since this topic is covered in detail in numerous other sources - see references at the end of this chapter.)

(a)  Substituting equations (6.4.1), (6.4.2) and (6.5.10) into (6.5.8) we get

$$
\delta e[k] = (A_\delta - H_\delta C_\delta) e[k] + [I, -H_\delta] \begin{bmatrix} v_\delta[k] \\ \omega_\delta[k] \end{bmatrix} \qquad (6.5.16)
$$

Both $v_\delta[k]$ and $\omega_\delta[k]$ have zero means and are i.i.d. sequences. As a result, $e[k]$ is independent of both $v_\delta[k]$ and $\omega_\delta[k]$ . Hence,

$$
\text{cov}\{v_\delta[k], e[k]\} = 0
$$
$$
\text{cov}\{\omega_\delta[k], e[k]\} = 0
$$

and, using (6.4.6) with $S_\delta = 0$ , (6.5.9), (6.5.10) and (6.5.16) we get

$$
\delta P[k] = \Delta A_\delta P[k] A_\delta^T + A_\delta P[k] + P[k] A_\delta^T + \Omega_\delta \\
- H_\delta C_\delta P[k] (I + \Delta A_\delta^T) - (I + \Delta A_\delta) P[k] C_\delta^T H_\delta^T \\
+ H_\delta (\Gamma_\delta + \Delta C_\delta P[k] C_\delta^T) H_\delta^T \qquad (6.5.17)
$$

The right hand side of equation (6.5.17) can be rearranged as follows:

$$\delta P[k] = \Delta A_\delta P[k] A_\delta^T + A_\delta P[k] + P[k] A_\delta^T + \Omega_\delta$$

$$- (I + \Delta A_\delta) P[k] C_\delta^T R[k]^{-1} C_\delta P[k] (I + \Delta A_\delta^T)$$

$$+ \left( H_\delta - (I + \Delta A_\delta) P[k] C_\delta^T R[k]^{-1} \right)$$

$$\cdot R[k] \left( H_\delta^T - R[k]^{-1} C_\delta P[k] (I + \Delta A_\delta^T) \right) \qquad (6.5.18)$$

where

$$R[k] = \Gamma_\delta + \Delta C_\delta P[k] C_\delta^T \qquad (6.5.19)$$

Since $\delta P[k] = \left( (P[k+1] - P[k]) \right) / \Delta$ it is clear from (6.5.18) that given $P[k]$, then $P[k+1]$ is minimized by choosing $H_\delta$ at each time instant so that the last term in (6.5.18) is set to zero. Thus, at each time instant, we get

$$H_\delta[k] = (I + \Delta A_\delta) P[k] C_\delta^T (\Gamma_\delta + \Delta C_\delta P[k] C_\delta^T)^{-1} \qquad (6.5.20)$$

Any other choice for $H_\delta$ will result in a different matrix sequence $\tilde{P}[k] \geq P[k]$ for every $k \geq 0$.

Substituting (6.5.20) into (6.5.18) leads to the following equation for the optimal estimation error covariance:

$$\delta P[k] = \Delta A_\delta P[k] A_\delta^T + A_\delta P[k] + P[k] A_\delta^T$$

$$+ \Omega_\delta - (I + \Delta A_\delta) P[k] C_\delta^T R[k]^{-1} C_\delta P[k] (I + \Delta A_\delta^T) \qquad (6.5.21)$$

We call this equation a (discrete-time) matrix Riccati equation.

With $P[k_0] \stackrel{\Delta}{=} P_0 = E\{e[k_0]e[k_0]^T\}$ assumed to be known, the sequence $\{P[k]\}$ and hence $\{H_\delta[k]\}$ can be calculated off-line and used in the observer (6.5.11). This constitutes the well known Kalman filter.

(b) For the stationary case, we refer to other literature for details. It can be shown (Gelb (1974), Goodwin and Sin (1984), Middleton and Goodwin (1990)) that, if $\left(A_\delta, \Omega^{1/2}\right)$ is a stabilizable pair and $(C_\delta, A_\delta)$ a detectable pair the sequence $\{P[k]\}$ converges to a constant matrix $P_\infty > 0$. This leads to the stationary Kalman filter. By setting $\delta P_\infty = 0$ in (6.5.13) we see that $P_\infty$ will then be the solution of the algebraic Riccati equation

$$\Delta A_\delta P_\infty A_\delta^T + A_\delta P_\infty + P_\infty A_\delta^T + \Omega_\delta$$

$$- (I + \Delta A_\delta) P_\infty C_\delta^T R^{-1} C_\delta P_\infty (I + \Delta A_\delta^T) = 0 \qquad (6.5.22)$$

where

$$R = \Gamma_\delta + \Delta C_\delta P_\infty C_\delta^T$$

and the Kalman gain becomes also constant

$$H_\delta = (I + \Delta A_\delta) P_\infty C_\delta^T \left(\Gamma_\delta + \Delta C_\delta P_\infty C_\delta^T\right)^{-1} \qquad (6.5.23)$$

It is also shown in the references that this choice for the optimal steady state filter gain leads to a stable filter (subject to the stabilizability and detectability assumptions referred to above).

$$\wedge\wedge\wedge$$

We illustrate the above result by a simple example.

Example 6.5.1 : Consider the problem of estimating a sinewave of frequency $\omega_0 = \pi$ rad sec$^{-1}$ measured in white noise. Sampling with period $\Delta = 0.1$ sec. results in the discrete-time model

$$y[k] = s[k] + \omega[k]$$

where $\omega[k]$ is an i.i.d. stationary sequence, zero mean and variance 10.

(a) Design a steady-state Kalman filter to recover the sinewave $s[k]$.

(b) Simulate the resultant filter.

## Solution

(a) The state-space model for the sinewave can be written as

$$\delta x[k] = \begin{bmatrix} \dfrac{\cos \omega_0 \Delta - 1}{\Delta} & \dfrac{1}{\omega_0 \Delta} \sin \omega_0 \Delta \\ \dfrac{-\omega_0 \sin \omega_0 \Delta}{\Delta} & \dfrac{\cos \omega_0 \Delta - 1}{\Delta} \end{bmatrix} x[k]$$

$$= A_\delta x[k] \tag{6.5.24}$$

$$y[k] = \begin{bmatrix} 1 & 0 \end{bmatrix} x[k] + \omega[k] \tag{6.5.25}$$

The Kalman filter is then given by

$$\delta \hat{x}[k] = A_\delta \hat{x}[k] + H_\delta[y[k] - C\hat{x}[k]]$$

and the sinewave estimate is

$$\hat{s}[k] = C\hat{x}[k]$$

(b) A picture of the data used for this example is shown in Figure 6.5.1. It can be seen that the underlying sinewave has been all but obliterated by the noise.

A difficulty with the system (6.5.24), (6.5.25) is that it is not stabilizable from the process noise sequence since the model has eigenvalues on the stability boundary and there is no process noise. Thus, we assume the presence of a little process noise by adding $v_\delta[k]$ to the right hand side of (6.5.24) having different values of the spectral density $\Omega_d$.

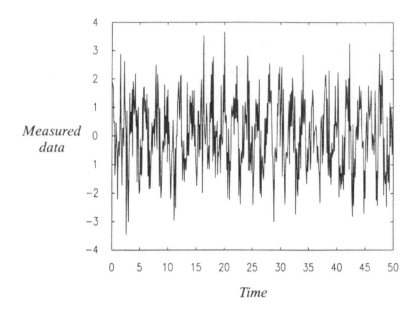

*Figure 6.5.1 : Noisy data for Example 6.5.1.*

We try three cases: $\Omega_\delta = 10^{-2}I$, $10^{-3}I$, $10^{-4}I$ and compare the results.

Solving the Riccati equation (6.5.15) for each case (with $\Gamma_\delta = 10\Delta = 1$) and calculating the corresponding $H_\delta$ using (6.5.23) we obtain

$$H_\delta = \begin{bmatrix} 0.9957 \\ 0.0463 \end{bmatrix} ; \begin{bmatrix} 0.3264 \\ 0.0048 \end{bmatrix} ; \begin{bmatrix} 0.1044 \\ 0.0005 \end{bmatrix}$$

Figure 6.5.2 shows the frequency response of resulting optimal filters linking $\bar{y}[k]$ to $\hat{x}_1[k]$ correspond to $\Omega_\delta = 10^{-2}I$, $10^{-3}I$ and $10^{-4}I$ respectively. We see that each filter is a bandpass filter centred on the signal frequency. As we hypothesised, with less process noise (i.e. $\Omega_\delta$ decreases) we see that the bandwidth

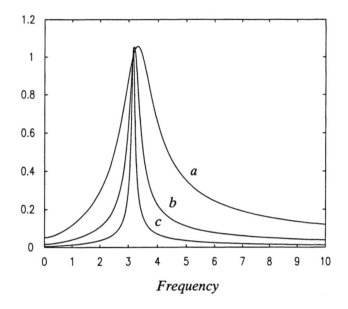

*Figure 6.5.2 :  Frequency responses of different filters :*
*(a) $\Omega_\delta = 10^{-2}I$,  (b) $\Omega_\delta = 10^{-3}I$,*
*(c) $\Omega_\delta = 10^{-4}I$.*

of the filter also decreases. This will mean that the filter transient time will increase but once locked onto the signal then the
rejection of the measurement noise will be better. We see that
we have a clear trade-off between responsiveness to real state
changes due to process noise and final ability to discriminate
against measurement noise.

The three filters corresponding to  $\Omega_\delta = 10^{-2}I$,  $10^{-3}I$  and
$10^{-4}I$  were simulated and the results appear in Figures 6.5.3,
6.5.4, 6.5.5 respectively. Note that as we reduce $\Omega_\delta$ the initial
transient time increases but the final sensitivity to measurement
noise decreases.

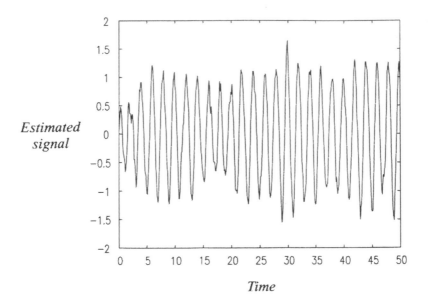

*Figure 6.5.3 :   Filter Response for* $\Omega_\delta = 10^{-2}I$.

The performance of all the filters is actually remarkably good
if we compare their outputs with the original measured data
shown in Figure 6.5.1. Of course, it should be remembered that
we assume perfect model of the sinewave frequency in this example.

△△△

## 6.5.3   Relationship to Finite Impulse Response Filters

We note that the optimal estimator (6.5.11), (6.5.12) expresses the state estimate $\hat{x}[k_1]$ as a linear function of the past data, $\delta y[k_0], \delta y[k_0 + 1]$,   …,
$\delta y[k_1 - 1]$ and the initial conditions $\hat{x}[k_0] = \hat{x}_0$. Thus, if we focus on a particular
linear combination of the states, say

$$s[k_1] = f^T x[k_1]$$

(6.5.26)

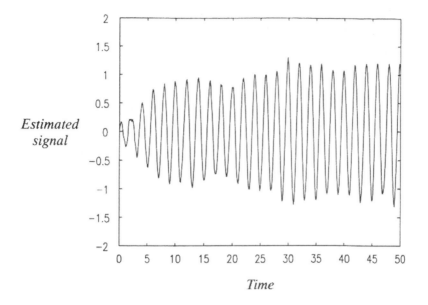

*Figure 6.5.4 : Filter Response for $\Omega_\delta = 10^{-3}I$.*

then the optimal estimator for $s[k_1]$ takes the form:

$$\hat{s}[k_1, N] = \Delta \sum_{i=1}^{N} h[i, N]\delta y[k_1 - i] + g^T \hat{x}_0 \qquad (6.5.27)$$

where $N$ is the number of data points i.e.

$$N = k_1 - k_0 \qquad (6.5.28)$$

and where the notation $\hat{s}[k_1, N]$ indicates that the estimate of $s[k_1]$ is based on the past $N$ data points. We have used the notation $\{h[i, N]\}$ to specifically indicate that the weighting function in (6.5.25) depend upon $N$ and hence $k_1$.

The reader will note the connection between (6.5.27) and the estimator described in equation (5.4.1). The differences are that here we include the initial condition term $\hat{x}_0$ and we assume that the description of $\{s[k]\}$ has a finite state

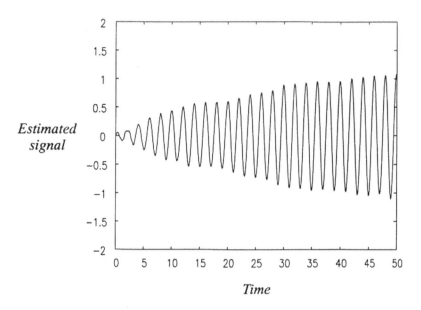

Estimated
signal

Time

*Figure 6.5.5 : Filter Response for $\Omega_\delta = 10^{-4}I$.*

model. Thus the Kalman filter is an appropriate generalization of the results in Chapter 5 to the situation where the signal model is described in state-space form. An important consequence of having a finite state-space model is that the filter impulse response in (6.5.27) can be expressed in terms of a finite dimensional model. In particular, it follows from Theorem 6.5.1 that the optimal sequence $\{h(i, N)\}$ can be expressed as

$$h[i, N] = f^T \psi[k_1, k - i + 1] H_\delta[k - i] \tag{6.5.29}$$

where $\{H_\delta[k]\}$ is the optimal filter gain sequence, and $\psi(k_1, i)$ is the state transition matrix of the optimal filter, i.e. $\psi[\cdot, \cdot]$ satisfies:

$$\delta\psi[k, k_0] = (A_\delta - H_\delta[k]C_\delta)\psi[k, k_0] \tag{6.5.30}$$

$$\psi[k_0, k_0] = I \tag{6.5.31}$$

The reverse procedure of converting the linear estimator (6.5.27) back to state-space form is explored in Problem 8.26.

## 6.6 Continuous-Time State Estimation

In this section we turn to the problem of estimating the *continuous-time* state of the original system (6.2.3), (6.2.4). We will study two cases: (a) when continuous data is available, and (b) when only sampled data is available.

### 6.6.1 Continuous-Time Data

Though it might be argued that one always uses sampled-data in practice, it is interesting to consider the limiting case of the filter in Theorem 6.5.1 as the sampling period goes to zero. This leads (formally) to the following result:

**Theorem 6.6.1** *The continuous-time Kalman filter*

Consider the continuous-time signal model (6.2.1), (6.2.2). The optimal estimator for the continuous-time state is

$$d\hat{x} = A\hat{x}dt + H(t)(dy - C\hat{x}dt) \qquad (6.6.1)$$

where

$$H(t) = P(t)C^T\Gamma^{-1} \qquad (6.6.2)$$

and *P(t)* satisfying the Riccati differential equation

$$\dot{P}(t) = AP(t) + P(t)A^T + \Omega - P(t)C^T\Gamma^{-1}CP(t) \qquad (6.6.3)$$

Proof : Equations (6.6.1) to (6.6.3) can be obtained formally by simply letting $\Delta \to 0$ in Theorem 6.5.1. Actually (6.6.1) to (6.6.3) can be rigorously derived directly from (6.2.1), (6.2.2) but this lies beyond the scope of this book.

$\wedge\!\wedge\!\wedge$

## 6.6.2   Sampled Data

A more practical question is what can be said about the *continuous-time* state given the discrete-time data $\{\delta y[k]\}$ .

The best estimate of the state at the sample points is given by the discrete Kalman filter. We want to know, however, something about the state between samples.

We recall that in the definition of the discrete-time model (6.4.1), (6.4.2) we actually defined $\delta y[k] = \bar{y}[k+1]$ . Hence equation (6.5.11) for the Kalman filter expresses $\hat{x}[k+1]$ in terms of $\bar{y}[k+1]$ . Thus the filter is proper but not strictly proper. Now, between samples the best estimate of the state is obtained by simply running the continuous-time state equation in open loop, beginning from the best available state estimate. This is true because no extra measurements are made and because the noise is white (i.e. uncorrelated with its past values). Thus we simply propagate the state estimate between samples using the open loop model. Hence the *continuous-time* state estimate given the sampled data can be expressed as follows:

$$\hat{x}(t) = e^{A(t-k\Delta)}\hat{x}[k] \; ; \; k\Delta \; \leq \; t < (k+1)\Delta \qquad (6.6.4)$$

$$\hat{x}[k+1] = e^{A\Delta}\hat{x}[k] + \Delta H_\delta[\bar{y}[k+1] - C_\delta\hat{x}[k]] \qquad (6.6.5)$$

where

$$C_\delta = \frac{1}{\Delta}C\int_0^\Delta e^{A\sigma}d\sigma \qquad (6.6.6)$$

Note that at the sampling instants (6.6.5) is equivalent to (6.5.11).

Actually these equations can be written in a more compact way as described below.

**Lemma 6.6.1** The best estimate of the *continuous-time* state $x(t)$ given the sampled data $\{\bar{y}[k+1]\} = \{\delta y[k]\}$ is given by

$$\boxed{\frac{d}{dt}\hat{x} = A\hat{x} + H_\delta v_f^s}$$   (6.6.7)

where $v_f^s$ is the $\Delta$ -pulse sampled version of $v_f$ and where $v_f$ is obtained by passing $v$ through an anti-aliasing filter having transfer function

$$H_0(s) = \left(1 - e^{-s\Delta}\right)/(s\Delta)$$

Finally, $v$ is the continuous innovations process defined by

$$\boxed{v = y' - C\hat{x}}$$   (6.6.8)

**Proof :** We define a continuous-time filter with sampled observations by the equation

$$\frac{d}{dt}\hat{x} = A\hat{x} + H_\delta v_f^s$$   (6.6.9)

where $v_f^s$ is a $\Delta$ -pulse sampled version of filtered innovations. The innovations are defined by (6.6.8).

If $v(t)$ is passed through the anti-aliasing filter having transfer function

$$H_0(s) = \frac{1 - e^{-s\Delta}}{s\Delta}$$   (6.6.10)

then the result at time $t = (k+1)\Delta$ is $v_f[k+1]$ where

$$v_f[k+1] = \frac{1}{\Delta} \int_{k\Delta}^{(k+1)\Delta} y'(\tau) \; d\tau - \frac{1}{\Delta} C \int_{k\Delta}^{(k+1)\Delta} \dot{x}(\tau) \; d\tau$$

$$= \bar{y}[k+1] - C_\delta \hat{x}[k] \qquad\qquad (6.6.11)$$

When this is $\Delta$-pulse sampled we obtain

$$v_f^s(t) = \sum_{k=-\infty}^{\infty} \Delta\delta(t-(k+1)\Delta)\{\bar{y}[k+1] - C_\delta \hat{x}[k]\}$$

$$= \bar{y}^s(t) - C_\delta \hat{x}^s(t-\Delta) \qquad\qquad (6.6.12)$$

Finally, substituting (6.6.12) into (6.6.7) leads to the result (6.6.4), (6.6.5).

$$\wedge\!\wedge\!\wedge$$

### 6.6.3   Relationships to Shannon Reconstruction Theorem

We have seen in Section 6.6.2 that it is possible, under certain circumstances, to use the Kalman filter to estimate a continuous-time signal from sampled data. The reader might well wonder if there is any connection between this result and the Shannon reconstruction theorem given in Chapter 2.

We recall that this latter theorem states that a signal can be perfectly recovered by the (infinite dimensional and noncausal) filter of equation (2.4.1) provided the signal is bandlimited to the range $(0, \omega_s/2)$. To put this problem into the framework of the Kalman filter we need a model having a finite state dimension. Hence we will consider a special case where the signal comprises a finite number of sinusoids whose frequencies lie in the given range $(0, \omega_s/2)$.

We will show that perfect signal reconstruction is possible using a *finite-dimensional and causal* linear filter when the signal comprises this finite combination of sinewave components. Given the sample values of the output, one can use a Kalman filter to estimate the state at each sample point. If one then propagates the continuous-time model between samples, then perfect signal recovery is achieved.

A key factor required for the success of this scheme is that the signal model be observable. A very interesting fact is that observability of this model is assured if the sinewave frequencies $\omega_i$ all lie in the range $0 \le \omega_i < \omega_s/2$ where $\omega_s$ is the sampling frequency. This result establishes an interesting link between the Shannon reconstruction theorem (which applies to arbitrary bandlimited signals) and Kalman observers (which apply to finite dimensional models).

The observability result can be easily argued in words as follows: Assume that the signal is composed of a finite number of sinewaves in the range $0 \le \omega_i < \omega_s/2$. We know from the Shannon reconstruction theorem that the sinewaves can be uniquely recovered from the sampled data. However, this implies that there does not exist a non-zero initial state which produces a zero output. This immediately implies observability.

To formally develop this idea, let the system be described by the following continuous-time state-space model for a finite sum of sinusoids:

$$\frac{d}{dt}x = A_c x + \dot{v} \tag{6.6.13}$$

$$y = Cx + \dot{\omega} \tag{6.6.14}$$

where $\dot{v}$, $\dot{\omega}$ are noise sources,

$$A_c = \text{diag}\left[A_{c_1}, A_{c_2} \ \cdots \ A_{c_n}\right] \tag{6.6.15}$$

$$A_{c_i} = \begin{bmatrix} 0 & 1 \\ -\omega_i^2 & 0 \end{bmatrix} \text{ for } i = 1, 2, \ \cdots \ n \tag{6.6.16}$$

$$C = \begin{bmatrix} 1 & 0 & 1 & 0 & \cdots & 0 \end{bmatrix} \tag{6.6.17}$$

Let $y(t)$ be sampled at period $\Delta$, then the corresponding sampled system satisfies

$$\delta x[k] = A_d x[k] \tag{6.6.18}$$

$$y[k] = C x[k] \tag{6.6.19}$$

where

$$A_d = \text{diag} \left[ A_{d_1}, A_{d_2} \ \cdots \ A_{d_n} \right] \tag{6.6.20}$$

$$A_{d_i} = \begin{bmatrix} \dfrac{\cos \omega_i \Delta - 1}{\Delta} & \dfrac{1}{\omega_i \Delta} \sin \omega_i \Delta \\[3mm] \dfrac{-\omega_i \sin \omega_i \Delta}{\Delta} & \dfrac{\cos \omega_i \Delta - 1}{\Delta} \end{bmatrix} \tag{6.6.21}$$

We then have the following result.

**Lemma 6.6.2** Consider the signal model described in (6.6.13) to (6.6.21). Provided the sampling period $\Delta$ is chosen so that $0 \leq \omega_i < \pi/\Delta$; $i = 1, \ldots,$ $n$, then the discrete model is observable.

**Proof :** Straightforward using the PBH test for observability – see Problem 6.6.

$$\wedge\wedge\wedge$$

Subject to observability, we can then construct an exponentially stable Kalman filter to recover the state *at the sample points* of the form:

$$\delta \hat{x} = A_\delta \hat{x} + H(y - C\hat{x}) \tag{6.6.22}$$

The inter-sample state estimates are finally obtained by propagating the continuous model (6.6.13) with the noise removed. Actually, this can be incorporated into the one state estimator by simply implementing (6.6.22) as in (6.6.7).

In summary, the Shannon reconstruction theorem shows how we can recover an arbitrary bandlimited signal using an infinite-dimensional non-causal filter whilst the Kalman filter shows how we can recover a finite set of sinusoids using a finite-dimensional causal filter provided the sinewave frequencies are known.

## 6.7    Further Reading and Discussion

There are many books on stochastic systems including Cox and Miller (1965), Caines (1988), Bhattacharya and Waymire (1990) and Doob (1953).

There are also many books dealing with optimal linear estimation – see for example Anderson and Moore (1979), Åström (1970), Goodwin and Sin (1984), Gelb (1974). See also Antoulas (1991).

Recent papers giving properties of the Riccati equation of optimal filtering are Chan et. al (1984) and de Souza et. al (1986).

The delta form of the Kalman filter is based on Salgado, Middleton and Goodwin (1986).

## 6.8    Problems

6.1    Derive equation (6.2.8).

6.2    Let the matrices $A$, $C$, $\Omega$ and $\Gamma$ in equations (6.2.3) – (6.2.4) be

$$A = \begin{bmatrix} 0 & 1 \\ -1 & -2 \end{bmatrix}, \quad C = [0, 1], \quad \Omega = 3I, \quad \Gamma = \begin{bmatrix} 1 & 1 \\ 1 & 3 \end{bmatrix}$$

(a)    Choose the anti-aliasing filter $h(t) = \begin{cases} e^{-at} & \text{for } t \geq 0 \text{ with } a > 0. \\ 0 & \text{for } t < 0 \end{cases}$

Find $A_\delta$, $C_\delta$, $\text{cov}\{v_\delta[k]\}$ and $\text{cov}\{\omega_\delta[k]\}$ for $\Delta = 1$, 0.5, 0.1 .

(b)    Repeat (a) with $h(t) = \begin{cases} \dfrac{1}{\Delta} & \text{for } t \in [0, \Delta) \\ 0 & \text{elsewhere} \end{cases}$ .

6.3   In an attempt to simulate the continuous process described in this chapter
      we do the following: For the sampling interval $\Delta$ , we generate two inde-
      pendent i.i.d sequences each with zero mean and covariance $1/\Delta_1$ ,
      $\{v_1[k]\}$ and $\{\omega_1[k]\}$ . Let $y'[k]$ be defined now by

$$\delta x[k] = \frac{1}{\Delta_1}\left(e^{-a\Delta_1} - 1\right)x[k] + v_1[k]$$

$$y'[k] = x[k] + \omega_1[k]$$

where $a > 0$ and $\delta = q - 1/\Delta_1$ is the delta operator. The continuous process $y'(t)$ is now generated by passing $y'[k]$ through a ZOH.

(a)   Express $\text{cov}\{y'(t)\} = E\{y'(t+\tau)y'(\tau)\}$ (assume steady state solu-
      tions in terms of $a$ and $\Delta_1$ ).

(b)   Find what happens to your result in (a) when $\Delta_1 \to 0$. Compare
      with the result in equation (6.2.8).

6.4   (a)   Show that over the interval $k\Delta < t \le (k+1)\Delta$ the system in Figure
            6.3.1 can also be described in state-space form

$$\frac{d}{dt}\begin{bmatrix} x \\ \bar{y} \end{bmatrix} = \begin{bmatrix} A & 0 \\ \frac{1}{\Delta}C & 0 \end{bmatrix}\begin{bmatrix} x \\ \bar{y} \end{bmatrix} + \begin{bmatrix} I & 0 \\ 0 & \frac{1}{\Delta}I \end{bmatrix}\begin{bmatrix} \dot{v} \\ \dot{\omega} \end{bmatrix}$$

            where $\bar{y}(k\Delta) = 0$;   $\bar{y}[k+1] = \bar{y}((k+1)\Delta)$ .

(b)   Hence show that the model can be expressed as

$$\begin{bmatrix} x[k+1] \\ \bar{y}[k+1] \end{bmatrix} = e^{\tilde{A}\Delta}\begin{bmatrix} x[k] \\ 0 \end{bmatrix} + \int_0^\Delta e^{\tilde{A}(\Delta-\tau)}\begin{bmatrix} \dot{v}(\tau + k\Delta) \\ \frac{1}{\Delta}\dot{\omega}(\tau + k\Delta) \end{bmatrix} d\tau$$

      where

$$\tilde{A} = \begin{bmatrix} A & 0 \\ \frac{1}{\Delta}C & 0 \end{bmatrix}$$

(c)     Show that the result in part (b) can be expressed as

$$\delta x[k] = A_\delta x[k] + v_\delta[k]$$

$$y_0 = \bar{y}[k+1] = C_\delta x[k] + \omega_\delta[k]$$

where

$$A_\delta = \frac{e^{A\Delta} - I}{\Delta}$$

$$C_\delta = \frac{1}{\Delta} C \int_0^\Delta e^{A(\Delta - \tau)} d\tau$$

$$\begin{bmatrix} v_\delta[k] \\ \omega_\delta[k] \end{bmatrix} = \begin{bmatrix} \frac{1}{\Delta}I & 0 \\ 0 & I \end{bmatrix} \int_0^\Delta e^{\tilde{A}(\Delta - \tau)} \begin{bmatrix} \dot{v}(\tau + k\Delta) \\ \frac{1}{\Delta}\dot{\omega}(\tau + k\Delta) \end{bmatrix} d\tau$$

(d)     Hence show that an alternative expression for the covariance of

$$\begin{bmatrix} v_\delta \\ \omega_\delta \end{bmatrix} \text{ is } E\left\{ \begin{bmatrix} v_\delta \\ \omega_\delta \end{bmatrix} \begin{bmatrix} v_\delta^T & \omega_\delta^T \end{bmatrix} \right\} = \begin{bmatrix} \Omega_\delta & S_\delta \\ S_\delta^T & \Gamma_\delta \end{bmatrix} \delta_d[m]$$

where $\begin{bmatrix} \Omega_\delta & S_\delta \\ S_\delta^T & \Gamma_\delta \end{bmatrix}$

$$= \begin{bmatrix} \frac{1}{\Delta}I & 0 \\ 0 & I \end{bmatrix} \int_0^\Delta \left( e^{\tilde{A}\tau} \right) \begin{bmatrix} \Delta\Omega & 0 \\ 0 & \frac{1}{\Delta}\Gamma \end{bmatrix} \left( e^{\tilde{A}\tau} \right)^T d\tau \begin{bmatrix} \frac{1}{\Delta}I & 0 \\ 0 & I \end{bmatrix}$$

(e)     Show that the expression for $\begin{bmatrix} \Omega_\delta & S_\delta \\ S_\delta^T & \Gamma_\delta \end{bmatrix}$ in part (d) is equal to that

given in Lemma 6.4.1.

6.5     Say we are told that a signal is measured in wideband noise having spectral density $S$ and a sampling period of $\Delta$. We then have two people who design Kalman filters. One converts the spectral density into a variance by computing $S2\pi/\Delta$. The other uses $S$ directly via the model (6.4.1) – (6.4.7). We are subsequently told that the noise has a bandwidth $B_N$ which is greater than the signal bandwidth but less than $\pi/\Delta$. The person who used the model expressed in terms of spectral densities knows that she need make no changes to the design. However, the other person is stuck with a dilemma : should he adjust the noise variance to its real value of $2SB_n$ or not.

Discuss why the second person should use the incorrect variance $S2\pi/\Delta$.

6.6    Verify the observability result in Lemma 6.6.2.

6.7    (a)    Show that the system (6.5.24), (6.5.25) is not stabilizable from $\overset{\bullet}{v}$ unless we hypothesis that (the process noise spectral density) $\Omega$ is non–singular.

       (b)    Show that as $\Omega \to 0$, the Kalman filter given in Example 6.5.1 approaches a notch filter of diminishing bandwidth.

       (c)    Hence discuss why we should assume some process noise is present in (6.5.24) even if in fact it is not there.

6.8    Consider a Stochastic model comprising two subparts – a stabilizable stochastic process and a deterministic process. Thus we write

$$\delta x_1 = A_1 x_1 + v$$
$$\delta x_2 = A_2 x_2$$
$$y = C_1 x_1 + C_2 x_2 + \omega$$

Show that this model can be converted to the form given in (6.4.1), (6.4.2).

6.9    Use the optimal filter of Section 6.5 to derive a simple expression for an optimal $d$-step-ahead of predictor of $\bar{y}[k]$.

# Chapter 7

# Periodic and Multirate Filtering

## 7.1    Introduction

In this chapter we will consider more general filtering problems which apply to cases where the underlying system is inherently time-varying. We will give particular emphasis to the special case of periodic systems. Potential application areas of these results are extensive and include subband coding of speech and video, narrowband filtering, short-time spectral analysis and geophysical signal processing.

We will present both frequency domain and time domain methods for analyzing periodic filters. Our results will build on the results presented in earlier chapters.

Periodic filtering problems can arise in a variety of ways. For example, the underlying system under study may itself be inherently periodic. Examples of such systems are those involving daily or yearly cycles. Alternatively, the periodicity may arise from the imposition of various sampling strategies. For example, it may be that certain ecological data is collected only on weekdays, giving a $(1\,1\,1\,1\,1\,0\,0)$ type periodic sampling pattern. In other cases, the periodicity may be due to the deliberate introduction of decimators and filters (see Section 1.7.1) so as to achieve computational speed advantages over filters that run at a single sampling rate.

We will develop tools applicable to all of the above scenarios by considering general periodic systems.

## 7.2    Models for Periodic Linear Systems

We begin by showing that sampling a continuous-time periodic process at a rate which bears an integer relationship to the period results in a periodic discrete-time process.

Let us start with the continuous-time linear *periodic* system.

$$\dot{x}(t) = A(t)x(t) + B(t)u(t) \tag{7.2.1}$$

$$y(t) = C(t)x(t) \tag{7.2.2}$$

where

$$A(t+T) = A(t), \quad B(t+T) = B(t), \quad C(t+T) = C(t) \quad \text{for all } t \tag{7.2.3}$$

**Definition 7.2.1** For the system in (7.2.1) – (7.2.2) the monodromy matrix $\psi_A(t_0)$ is defined as

$$\psi_A(t_0) = \Phi(t_0 + T, t_0) \tag{7.2.4}$$

where $\Phi(t, \tau)$ is the transition matrix corresponding to $A(t)$.

$$\wedge\wedge\wedge$$

It can then be readily shown that

$$\Phi(t+T, \tau+T) = \Phi(t, \tau) \quad \text{for any } t, \tau \tag{7.2.5}$$

and that the eigenvalues of $\psi_A(t_0)$ are independent of $t_0$ (see Problem 7.1).

Then we have

**Lemma 7.2.1** The system (7.2.1) – (7.2.2) is asymptotically stable if and only if all eigenvalues of $\psi_A(t_0)$ are inside the unit circle.

**Proof :** From (7.2.1) we have for $t \geq kT$ and $u(t) \equiv 0$

$$x(t) = \Phi(t, kT)x(kT) \tag{7.2.6}$$

and for $t = (k + 1)T$

$$x((k + 1)T) = \psi_A(kT)x(kT)$$

$$= \psi_A(0)x(kT)$$

where we have used (7.2.5).

Clearly, if the eigenvalues of $\psi_A(0)$ are inside the unit circle, $\lim_{k \to \infty} x(kT) = 0$. Under very mild restrictions on $A(t)$, $\phi(t, kT)$ is bounded on the interval $(kT, (k + 1)T)$ hence, from (7.2.6) we also have $\lim_{t \to \infty} x(t) = 0$ for any initial condition. On the other hand, if $\psi_A(0)$ has an eigenvalue on or outside the unit circle, there exists initial conditions such that $\lim_{k \to \infty} x(kT) \neq 0$, which completes the proof.

$$\wedge\wedge\wedge$$

Next we recall some basic definitions from linear system theory:

**Definition 7.2.2** The system (7.2.1) - (7.2.2) is *reachable* on the interval $[t_0, t_0 + T)$ if for any vectors $x_0$ and $x_1$ there exists a function $u(\tau)$, $\tau \in [t_0, t_0 + T)$ which takes the system from state $x_0$ at time $t_0$ to state $x_1$ at time $t_0 + T$.

**Definition 7.2.3** The system (7.2.1) - (7.2.2) is *observable* in the interval $[t_0, t_0 + T)$ if any initial state $x(t_0) = x_0$ with $u(t) \equiv 0$ can be uniquely recovered from the measurements $y(\tau)$, $\tau \in [t_0, t_0 + T)$.

We also state the following linear system theory results without proof.

**Lemma 7.2.2** The system (7.2.1) - (7.2.2) is reachable (observable) on the interval $[t_0, t_0 + T)$ if and only if the controllability grammian C (observability grammian O) is nonsingular, where

$$C = \int_{t_0}^{t_0+T} \Phi(t_0 + T, \tau) \; B(\tau) \; B(\tau)^T \; \Phi^T(t_0 + T, \tau) \; d\tau \qquad (7.2.7)$$

$$O = \int_{0}^{t_0+T} \Phi^T(\tau, t_0) \; C^T(\tau) \; C(\tau) \; \Phi(\tau, t_0) \; d\tau \qquad (7.2.8)$$

## 7.3 The Raising Procedure

An idea that can be used to great advantage is the notion that each periodic system is equivalent to a linear time-invariant system if the inputs and outputs are blocked together in vectors of length equal to the period of the periodic system. This idea is commonly called 'raising' and is discussed below.

Consider a discrete-time periodic system (say a filter) expressed in state-space form as follows:

$$x^c[k + 1] = F[k]x^c[k] + G[k]e[k] \qquad (7.3.1)$$

$$u[k] = J[k]x^c[k] + K[k]e[k] \qquad (7.3.2)$$

where $\{F[k]\}$, $\{G[k]\}$, $\{J[k]\}$ and $\{K[k]\}$ are periodic sequences of period $N$.

Assuming all sequences in these equations start at $k = 0$ we define the following 'raised' vectors:

$$e_R[m] = \Big[e[mN], \; e[mN + 1], \; \ldots, \; e[mN + N - 1]\Big]^T \in \mathbf{R}^N \qquad (7.3.3)$$

$$u_R[m] = \Big[u[mN], \; u[mN + 1], \; \ldots, \; u[mN + N - 1]\Big]^T \in \mathbf{R}^N \qquad (7.3.4)$$

Then, the following relationship between $e_R[m]$ and $u_R[m]$ holds.

**Lemma 7.3.1** With equations (7.3.1), (7.3.2) and definitions (7.3.3), (7.3.4), the response of the periodic system (7.3.1), (7.3.2) can be described by the following *time-invariant* raised system:

$$x_R^c[m+1] = F_R x_R^c[m] + G_R e_R[m] \qquad (7.3.5)$$

$$u_R[m] = J_R x_R^c[m] + K_R e_R[m] \qquad (7.3.6)$$

where

$$F_R = F[N-1]\, F[N-2]\, \dots\, F[0] \qquad (7.3.7)$$

$$G_R = \Big[F[N-1]\, \dots\, F[1]\, G[0],\quad F[N-1]\, \dots\, F[2]\, G[1],$$
$$\dots\,,\quad F[N-1]\, G[N-2],\quad G[N-1]\Big] \qquad (7.3.8)$$

$$J_R = \begin{bmatrix} J[0] \\ J[1]F[0] \\ J[2]F[1]F[0] \\ \cdot \\ \cdot \\ \cdot \\ J[N-1]F[N-2]\, \cdots\, F[0] \end{bmatrix} \qquad (7.3.9)$$

$$
K_R = \begin{bmatrix} K[0] \\ J[1]G[0] \\ J[2]F[1]G[0] \\ \cdot \\ \cdot \\ \cdot \\ J[N-1]F[N-2] \cdots F[1]G[0] \end{bmatrix},
$$

$$
\begin{bmatrix} 0 & & 0 \\ K[1] & & 0 \\ J[2]G[1] & & 0 \\ \cdot & , \cdots , & \cdot \\ \cdot & & \cdot \\ \cdot & & \cdot \\ J[N-1]F[N-2] \cdots F[2]G[1] & & K[N-1] \end{bmatrix} \qquad (7.3.10)
$$

$$
x_R^c[m] = x^c[mN] \qquad (7.3.11)
$$

Proof :   From (7.3.1) we have

$$
u[mN+l] = J[mN+l]x^c[mN+l] + K[mN+l]e[mN+l]
$$

$$
= J[l]x^c[mN+l] + K[l]e[mN+l] \qquad (7.3.12)
$$

and from (7.3.1) we have

$$
x^c[mN+l] = F[mN+l-1] \cdots F[mN]x^c[mN]
$$

$$
+ \sum_{p=0}^{l-2} F[mN+l-1] \cdots F[mN+p+1]G[mN+p]e[mN+p]
$$

$$
+ G[mN+l-1]e[mN+l-1] \qquad 1 \le l \le N
$$

$$= F[l-1] \ \cdots \ F[0]x^c[mN]$$

$$+ \sum_{p=0}^{l-2} F[l-1] \ \cdots \ F[p+1]G[p]e[mN+p]$$

$$+ G[l-1]e[mN+l-1] \qquad 1 \le l \le N \qquad (7.3.13)$$

Denoting $x_R^c[m] = x^c[mN]$ and writing (7.3.13) for $l = N$ establishes (7.3.5), (7.3.7) and (7.3.8). Substituting (7.3.13) into (7.3.12) for $l = 0, 1,$ ..., $N-1$ establishes (7.3.6), (7.3.9) and (7.3.10), which completes the proof of the lemma.

$$\wedge\wedge\wedge$$

We illustrate the raising mechanism described above by a simple example.

**Example 7.3.1 :** Suppose we are given the system in equations (7.3.1) – (7.3.2) where

$$F[k] = (-1)^k F, \quad G[k] = G, \quad J[k] = (1 + \cos(k\pi))J, \quad K[k] = 0$$

Develop the raised system and calculate its transfer matrix.

**Solution** We first observe that indeed this is a periodic linear system with periodicity $N = 2$. Then, using Lemma 7.3.1, we assign

$$F_R = -F^2$$

$$G_R = [-F, I]G$$

$$J_R = \begin{bmatrix} 2J \\ 0 \end{bmatrix}$$

$$K_R = 0$$

The resultant $2 \times 2$ linear system has the following transfer function (in the shift domain):

$$\tilde{C}_R(z) = J_R(zI - F_R)^{-1}G_R + K_R$$

$$= \begin{bmatrix} 2J \\ 0 \end{bmatrix} (zI + F^2)^{-1}[-F, I]G$$

$$= \begin{bmatrix} -2J(zI + F^2)^{-1}FG & 2J(zI + F^2)^{-1}G \\ 0 & 0 \end{bmatrix}$$

$\wedge\wedge\wedge$

We have thus seen that a periodic controller (or indeed any periodic system) of period $N$ can be converted into an $N$-input, $N$-output linear time-invariant system.

**Remark 7.3.1** An obvious question is "should we implement periodic controllers in their raised form?" The answer is quite clear – in its raised form the controller does block processing. It takes a block of its inputs $\{e[mN], \ldots,$ $e[mN + N - 1]\}$ at time $k = mN + N - 1$ and produces the block $\{u[mN], \ldots,$ $u[mN + N - 1]\}$. But the system expects to get $u[mN]$ at time $k = mN$ and so on. Thus direct implementation in the 'raised' form is not possible due to causality issues. However, subject to structural constraints on the raised controller, the output response will indeed be causally related to the input and hence can be implemented sequentially, which, as we will see, results (generally) in a periodic system. Thus, the raised form is convenient for analysis and we need only remember to take the outputs sequentially to achieve a causal implementation.

$\wedge\wedge\wedge$

In this regard, we claim the following converse to Lemma 7.3.1:

**Lemma 7.3.2** Every $N$-input $N$-output linear time-invariant system with a transfer matrix $\tilde{C}_R(z)$ such that $\tilde{C}_R(\infty)$ is a lower triangular matrix, and such that the inputs and outputs are as in (7.3.3), (7.3.4), can be implemented sequentially as a linear SISO $N$-periodic system.

**Proof :** The proof will be given by constructing the sequential implementation.

Given the $N \times N$ proper transfer matrix

$$
\tilde{C}_R(z) = \begin{bmatrix}
\tilde{C}_{11}(z) & \tilde{C}_{12}(z) & \cdots & \tilde{C}_{1N}(z) \\
\tilde{C}_{21}(z) & \tilde{C}_{22}(z) & \cdots & \tilde{C}_{2N}(z) \\
\cdot & \cdot & & \cdot \\
\cdot & \cdot & & \cdot \\
\cdot & \cdot & & \cdot \\
\tilde{C}_{N1}(z) & \tilde{C}_{N2}(z) & \cdots & \tilde{C}_{NN}(z)
\end{bmatrix} \tag{7.3.14}
$$

with all $\tilde{C}_{ij}(z)$, $i < j$, strictly proper, define the following $N$ transfer functions:

$$
\tilde{C}_l(z) = \sum_{j=1}^{N} z^{j-l-1} \tilde{C}_{l+1,j}(z^N) \qquad l = 0, 1, \ldots, N-1 \tag{7.3.15}
$$

Note that since $\tilde{C}_{ij}(z)$ for $i < j$ are strictly proper so are $z^{N-1}\tilde{C}_{ij}(z^N)$ for $i < j$. Hence, all transfer functions defined in (7.3.15) are proper.

Let $(F, G)$ be a reachable pair such that the characteristic polynomial of $F$ is the least common denominator of $\tilde{C}_l(z)$ and $J[l]$, $K[l]$ $l = 0, 1, \ldots,$ $N-1$ such that $\left( J[l], F, G, K[l] \right)$ is a realization of $\tilde{C}_l(z)$. Then we claim that

$$
x^c[k+1] = Fx^c[k] + Ge[k] \tag{7.3.16}
$$

$$
u[k] = J[l]x^c[k] + K[l]e[k] \tag{7.3.17}
$$

with $l = k \bmod N$, is a sequential realization of $\tilde{C}_R(z)$ (clearly periodic).

To verify this claim we raise the system $(7.3.16) - (7.3.17)$ and show that the transfer matrix of the raised system is indeed $\tilde{C}_R(z)$.

From $(7.3.7) - (7.3.10)$ and $(7.3.16) - (7.3.17)$ we get

$$
J_R = \begin{bmatrix} J[0] \\ J[1]F \\ \cdot \\ \cdot \\ \cdot \\ J[N-1]F^{N-1} \end{bmatrix}
$$

$$
K_R(z) = \begin{bmatrix} K[0] & 0 & - - - - & 0 \\ J[1]G & K[1] & - - - - & 0 \\ \cdot & \cdot & & \cdot \\ \cdot & \cdot & & \cdot \\ \cdot & \cdot & & \cdot \\ J[N-1]F^{N-2}G & J[N-1]F^{N-3}G & - - - - & K[N-1] \end{bmatrix}
$$

The corresponding transfer matrix is

$$
\tilde{C}^1(z) = J_R(zI - F_R)^{-1}G_R + K_R
$$

$$
= \begin{bmatrix} J[0] \\ J[1]F \\ \cdot \\ \cdot \\ \cdot \\ J[N-1]F^{N-1} \end{bmatrix} (zI - F^N)^{-1} \left[ F^{N-1}G, \ F^{N-2}G, \ ..., \ G \right]
$$

$$+ \begin{bmatrix} K[0] & 0 & - - - - & 0 \\ J[1]G & K[1] & - - - - & 0 \\ \cdot & \cdot & & \cdot \\ \cdot & \cdot & & \cdot \\ \cdot & \cdot & & \cdot \\ J[N-1]F^{N-2}G & J[N-1]F^{N-3}G & - - - - & K[N-1] \end{bmatrix}$$

Hence

$$\tilde{C}^1_{ij}(z) = J[i-1]F^{i-1}(zI-F^N)^{-1}F^{N-j}G + L_{ij}$$

where

$$L_{ij} = \begin{cases} J[i-1]F^{i-j-1}G & \text{for } i > j \\ K[i] & \text{for } i = j \\ 0 & \text{for } i < j \end{cases}$$

We then calculate the following sums

$$\sum_{j=1}^{N} z^{-i}\tilde{C}^1_{ij}(z^N) = J[i-1]F^{i-1}(z^N I - F^N)^{-1}\left[F^{N-1}z + F^{N-2}z^2 + \dots + z^N I\right]Gz^{-i}$$
$$+ J[i-1]\left[F^{i-2}z + F^{i-3}z^2 + \dots + z^{i-1}I\right]Gz^{-i} + K[i-1]$$

$$= J[i-1]F^{i-1}(z^N I - F^N)^{-1}(z^N I - F^N)(I - z^{-1}F)^{-1}Gz^{-i}$$
$$+ J[i-1](z^{i-1}I - F^{i-1})(I - z^{-1}F)^{-1}Gz^{-i} + K[i-1]$$

$$= J[i-1](zI-F)^{-1}G + K[i-1]$$

From the definition of $\left(J[l], \ F, \ G, \ K[l]\right)$ and (7.3.15) we have

$$\sum_{j=1}^{N} z^{j-i} \tilde{C}_{ij}^1(z^N) = \tilde{C}_{i-1}(z)$$

$$= \sum_{j=1}^{N} z^{j-i} \tilde{C}_{ij}(z^N)$$

Equating the coefficients in the numerator and denominator of the above expression.

$$\tilde{C}_{ij}^1(z) = \tilde{C}_{ij}(z)$$

which completes the proof of the lemma.

$$\wedge\wedge\wedge$$

**Example 7.3.2 :** Let

$$\tilde{C}_R(z) = \begin{bmatrix} \dfrac{z+0.2}{z-0.5} & \dfrac{1}{z-0.5} \\[3mm] \dfrac{z}{z-0.5} & \dfrac{5z-1}{z-0.5} \end{bmatrix}$$

Find the corresponding sequential implementation.

**Solution**

$$\tilde{C}_0(z) = \frac{z^2+0.2}{z^2-0.5} + \frac{z}{z^2-0.5} = \frac{z^2+z+0.2}{z^2-0.5} = \frac{z+0.7}{z^2-0.5} + 1$$

$$\tilde{C}_1(z) = \frac{z}{z^2-0.5} + \frac{5z^2-1}{z^2-0.5} = \frac{5z^2+z-1}{z^2-0.5} = \frac{z+1.5}{z^2-0.5} + 5$$

Let

$$F = \begin{bmatrix} 0 & 1 \\ 0.5 & 0 \end{bmatrix}, \qquad G = \begin{bmatrix} 0 \\ 1 \end{bmatrix}$$

$$J[0] = [0.7 , 1] \qquad J[1] = [1.5 , 1]$$

$$K[0] = 1 \qquad K[1] = 5$$

So, the sequential implementation of the given $\tilde{C}_R(z)$ will be

$$x^c[k+1] = Fx^c[k] + Ge[k]$$

$$u[k] = J[l]x^c[k] + K[l]e[k]$$

where $l = k \bmod 2$.

Note that this realization is not necessarily minimal.

$\wedge\wedge\wedge$

The result of the implementation presented in the proof of the Lemma 7.3.2 immediately suggests the following:

**Remark 7.3.2** Every linear periodic system can be realized so that the state equation is time-invariant but where the output map is periodic.

$\wedge\wedge\wedge$

The power of the raising technique discussed above is that the periodic filter design can be carried out using the well established tools used for time-invariant systems. However, this time domain approach can hide some subtle features of the response within a period. These issues are most easily seen in the frequency domain. This alternative viewpoint is taken up in the next section.

## 7.4    Frequency Domain Analysis of Periodic Filters

To obtain a frequency domain description of periodic filters we will consider a system that is time-invariant, save for a periodic outpit map. Note that, in view of Remark 7.3.2, this is without loss of generality. Thus we write the filter as

$$\dot{x}_{[k+1]} = Fx[k] + Ge[k] \tag{7.4.1}$$

$$u[k] = J[k]x[k] + K[k]e[k] \tag{7.4.2}$$

where $\{e[k]\}$, $\{u[k]\}$ are the filter input and output respectively, and $\{J[k]\}$, $\{K[k]\}$ are assumed to be $N$-periodic.

Applying the Fourier transform to (7.4.1), (7.4.2) and using (1.4.10) we obtain

$$e^{j\omega\Delta}X^d(\omega) = FX^d(\omega) + GE^d(\omega) \tag{7.4.3}$$

$$U^d(\omega) = \frac{1}{2\pi} \int_{-\frac{\pi}{\Delta}}^{\frac{\pi}{\Delta}} J^d(\eta)X^d(\omega-\eta)\, d\eta + \frac{1}{2\pi} \int_{-\frac{\pi}{\Delta}}^{\frac{\pi}{\Delta}} K^d(\eta)E^d(\omega-\eta)\, d\eta \tag{7.4.4}$$

where $E^d(\omega)$, $U^d(\omega)$, $J^d(\omega)$ and $K^d(\omega)$ are the transforms of the sequences $e[k]$, $u[k]$, $J[k]$ and $K[k]$ respectively.

To calculate $J^d(\omega)$ and $K^d(\omega)$ we exploit the fact that the corresponding sequences are $N$-periodic and hence have Fourier series representations :

$$\bar{J}[k] = \frac{1}{N}\sum_{i=0}^{N-1} J_i\, e^{j\omega_N\Delta ki} \tag{7.4.5}$$

$$\bar{K}[k] = \frac{1}{N}\sum_{i=0}^{N-1} K_i\, e^{j\omega_N\Delta ki} \tag{7.4.6}$$

where

$$J_i = \sum_{k=0}^{N-1} \bar{J}[k]\, e^{-j\omega_N\Delta ki} \tag{7.4.7}$$

$$K_i = \sum_{k=0}^{N-1} \overline{K}[k] \; e^{-j\omega_N k \Delta i} \qquad (7.4.8)$$

and

$$\omega_s = \frac{2\pi}{\Delta}, \quad \omega_N = \frac{\omega_s}{N} \qquad (7.4.9)$$

Hence, the DTFT's $J^d(\omega)$ and $K^d(\omega)$ take the form of impulse-sampled line spectra, i.e.

$$J^d(\omega) = \frac{2\pi}{N} \sum_{i=-\infty}^{\infty} J_i \; \delta(\omega - i\omega_N) \qquad (7.4.10)$$

$$K^d(\omega) = \frac{2\pi}{N} \sum_{i=-\infty}^{\infty} K_i \; \delta(\omega - i\omega_N) \qquad (7.4.11)$$

Substituting into (7.4.4) gives

$$U^d(\omega) = \frac{1}{N} \sum_{i=0}^{N-1} \left[ J_i \; X^d(\omega - i\omega_N) + K_i \; E^d(\omega - i\omega_N) \right] \qquad (7.4.12)$$

Since the above expression involves shifted versions of $X^d(\omega)$ and $E^d(\omega)$ it will be convenient to use the modulation representation in equations (1.7.39) and (2.6.7), i.e.

$$\tilde{X}^d(\omega) = \begin{bmatrix} X^d(\omega) \\ X^d(\omega - \omega_N) \\ \cdot \\ \cdot \\ \cdot \\ X^d(\omega - (N-1)\omega_N) \end{bmatrix} \qquad (7.4.13)$$

and similarly for $\tilde{E}^d(\omega)$ and $\tilde{U}^d(\omega)$. Equations (7.4.3) and (7.4.12) can then be written as

$$\tilde{U}^d(\omega) = \tilde{H}(\omega)\tilde{E}^d(\omega) \qquad (7.4.14)$$

where $\tilde{H}(\omega)$ is an $N$ by $N$ matrix – sometimes called the *alias component matrix*. Here $\tilde{H}(\omega)$ is given by

$$\tilde{H}(\omega) = \tilde{J}\left(e^{j\omega\Delta}\tilde{W} - \tilde{F}\right)^{-1}\tilde{G} + \tilde{K}$$

where

$$\tilde{J} = \frac{1}{N}\begin{bmatrix} J_0 & J_1 & & & J_{N-1} \\ J_{N-1} & J_0 & & & J_{N-2} \\ J_{N-2} & J_{N-1} & & & J_{N-3} \\ \cdot & \cdot & & - - - - & \cdot \\ \cdot & \cdot & & & \cdot \\ J_1 & J_2 & & & J_0 \end{bmatrix} \qquad (7.4.15)$$

$$\tilde{K} = \frac{1}{N}\begin{bmatrix} K_0 & K_1 & & & K_{N-1} \\ K_{N-1} & K_0 & & & K_{N-2} \\ K_{N-2} & K_{N-1} & & & K_{N-3} \\ \cdot & \cdot & & - - - - & \cdot \\ \cdot & \cdot & & & \cdot \\ K_1 & K_2 & & & K_0 \end{bmatrix} \qquad (7.4.16)$$

$$\tilde{F} = \begin{bmatrix} F & 0 & & & 0 \\ 0 & F & & & 0 \\ 0 & 0 & & & 0 \\ \cdot & \cdot & & - - - - & \cdot \\ \cdot & \cdot & & & \cdot \\ 0 & 0 & & & F \end{bmatrix} \qquad (7.4.17)$$

$$\tilde{G} \;=\; \begin{bmatrix} G & 0 & & & 0 \\ 0 & G & & & 0 \\ 0 & 0 & & & 0 \\ \cdot & \cdot & & \cdots & \cdot \\ \cdot & \cdot & & & \cdot \\ 0 & 0 & & & G \end{bmatrix} \qquad\qquad (7.4.18)$$

$$\tilde{W} \;=\; \begin{bmatrix} I & 0 & & & 0 \\ 0 & e^{-j\omega_N\Delta}\,I & & & 0 \\ 0 & 0 & & & 0 \\ \cdot & \cdot & & \cdots & \cdot \\ \cdot & \cdot & & & \cdot \\ 0 & 0 & & & e^{-j(N-1)\omega_N\Delta}\,I \end{bmatrix} \qquad\qquad (7.4.19)$$

The representation in (7.4.14) allows one to analyze $N$-periodic systems in the frequency domain. We illustrate by a simple example.

Example 7.4.1 :

Say that for computational reasons we wish to implement a filter using a decimator and filler in series as shown in Figure 7.4.1. Here, $H(\omega)$, $G(\omega)$ are two transfer functions, and $D_2$, $F_2$ denote a two sample decimator and filler respectively. Note that every second sample has been deleted from the input to the block denoted $G(\omega)$.

Develop an expression for the corresponding alias component matrix.

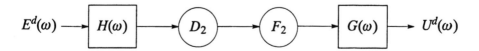

$E^d(\omega) \longrightarrow \boxed{H(\omega)} \longrightarrow (D_2) \longrightarrow (F_2) \longrightarrow \boxed{G(\omega)} \longrightarrow U^d(\omega)$

*Figure 7.4.1 : Signal processing using multirate sampling.*

Solution  It can be seen that the system in Figure 7.4.1 is a 2-periodic system. Its $2 \times 2$ alias component matrix is given as in (7.4.14) by

$$
\tilde{H}(\omega) = \begin{bmatrix} G(\omega) & 0 \\ 0 & G\left(\omega - \dfrac{\pi}{2}\right) \end{bmatrix}
$$

$$
\times \begin{bmatrix} \dfrac{1}{2} & \dfrac{1}{2} \\ \dfrac{1}{2} & \dfrac{1}{2} \end{bmatrix}
$$

$$
\times \begin{bmatrix} H(\omega) & 0 \\ 0 & H\left(\omega - \dfrac{\pi}{2}\right) \end{bmatrix}
$$

$$
= \begin{bmatrix} \dfrac{1}{2}G(\omega)H(\omega) & \dfrac{1}{2}G(\omega)H\left(\omega - \dfrac{\pi}{\Delta}\right) \\ \dfrac{1}{2}G\left(\omega - \dfrac{\pi}{\Delta}\right)H(\omega) & \dfrac{1}{2}G\left(\omega - \dfrac{\pi}{\Delta}\right)H\left(\omega - \dfrac{\pi}{2}\right) \end{bmatrix} \qquad (7.4.20)
$$

We will return to the above frequency domain tools when we consider periodic control of linear time-invariant systems in Chapter 11.

In the remaining sections of this chapter we will develop optimal filtering theory for sampled periodic systems.

## 7.5    Models for Sampled Periodic Stochastic Systems

We next turn to the problem of optimal state estimation in periodic stochastic systems.

Our starting point is the following continuous-time model (see (6.2.3), (6.2.4))

$$
\frac{dx(t)}{dt} = A(t)x(t) + \dot{v}(t) \qquad (7.5.1)
$$

$$y'(t) = \frac{dy(t)}{dt} = C(t)x(t) + \dot{\omega}(t) \tag{7.5.2}$$

where $\dot{v}(t)$ and $\dot{\omega}(t)$ denote independent continuous-time "white noise" processes with

$$E\left\{\dot{v}(t)\dot{v}(s)^T\right\} = \Omega(t)\delta(t-s) \tag{7.5.3}$$

$$E\left\{\dot{\omega}(t)\dot{\omega}(s)^T\right\} = \Gamma(t)\delta(t-s) \tag{7.5.4}$$

where $\Omega(t)$ and $\Gamma(t)$ are symmetric and positive semidefinite matrices. A key assumption here is that $A(t)$, $C(t)$, $\Omega(t)$ and $\Gamma(t)$ are all periodic with period $T$. The above process is called a *periodic stochastic process*, (or sometimes a *cyclostationary* stochastic process).

The following result shows that sampling the model (7.5.1) – (7.5.4) leads to a periodic discrete-time model provided the period is an integer multiple of sampling period.

**Lemma 7.5.1** Let the system of equations (7.5.1), (7.5.2) be sampled at a sampling interval $\Delta = T/N$ for some integer $N$ using the anti-aliasing filter with impulse response

$$h(t) = h_0(t) = \begin{cases} \dfrac{1}{\Delta} & \text{for } 0 \le t < \Delta \\ 0 & \text{otherwise} \end{cases} \tag{7.5.5}$$

Then the sampled output response can be described by the following discrete-time system ($\bar{y}(t)$ denotes the filtered output signal)

$$\delta x[k] = A_\delta[k]x[k] + v_\delta[k] \tag{7.5.6}$$

$$\bar{y}[k+1] = C_\delta[k]x[k] + \omega_\delta[k] \tag{7.5.7}$$

where

$$x[k] = x(k\Delta) \tag{7.5.8}$$

$$A_\delta[k] = \frac{1}{\Delta}\left(\Phi((k+1)\Delta, k\Delta) - I\right) \tag{7.5.9}$$

$$C_\delta[k] = \frac{1}{\Delta} \int\limits_{k\Delta}^{(k+1)\Delta} C(t)\Phi(t, k\Delta) \, dt \tag{7.5.10}$$

where $\phi(\cdot, \cdot)$ is the state transition matrix corresponding to $A(t)$ and $v_\delta[k]$, $\omega_\delta[k]$ are sequences with zero mean and the following covariances:

$$E\left\{\begin{bmatrix} v_\delta[k+m] \\ \omega_\delta[k+m] \end{bmatrix} \left[v_\delta[k]^T \omega_\delta[k]^T\right]\right\} = \begin{bmatrix} \Omega_\delta[k] & S_\delta[k] \\ S_\delta^T[k] & \Gamma_\delta[k] \end{bmatrix} \delta_d[m] \tag{7.5.11}$$

where $\delta_d[m]$ is the discrete $1/\Delta$ pulse and

$$\begin{bmatrix} \Omega_\delta[k] & S_\delta[k] \\ S_\delta^T[k] & \Gamma_\delta[k] \end{bmatrix} = \frac{1}{\Delta} \int\limits_{k\Delta}^{(k+1)\Delta} \overline{\Phi}(t, k\Delta) \begin{bmatrix} \Omega(t) & 0 \\ 0 & \Gamma(t) \end{bmatrix} \overline{\Phi}(t, k\Delta)^T \, dt \tag{7.5.12}$$

with $\overline{\Phi}(t, \tau)$ the transition matrix corresponding to $\overline{A}_f(t)$ where

$$\overline{A}_f(t) = \begin{bmatrix} A(t) & 0 \\ C(t) & 0 \end{bmatrix}$$

Proof :   Follows closely the development in Sections 6.4 and 6.5.

$$\triangle\triangle\triangle$$

It can be readily shown (see Problem 7.3) that the resulting discrete-time system – equations (7.5.6), (7.5.7) and (7.5.11) – is periodic with period $N$.

Periodic discrete-time models can also arise in many other ways. In the sequel we assume a discrete-time periodic model as in (7.5.6), (7.5.7), (7.5.11) without further comment. We next turn to the problem of estimating the state.

## 7.6 Periodic Optimal Filtering

An important observation is that the results of Section 6.5.2 apply to general time-varying state-space models and thus they are immediately applicable to the periodic case. However, by taking account of the periodic nature of the problem we can take the development further. Indeed, by use of the raising techniques, we can relate the solution of a time-varying periodic filtering problem to an associated time-invariant filtering problem. This will allow us to gain a comprehensive understanding of the periodic case by simply applying all of the known time-invariant theory.

Following the development in Section 6.5 we can show that the Kalman filter for the case of a linear periodic system takes the form:

$$\delta \hat{x}[k] = A_\delta[k]\hat{x}[k] + H_\delta[k](\delta y[k] - C_\delta[k]\hat{x}[k]) \tag{7.6.1}$$

where $\delta y[k] = \bar{y}[k+1]$. The Kalman gain $H_\delta[k]$ is given by

$$H_\delta[k] = (I + \Delta A_\delta[k])P[k]C_\delta[k]^T(\Gamma_\delta[k] + \Delta C_\delta[k]P[k]C_\delta[k]^T)^{-1} \tag{7.6.2}$$

and $P[k]$ is the solution of the discrete Riccati equation (in delta form)

$$
\begin{aligned}
\delta P[k] = {} & \Delta A_\delta[k]P[k]A_\delta[k]^T + A_\delta[k]P[k] + P[k]A_\delta[k]^T + \Omega_b[k] \\
& - \left((I + \Delta A_\delta[k])P[k]C_\delta^T[k] + S_\delta[k]\right) \\
& \quad \cdot \left(\Gamma_\delta[k] + \Delta C_\delta[k]P[k]C_\delta[k]^T\right)^{-1} \\
& \quad \cdot \left(C_\delta[k]^T P[k](I + \Delta A_\delta[k]^T) + S_\delta[k]^T\right)
\end{aligned}
\tag{7.6.3}
$$

with $P[0] = P_0 = E\{(x[0] - \hat{x}[0])(x[0] - \hat{x}[0])^T\}$.

We could leave the filter in this form. However, since the process we deal with is periodic we can use the ideas of the 'raised system' from Section 7.3 to simplify the solution of the above Riccati equation. It is somewhat easier to discuss the raised system in the shift operator form. Hence we rewrite equations (7.5.6), (7.5.7) and (7.6.3) as

$$x[k+1] = A_s[k]x[k] + v_s[k] \tag{7.6.4}$$

$$\delta y[k] = \bar{y}[k+1] = C_s[k]x[k] + \omega_s[k] \tag{7.6.5}$$

and

$$P[k+1] = \bar{\Omega}_s[k] + A_s[k]P[k]A_s[k]^T - \left(A_s[k]P[k]C_s[k]^T + S_s[k]\right)$$
$$\cdot \left(\bar{\Gamma}_s[k] + C_s[k]P[k]C_s[k]^T\right)^{-1}\left(C_s[k]P[k]A_s[k]^T + S_s[k]^T\right) \tag{7.6.6}$$

where

$$\left.\begin{aligned}
A_s[k] &= \Delta A_\delta[k] + I \\
C_s[k] &= C_\delta[k] \\
\bar{\Omega}_s[k] &= \Delta\Omega_\delta[k] \\
\bar{\Gamma}_s[k] &= 1/\Delta \cdot \Gamma_\delta[k] \\
\bar{S}_s[k] &= S_\delta[k] \\
v_s[k] &= \Delta v_\delta[k] \\
\omega_s[k] &= \omega_\delta[k]
\end{aligned}\right\} \tag{7.6.7}$$

Note that $\bar{\Omega}_s$ and $\bar{\Gamma}_s$ are variances as is traditional for the shift case.

We then have the following result which relates the periodic solution of the above filtering problem to the time-invariant solution of a non-periodic filtering problem associated with a raised system.

## Lemma 7.6.1

(a) We define the following raised quantities for $0 \le k_0 < N$ :

$$
\left.
\begin{aligned}
x_R[m] &= x[k_0 + mn] \\
v_R[m] &= \left[ v_s[k_0 + mn]^T, v_s[k_0 + mN + N]^T, \right. \\
&\qquad\qquad\qquad \left. \dots, \; v_s[k_0 + mN + N - 1]^T \right]^T \\
\omega_R[m] &= \left[ \omega_s[k_0 + mN]^T, \; \dots, \; \omega_s[k_0 + mN + N - 1]^T \right]^T \\
y_R^0[m] &= \left[ y_0[k_0 + mN]^T, \; \dots, \; y_0[k_0 + mN + N - 1]^T \right]^T
\end{aligned}
\right\}
\qquad (7.6.8)
$$

These quantities are related by the following time-invariant model:

$$
\boxed{x_R[m + 1] = A_R x_R[m] + E_R v_R[m]} \qquad (7.6.9)
$$

$$
\boxed{y_R^0[m] = C_R x_R[m] + \omega_R[m] + G_R v_R[m]} \qquad (7.6.10)
$$

where

$$
A_R = A_s[k_0 + N - 1] A_s[k_0 + N - 2] \; \dots \; A_s[k_0] \qquad (7.6.11)
$$

$$
\begin{aligned}
E_R = \bigl[ A_s[k_0 + N - 1] \; &\dots \; A_s[k_0 + 1], \\
A_s[k_0 + N - 1] \; &\dots \; A_s[k_0 + 2], \\
&\dots, \; A_s[k_0 + N - 1], \; I \bigr]
\end{aligned}
\qquad (7.6.12)
$$

$$
C_R = 
\begin{bmatrix}
C_s[k_0] \\
C_s[k_0 + 1] A_s[k_0] \\
\cdot \\
\cdot \\
\cdot \\
C_s[k_0 + N - 1] A_s[k_0 + N - 2] \; \dots \; A_s[k_0]
\end{bmatrix}
\qquad (7.6.13)
$$

and

$$G_R = \begin{bmatrix} G_{R,1} & G_{R,2} & \cdots & G_{R,N-1} & G_{R,N} \end{bmatrix} \qquad (7.6.14)$$

where

$$G_{R,1} = \begin{bmatrix} 0 \\ C_s[k_0 + 1] \\ C_s[k_0 + 2]A_s[k_0 + 1] \\ \cdot \\ \cdot \\ \cdot \\ C_s[k_0 + N - 1]A_s[k_0 + N - 2] \ \cdots \ A_s[k_0 + 1] \end{bmatrix}$$

$$G_{R,2} = \begin{bmatrix} 0 \\ 0 \\ C_s[k_0 + 2] \\ \cdot \\ \cdot \\ \cdot \\ C_s[k_0 + N - 1]A_s[k_0 + N - 2] \ \cdots \ A_s[k_0 + 2] \end{bmatrix}$$

$$G_{R,N-1} = \begin{bmatrix} 0 \\ 0 \\ 0 \\ \cdot \\ \cdot \\ \cdot \\ C_s[k_0 + N - 1] \end{bmatrix} \qquad G_{R,N} = \begin{bmatrix} 0 \\ 0 \\ 0 \\ \cdot \\ \cdot \\ \cdot \\ 0 \end{bmatrix}$$

(b)   The processes $\omega_R[m]$ and $v_R[m]$ defined in (7.6.8) are stationary processes;

(c)   The Kalman filter for with the *raised* system (which is linear *time-invariant* and *stationary*) is given by

$$\hat{x}_R[m+1] = A_R\hat{x}_R[m] + H_R[k]\left(y_R^0[m] - C_R\hat{x}_R[m]\right) \qquad (7.6.15)$$

The Kalman gain $H_R[k]$ is

$$H_R[k] = \left(A_R P_R[m]C_R^T + S_R\right)\left(\bar{\Gamma}_R + C_R P_R[m]C_R^T\right)^{-1} \qquad (7.6.16)$$

and $P_R$ is the solution of the following discrete-time (shift domain) Riccati equation

$$
\begin{aligned}
P_R[m+1] = {} & \bar{\Omega}_R + A_R P_R[m]A_R^T \\
& - \left(A_R P_R[m]C_R^T + \bar{S}_R\right)\left(\bar{\Gamma}_R + C_R P_R[m]\right)^{-1} \\
& \cdot \left(C_R P_R[m]A_R^T + \bar{S}_R^T\right)
\end{aligned}
\qquad (7.6.17)
$$

with

$$P_R[0] = P[k_0] \qquad (7.6.18)$$

where

$$
\begin{aligned}
\bar{\Omega}_R &\triangleq E\left\{ E_R v_R[m]v_R[m]^T E_R^T \right\} \\
&= E_R \begin{bmatrix} \bar{\Omega}_s[k_0] & & 0 \\ & \ddots & \\ 0 & & \bar{\Omega}_s[k_0+N-1] \end{bmatrix} E_R^T
\end{aligned}
\qquad (7.6.19)
$$

$$
\begin{aligned}
\bar{\Gamma}_R &\triangleq E\left\{ \left(\omega_R[m] + G_R v_R[m]\right)\left(\omega_R[m]^T + v_R[m]^T G_R^T\right) \right\} \\
&= \begin{bmatrix} \bar{\Gamma}_s[k_0] & & 0 \\ & \ddots & \\ 0 & & \bar{\Gamma}_s[k_0+N-1] \end{bmatrix}
\end{aligned}
$$

$$+ \begin{bmatrix} \bar{S}_s[k_0]^T & & 0 \\ & \ddots & \\ 0 & & \bar{S}_s[k_0+N-1\ ]^T \end{bmatrix} G_R^T$$

$$+ \; G_R \begin{bmatrix} \bar{S}_s[k_0] & & 0 \\ & \ddots & \\ 0 & & \bar{S}_s[k_0+N-1] \end{bmatrix}$$

$$+ \; G_R \begin{bmatrix} \bar{\Omega}_s[k_0] & & 0 \\ & \ddots & \\ 0 & & \bar{\Omega}_s[k_0+N-1] \end{bmatrix} G_R^T \qquad (7.6.20)$$

and

$$\bar{S}_R \triangleq E\left\{ E_R v_R[m]\left( \omega_R[m]^T + v_R[m]^T G_R^T \right) \right\}$$

$$= \; E_R \begin{bmatrix} \bar{S}_s[k_0] & & 0 \\ & \ddots & \\ 0 & & \bar{S}_s[k_0+N-1] \end{bmatrix}$$

$$+ \; E_R \begin{bmatrix} \bar{\Omega}_s[k_0] & & 0 \\ & \ddots & \\ 0 & & \bar{\Omega}_s[k_0+N-1] \end{bmatrix} G_R^T \qquad (7.6.21)$$

(d)  The optimal filter (7.6.1) to (7.6.3) for the periodic system and
     the optimal filter (7.6.15) to (7.6.18) for the raised system are
     related as follows:

$$\boxed{\hat{x}[k_0 + mN] = \hat{x}_R[m]} \qquad (7.6.22)$$

$$\boxed{P_R[m] = P[mN + k_0]} \tag{7.6.23}$$

**Proof :**  Parts (a) to (c) follow immediately by applying the results of Section 6.5 to the raised system.

Part (d) follows since both the original Kalman filter and the raised Kalman filter are *optimal* and the data is the *same* provided we set $\hat{x}[k_0] = \hat{x}_R[0]$ .

Alternatively, part (d) can be derived directly from the relationships in equations (7.6.4) - (7.6.21) by using (7.6.6) to express $P[mN + N]$ in terms of $P[mN]$ and showing that the result is identical to (7.6.17) .

$\wedge\wedge\wedge$

The significance of Lemma 7.6.1 is that the known results for stationary processes can be applied to the periodic case. In particular, it is known that, under mild conditions (e.g. $(C_R, A_R)$ detectable and $\left(A_R, \overline{\Omega}_R^{\frac{1}{2}}\right)$ stabilizable) the solution to equation (7.6.16) converges to a limit which satisfies the algebraic Riccati equation

$$P_R = \overline{\Omega}_R + A_R P_R A_R^T - \left(A_R P_R C_R^T + \overline{S}_R\right)\left(\overline{\Gamma}_R + C_R P_R C_R^T\right)^{-1}\left(C_R P_R A_R^T + \overline{S}_R^T\right) \tag{7.6.24}$$

Since the above holds for any $0 \le k_0 < N$ (7.6.23) implies that $P[k]$ converges to a *periodic* solution. Furthermore, the above argument suggests a simplified method of solving the periodic Riccati equation for its steady state solution. Solving (7.6.24) for $k_0 = 0$ provides $P[0] = P_R$ , and then sequential substitution in equation (7.6.6) for one period, i.e. $k = 0, 1, \ldots, N-2$ gives the steady-state (periodic) solution to (7.6.6).

Stability and other properties of the periodic Kalman filter also follow immediately from the associated time-invariant Kalman filter for the raised system.

We conclude this section with an illustrative example of periodic Kalman filtering.

**Example 7.6.1 :**   Consider the following scalar 2-periodic system

$$x[k+1] = a[k]x[k] + v_s[k]$$

$$y[k] = c[k]x[k] + \omega_s[k]$$

where

$$a[k] = 4.55 - 4.45\cos(\pi k)$$

$$c[k] = -\cos(\pi k)$$

The terms $v[k]$ and $\omega[k]$ are independent periodic sequences with the following variances

$$E\{v_s[k]v_s[m]\} \;=\; \begin{cases} 0.5(1 + \cos(\pi k)) & \text{for } k = m \\ 0 & \text{for } k \neq m \end{cases}$$

$$E\{\omega_s[k]\omega_s[m]\} \;=\; \begin{cases} 5.5 + 4.5\cos(\pi k) & \text{for } k = m \\ 0 & \text{for } k \neq m \end{cases}$$

Design the periodic Kalman filter for this system.

**Solution**  The steady state Riccati equation will be as in (7.6.6) i.e.

$$P[2m+1] \;=\; 1 + 0.01P[2m] - \frac{0.01P[2m]^2}{10 + P[2m]}$$

$$P[2m] \;=\; 81P[2m+1] - \frac{81P[2m+1]^2}{1 + P[2m+1]}$$

These equations are relatively simple to solve directly. However, following the methodology outlined in this section we find the raised system

$$A_R = 0.9$$

$$E_R = [9, 1]$$

$$C_R = \begin{bmatrix} -1 \\ 0.1 \end{bmatrix}$$

$$G_R = \begin{bmatrix} 0 & 0 \\ 1 & 0 \end{bmatrix}$$

$$\overline{\Omega}_R = [9, 1] \begin{bmatrix} 1 & 0 \\ 0 & 0 \end{bmatrix} \begin{bmatrix} 9 \\ 1 \end{bmatrix} = 81$$

$$\overline{\Gamma}_R = \begin{bmatrix} 10 & 0 \\ 0 & 1 \end{bmatrix} + \begin{bmatrix} 0 & 0 \\ 1 & 0 \end{bmatrix} \begin{bmatrix} 1 & 0 \\ 0 & 0 \end{bmatrix} \begin{bmatrix} 0 & 1 \\ 0 & 0 \end{bmatrix} = \begin{bmatrix} 10 & 0 \\ 0 & 2 \end{bmatrix}$$

$$\overline{S}_R = [9, 1] \begin{bmatrix} 1 & 0 \\ 0 & 0 \end{bmatrix} \begin{bmatrix} 0 & 1 \\ 0 & 0 \end{bmatrix} = [0, \ 9]$$

The steady-state Riccati equation corresponding to the raised system is then

$$P_R = 81 + (0.9)^2 P_R$$

$$- \begin{bmatrix} -81P_R, & 9 + 81P_R \end{bmatrix} \begin{bmatrix} 10 + P_R & -0.1P_R \\ -0.1P_R & 2 + 0.01P_R \end{bmatrix} \begin{bmatrix} -81P_R \\ 9 + 81P_R \end{bmatrix}$$

The positive solution to this equation is found to be

$$P_R = 42.0726$$

Using (7.6.23), we then have

$$P[2m] = P_R = 42.0726$$

Substituting this in equation (7.6.17) we get

$$P[2m + 1] = 1.0808$$

Hence, the corresponding Kalman gains (7.6.16) will be

$$H[2m] = -0.0808$$

$$H[2m + 1] = 4.6747$$

or

$$H[k] = 2.297 - 2.3778 \cos(\pi k)$$

The above periodic system and Kalman filter were simulated and the results are presented in Figure 7.6.1. It is quite clear that $\hat{x}[k]$ resembles quite
closely $x[k]$.

## 7.7   Further Reading and Discussion

Frequency domain analysis of time-varying systems has a rich history in
the area of signal processing. A pioneering paper describing frequency domain
tools for analyzing time-varying systems is Zadeh (1950). Application of these
ideas to periodic and and multirate filters appears in Vetterli (1987), (1989) and
Shenoy, Burnside and Parks (1994).

Connections between discrete periodic systems and discrete-time-invari-
ant systems are explored in Colanerim and Longhi (1995), Flamm (1991), Lin
and King (1993), Van Dooren and Sreedhar (1994) and Misra (1996). Colanerim
and Longhi (1995) deal with the question of how to realize a discrete-time transfer function matrix as a discrete-time periodic system using raising. Flamm
(1991) proposes a new family of time-invariant representations for periodic discrete systems. Lin and King (1993) studied the inverse problem of raising, that
is, given a transfer matrix find the corresponding periodic system. Van Dooren
and Sreedhar (1994) provides conditions under which a periodic discrete-time
linear state-space system can be transformed into a time-invariant one by change
of basis thus extending Floquet and Lyapunov results from continuous to discrete

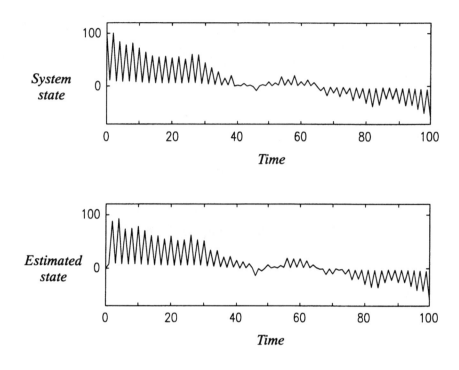

*Figure 7.6.1 :   Periodic Kalman filter response
from Example 7.6.1.*

time. Misra (1996) presents a further representation of periodic discrete-time
systems which preserves the time step.

Application areas of multirate and periodic filters are extensive – see for
example Vaidyanathan (1990), (1993), Vetterli (1987), Crochiere and Rabiner
(1983), Pridham and Mucci (1979), Scheibner and Parks (1984) and Mayne
(1962).

Optimal periodic filters are discussed in Bittanti, Laub and Willems (1991)
and Bittanti, Colaneri and De Nicolas (1991). The central idea in understanding
these developments is to relate the periodic filter to an underlying time-invariant
filter by use of raising or block processing. Early references on this technique are
Kranc (1957) and Meyer and Burrus (1975).

Example 7.4.1 is based on Shenoy, Burnside and Parks (1994) and, in turn, on Mintzer and Liu (1978).

## 7.8   Problems

7.1  (a)   Given that $A(t) = A(t + T)$ for all $t$ show that

$$\Phi(t + T, \tau + T) = \Phi(t, \tau)$$

(b)   Show that the eigenvalues of $\psi_A(t_0)$ are independent of $t_0$.

Hint: Use the property $\Phi(t_1, t_2)\Phi(t_2, t_3) = \Phi(t_1, t_3)$.

7.2  Show that every continuous time periodic system as in equations (7.2.1) - (7.2.2) can be directly transformed into the system

$$\dot{\tilde{x}}(t) = \tilde{A}\tilde{x}(t) + \tilde{B}(t)u(t)$$

$$y(t) = \tilde{C}(t)\tilde{x}(t)$$

where $\tilde{A}$ is time-invariant and the dimension of $x$ and $\tilde{x}$ are equal. To do this, follow the steps:

(a)   Find a constant nonsingular matrix $L$ so that

$$\Phi(t + T, 0) = \Phi(t, 0)L$$

(b)   Define the matrix $V$ so that $L = \exp(VT)$ and

$$R(t) = \Phi(t, 0)\exp(-Vt)$$

Show that $R(t + T) = R(t)$ for all $t$.

(c)   Define $\tilde{x}(t) = R^{-1}(t)x(t)$ and derive the state-space equations for $\tilde{x}(t)$. Verify that the resulting $\tilde{A}$ is indeed time invariant.

7.3  Show that $A_\delta[k]$, $C_\delta[k]$, $\Omega_\delta[k]$, $S_\delta[k]$ and $\Gamma_\delta[k]$ in Lemma 7.5.1 are $N$-periodic (recall that $T = N\Delta$ and that if $A(t)$ is $T$-periodic its corresponding transition matrix $\phi(t, \tau) = \phi(t + T, \tau + T)$).

7.4 Derive the shift version of the Riccati equation (7.6.3) using the notation in (7.6.7).

7.5 (a) Show that by defining $\omega[k] = \bar{y}[k] - C_\delta \dot{x}[k]$ the discrete Kalman filter can be expressed as

$$\delta \dot{x}[k] = A_\delta \dot{x}[k] + B_\delta u[k] + H_\delta \omega[k]$$

$$\hat{y}^0[k] = C_\delta \dot{x}[k]$$

$$y[k] = \hat{y}^0[k] + \omega[k]$$

(b) By assuming $A_\delta$, $B_\delta$, $H_\delta$ to have the following form:

$$A_\delta = \begin{bmatrix} -a_{n-1} & 1 & \cdot & \cdot & \cdot \\ a_{n-2} & 0 & \cdot & \cdot & \cdot \\ \cdot & \cdot & \cdot & \cdot & \cdot \\ \cdot & \cdot & \cdot & \cdot & 1 \\ -a_0 & \cdot & \cdot & \cdot & 0 \end{bmatrix} \qquad B_\delta = \begin{bmatrix} b_{n-1} \\ \cdot \\ \cdot \\ \cdot \\ b_0 \end{bmatrix} \qquad H_\delta = \begin{bmatrix} h_{n-1} \\ \cdot \\ \cdot \\ \cdot \\ h_0 \end{bmatrix}$$

$$C_\delta = \begin{bmatrix} 1 & 0 & \cdot & \cdot & \cdot & 0 \end{bmatrix}$$

Show that the input-output behaviour of a stationary linear stochastic process can be expressed in one of the following equivalent difference equation forms:

(i)

$$\bar{A}(\delta)\hat{y}^0[k] = \bar{B}(\delta)u[k] + \bar{H}(\delta)\omega[k]$$

$$y[k] = \hat{y}^0[k] + \omega[k]$$

where

$$\bar{A}(\delta) = \delta^n + a_{n-1}\delta^{n-1} + \ldots a_0$$

$$\bar{B}(\delta) = b_{n-1}\delta^{n-1} + \ldots + b_0$$

$$\bar{H}(\delta) = h_{n-1}\delta^{n-1} + \ldots + h_0$$

(ii)

$$\bar{A}(\delta)y[k] = \bar{B}(\delta)u[k] + \bar{C}(\delta)\omega[k]$$

where $\bar{C}(\delta) = \bar{A}(\delta) + \bar{H}(\delta)$ is a stable polynomial;

(iii)

$$\bar{C}(\delta)\mathring{y}^0[k] = \bar{B}(\delta)u[k] + \bar{H}(\delta)y[k]$$

# Chapter 8

# Discrete-Time Control

## 8.1    Introduction

In the remaining chapters of the book we will turn our attention to feedback control problems. In solving these problems, we will find that we will frequently call on the results of Chapters 1 to 7.

In Chapter 3 we saw how to analyse the characteristics of discrete-time systems and in Chapter 4 we saw how to obtain discrete-time models of sampled continuous-time systems In this chapter we discuss the problem of discrete-time (i.e. computer based) control.

The basic idea involved in control system design is to determine how best to adjust the input to a system so that the output responds in some desirable fashion. For example, in a robot one might manipulate the voltages applied to the various motors so that the end-effector moves along some desirable path, or in a chemical plant one might manipulate the feed rate of various reagents to produce a desirable end product.

If the model of the system is known, then control design can, in principle, be achieved by simply inverting the model so as to calculate that input sequence which produces the desired output behaviour. We shall indeed find that some form of inversion is inescapable in control design. However, it is usually desirable to carry out the inverse operation using feedback rather than by direct calculation.

In simple terms, feedback is a mechanism wherein the error between the measured output of the system and the desired output is used to adjust the input so as to cause the output to align itself with the desired output trajectory. We will show that this has significant advantages over the straightforward inverse solution.

An issue of paramount importance in feedback systems is that of stability of the overall system. In particular, it is obviously desirable that all signals in the system remain bounded. If this is true we say that the system is *internally* stable.

Another issue of importance is the sensitivity of the resultant control system to various inperfections including noise, disturbances and model errors. It is clearly desirable to have low sensitivity of the system to these kinds of imperfections. However, there are fundamental constraints on the capacity of feedback to reduce sensitivity. For example, we shall see that an unavoidable constraint is that the sum of the sensitivity to output noise and the sensitivity to output disturbances is equal to one at all frequencies. Also, it is usually true that reduction of sensitivity in one frequency range leads to an increase in sensitivity in some other frequency range. Thus control system design inevitably involves trade-offs of various types.

In determining how best to assign these trade-offs it would be helpful if one could have a simple characterization of all control laws which, at least, achieved stability for the system. We shall see that this is possible using the so-called *Youla parameterization* of all stabilizing controllers. An added advantage of this parameterization is that many transfer functions of interest, including the various sensitivity functions, are linear in the parameter expressing the available design freedom.

We shall give both transfer-function and state-space interpretations of the results. For simplicity we mainly concentrate on the single-input single-output case, although we will, where appropriate, refer to the multivariable generalization of the results. Our main objective in this chapter is to lay the foundation for our subsequent treatment of other problems including inter-sample response, generalized holds and periodic systems.

A typical feedback control loop using computer implementation is shown in Figure 8.1.1. Note that the input is applied to the system via a zero-order-hold and that an anti-aliasing filter is applied to the output signal before samples are taken.

A reasonable question which might immediately occur to the reader is whether or not it is really necessary to worry about the discrete-time nature of the control law. Why not simply carry out a continuous-time design and then map this in some reasonable fashion into a discrete-time law? If this were as easy as it sounds, then there would be no need to go beyond continuous-time control theory and it would hardly be worth worrying the reader with a special topic on digital control. Unfortunately, however, this ad-hoc approach often leads to un-acceptable results and this has motivated the development of discrete-time and sampled-data control theory as presented below.

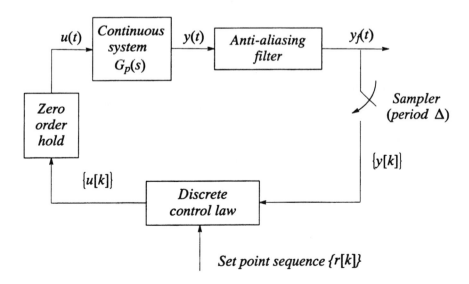

*Figure 8.1.1 : Discrete control system.*

To illustrate the effect of simple continuous to discrete mappings in control system design, we mention three methods drawn from the digital signal processing literature. These are:

(1)  Simply replace $s$ by $\gamma$ (this is the simplest transformation and might be the one we would first think of). This leads to

$$\overline{C}_1(\gamma) = C(s) \Big|_{s = \gamma} \qquad\qquad (8.1.1)$$

where $C(s)$ is the transfer function of the continuous-time controller and where $\overline{C}_1(\gamma)$ is the resultant transfer function of the discrete-time controller.

(2)  Convert the controller to a zero-order-hold discrete equivalent as in Table 4.3.1 of Chapter 4. (This is called a *step invariant transformation*.) Using the results in Section 4.3 this leads to

$$\overline{C}_2(\gamma) = D\left\{S_\Delta\left\{L^{-1}\{C(s)H_0(s)\}\right\}\right\} \qquad\qquad (8.1.2)$$

where $C(s)$, $H_0(s)$ and $\overline{C}_2(\gamma)$ are the transfer functions of the continuous-time controller, zero-order-hold and resultant discrete-time controller respectively.

(3)  We could use a more sophisticated mapping from $s$ to $\gamma$. For example, we could carry out the following transformation (which is commonly called a *bilinear transformation with pre-warping*). We first let

$$s = \frac{\alpha\gamma}{\frac{\Delta}{2}\gamma + 1} \qquad\qquad (8.1.3)$$

or

$$\gamma = \frac{s}{\alpha - \frac{\Delta}{2}s} \qquad\qquad (8.1.4)$$

The discrete controller is then defined by

$$\bar{C}_3(\gamma) = C(s)\Big|_{s = \dfrac{\alpha\gamma}{\frac{\Delta}{2}\gamma + 1}} \tag{8.1.5}$$

We next choose $\alpha$ so as to match the frequency responses of the two controllers at some desired frequency, say $\omega^*$. For example, one might choose $\omega^*$ as the frequency at which the continuous-time sensitivity function has its maximum value.

Now we recall from Section 3.11 that the discrete frequency response at $\omega^*$ is evaluated using $\gamma = \left(e^{j\omega^*\Delta} - 1\right)/\Delta$ and the continuous frequency response is obtained by replacing $s$ by $j\omega^*$. Hence to equate the discrete and continuous frequency responses at $\omega^*$ we require that $\alpha$ satisfy

$$j\omega^* = \dfrac{\alpha\left[\dfrac{e^{j\omega^*\Delta} - 1}{\Delta}\right]}{\dfrac{\Delta}{2}\left[\dfrac{e^{j\omega^*\Delta} - 1}{\Delta}\right] + 1} \tag{8.1.6}$$

The solution is

$$\alpha = \dfrac{\omega^*\Delta}{2}\left[\dfrac{\sin\omega^*\Delta}{1 - \cos\omega^*\Delta}\right] = \dfrac{\omega^*\Delta}{2}\tan\left[\dfrac{\omega^*\Delta}{2}\right] \tag{8.1.7}$$

The fact that none of these ad-hoc transformations is entirely satisfactory is illustrated in the following example.

**Example 8.1.1** : This example has been used in the literature (see Rattan (1984) and Keller et. el (1992)).

The system is given by

$$G(s) = \frac{10}{s(s+1)}$$

and the continuous-time controller is

$$C(s) = \frac{0.416s + 1}{0.139s + 1}$$

Replace this controller by a digital controller with $\Delta = 0.157$ sec. preceded by a sampler and followed by a ZOH using the three approximations outlined above. Test the unit step response for each such approximation.

## Solution

(1)  Replacing $s$ by $\gamma$ in $C(s)$ we get

$$\bar{C}_1(\gamma) = \frac{0.416\gamma + 1}{0.139\gamma + 1}$$

(2)  The ZOH equivalent of $C(s)$ (using Table 4.3.1) is simply

$$\bar{C}_2(\gamma) = \frac{0.916\gamma + 1}{0.3061\gamma + 1}$$

(3)  Finally, for the bilinear pre-warping we first look at the continuous-time sensitivity function

$$S(\omega) = \frac{1}{1 + C(j\omega)G(j\omega)}$$

We find that $|S(\omega)|$ has a maximum at

$$\omega^* = 5.48 \text{ rad sec}^{-1}$$

Now using the formula

$$s = \frac{\alpha\gamma}{\frac{\Delta}{2}\gamma + 1}$$

$\alpha$ is determined so that

$$j\omega^* = \left. \frac{\alpha\gamma}{\frac{\Delta}{2}\gamma + 1} \right|_{\gamma = \frac{e^{j\omega^*\Delta} - 1}{\Delta}}$$

and we get

$$\alpha = 0.9375$$

With this value of $\alpha$ we find the approximation

$$\overline{C}_3(\gamma) = C(s) \left|_{s = \frac{\alpha\gamma}{\frac{\Delta}{2}\gamma + 1}} \right. = \frac{0.4685\gamma + 1}{0.2088\gamma + 1}$$

The closed loop unit step responses obtained with the continuous-time controller $C(s)$ and the three discrete-time controllers $\overline{C}_1(\gamma)$, $\overline{C}_2(\gamma)$ and $\overline{C}_3(\gamma)$ are presented in Figure 8.1.2.

We see from the figure that none of the approximations adequately reproduces the closed-loop response obtained with the continuous-time controller.

△△△

The above example shows the difficulty of obtaining discrete-time control laws by ad-hoc means. We therefore proceed in this and subsequent chapters to develop rigorous theory for discrete-time and sampled-data control system design.

A key feature of our treatment in this chapter is that we restrict ourselves to the *at-sample response* : we ignore what happens *between* samples. One might suspect that this will not make much difference provided the sampling rate is sufficiently high and provided one does not deliberately try to force the sampled response to behave in ways which are unnatural for the underlying continuous-time system. This is indeed the case as we shall see in subsequent chapters.

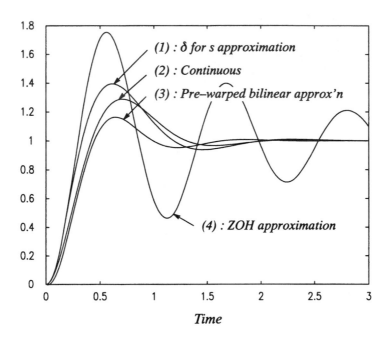

*Figure 8.1.2 :  Step responses for Example 8.1.1.*

Since for the moment we are only concerned with the at-sample response
it suffices to represent the combination of the hold, plant and anti-aliasing filter
by a single discrete transfer function. The reader is reminded of the methods used
in Chapter 4 to obtain the discrete transfer function from the continuous-time
transfer function of the system. We will denote the resultant discrete transfer
function as $\overline{G}(\gamma)$. We also use $\{y[k]\}$ to denote the sampled (and filtered) output
sequence.

## 8.2    Closed-Loop Stability and Pole Assignment

As stated in the introduction to this chapter, stability is of paramount importance in control system design. Thus, we will next study how we can ensure
stability of a feedback control loop. We assume that the discrete plant transfer-

function (including hold and anti-aliasing filter) is expressed as the ratio of two polynomials as follows:

$$\bar{G}(\gamma) = \frac{\bar{B}(\gamma)}{\bar{A}(\gamma)} \qquad (8.2.1)$$

where $\bar{B}(\gamma)$, $\bar{A}(\gamma)$ are polynomials in $\gamma$.

Similarly we will assume that the controller is also described by a ratio of polynomials : we write

$$\bar{C}(\gamma) = \frac{\bar{P}(\gamma)}{\bar{L}(\gamma)} \qquad (8.2.2)$$

Equation (8.2.1) implies that the plant input-output behaviour is described by a difference equation model of the form:

$$\bar{A}(\delta)y[k] = \bar{B}(\delta)u[k] \qquad (8.2.3)$$

Similarly, the control law (8.2.2) when implemented as in Figure 8.1.1 is equivalent to:

$$\bar{L}(\delta)u[k] = \bar{P}(\delta)(r[k] - y[k]) \qquad (8.2.4)$$

In the sequel it will also be useful to generalize (8.2.4) by adding an extra degree of freedom to the control law (8.2.4) leading to

$$\bar{L}(\delta)u[k] = -\bar{P}(\delta)y[k] + \bar{Q}(\delta)r[k] \qquad (8.2.5)$$

Now, if we operate on (8.2.3) by $\bar{L}(\delta)$ and use (8.2.5) to eliminate $u[k]$ then this leads to the following closed loop equation:

$$\left(\bar{A}(\delta)\bar{L}(\delta) + \bar{B}(\delta)\bar{P}(\delta)\right)y[k] = \bar{B}(\delta)\bar{Q}(\delta)r[k] \qquad (8.2.6)$$

From (8.2.6) we see that the closed-loop transfer-function between $\{r[k]\}$ and $\{y[k]\}$ is

$$\overline{G}_c = \frac{\overline{B}\overline{Q}}{(\overline{A}\overline{L} + \overline{B}\overline{P})} \tag{8.2.7}$$

Similarly, we see that $\{u[k]\}$ satisfies

$$\left(\overline{A}(\delta)\overline{L}(\delta) + \overline{B}(\delta)\overline{P}(\delta)\right)u[k] = \overline{A}(\delta)\overline{Q}(\delta)r[k] \tag{8.2.8}$$

From the closed loop equations (8.2.6), (8.2.8) we see that closed loop stability (recall Section 3.10) is guaranteed provided the polynomial $\left(\overline{A}(\delta)\overline{L}(\delta) + \overline{B}(\delta)\overline{P}(\delta)\right)$ has all its zeros in the stability domain.

This leads us to formulate a design question as follows : *"Given the plant model $\overline{A}(\delta)$ and $\overline{B}(\delta)$, is it possible to choose $\overline{L}(\delta)$ and $\overline{P}(\delta)$ so that the closed loop is guaranteed to be stable?"* We shall see below that under rather general conditions we *can* always choose $\overline{L}(\delta)$ and $\overline{P}(\delta)$ to ensure closed loop stability.

Actually, much more can be said. It is possible (subject to mild restrictions on the plant) to choose $\overline{L}(\delta)$, $\overline{P}(\delta)$ to place the closed loop poles at arbitrary locations. To see why this is so, let $\overline{A}(\delta)$ have degree $n$ and let $\overline{A}^*(\delta)$ be any desired closed loop characteristic polynomial of degree $2n$. Then to assign the closed loop poles we need to be able to solve the following polynomial identity for $\overline{L}(\delta)$, $\overline{P}(\delta)$ :

$$\overline{A}(\gamma)\overline{L}(\gamma) + \overline{B}(\gamma)\overline{P}(\gamma) = \overline{A}^*(\gamma) \tag{8.2.9}$$

Equating like powers of $\gamma$ on both sides of (8.2.9) leads to the following equation:

$$
\begin{bmatrix}
a_n & & 0 & & \\
a_{n-1} & \cdot & b_n & \cdot & \cdot \\
\cdot & \cdot & & \cdot & \cdot \\
\cdot & \cdot & & \cdot & \cdot \\
\cdot & \cdot & a_n & \cdot & \cdot & \cdot \\
\cdot & & a_{n-1} & b_0 & \cdot & \cdot & 0 \\
a_0 & \cdot & & \cdot & & b_n \\
& \cdot & & & \cdot & \cdot \\
& \cdot & & & \cdot & \cdot \\
& \cdot & & & \cdot & \cdot \\
& & a_0 & & & b_0
\end{bmatrix}
\begin{bmatrix}
l_n \\
\cdot \\
\cdot \\
\cdot \\
l_0 \\
p_{n-1} \\
\cdot \\
\cdot \\
\cdot \\
p_0
\end{bmatrix}
=
\begin{bmatrix}
a_{2n}^* \\
\cdot \\
\cdot \\
\cdot \\
\cdot \\
\cdot \\
\cdot \\
\cdot \\
\cdot \\
a_0^*
\end{bmatrix}
\qquad (8.2.10)
$$

where $\quad \overline{A}^*(\gamma) = a_{2n}^* \gamma^{2n} + \ldots + a_0^*$

$\overline{A}(\gamma) = a_n \gamma^n + \ldots a_0$

$\overline{B}(\gamma) = b_n \gamma^n + b_{n-1} \gamma^{n-1} + \ldots b_0$

$\overline{L}(\gamma) = l_n \gamma^n + \ldots l_0$

$\overline{P}(\gamma) = p_{n-1} \gamma^{n-1} + \ldots p_0$

The matrix on the left hand side of (8.2.10) is known as the *Sylvester* or *eliminant matrix* and is denoted by $M_e$. Clearly, we need this matrix to be nonsingular if we are going to be able to solve for the coefficients of $\overline{L}(\gamma)$ and $\overline{P}(\gamma)$. However, nonsingularity is guaranteed if $\overline{A}(\gamma)$ and $\overline{B}(\gamma)$ have no common factors i.e. if $\overline{A}(\gamma)$ and $\overline{B}(\gamma)$ are relatively prime. This is established in the following result:

## Lemma 8.2.1 *Sylvester's Theorem*

Two polynomials of degree $n$ are relatively prime if and only if their eliminant matrix, $M_e$, is nonsingular.

Proof :

If:  Assume that there is a common root $r$, and write

$$\bar{A}(\gamma) = (\gamma - r)\left(a'_{n-1}\gamma^{n-1} + \ldots + a'_0\right) \tag{8.2.11}$$

$$\bar{B}(\gamma) = (\gamma - r)\left(b'_{n-1}\gamma^{n-1} + \ldots + b'_0\right) \tag{8.2.12}$$

Eliminating $(\gamma - r)$ gives

$$\bar{A}(\gamma)\left[b'_{n-1}\gamma^{n-1} + \ldots + b'_0\right]$$
$$-\bar{B}(\gamma)\left[a'_{n-1}\gamma^{n-1} + \ldots + a'_0\right] = 0 \tag{8.2.13}$$

Equating coefficients on both sides gives

$$M_e\theta' = 0 \tag{8.2.14}$$

where

$$\theta' = \left[0, \; b'_{n-1}, \; \ldots, \; b'_0, \; -a'_{n-1}, \; \ldots, \; -a'_o\right] \tag{8.2.15}$$

However, (8.2.14) has a nontrivial solution if and only if $\det M_e = 0$.

Only if:  Reverse the above argument.

$$\wedge\wedge\wedge$$

We see from the above result that, provided $\bar{A}(\gamma)$ and $\bar{B}(\gamma)$ are relatively prime then it is straightforward to choose $\bar{L}(\gamma)$ and $\bar{P}(\gamma)$ to achieve closed loop stability. Indeed, we have seen that $\bar{L}(\gamma)$, $\bar{P}(\gamma)$ can actually be chosen to arbitrarily assign the closed loop polynomial $\bar{A}(\gamma)\bar{L}(\gamma) + \bar{B}(\gamma)\bar{P}(\gamma)$. Of course, this raises the issue of what would be a desirable value for this polynomial. It turns out this is closely related to the location of the open-loop zeros (the zeros of

$\overline{B}(\gamma)$ ) and the open-loop poles (the zeros of $\overline{A}(\gamma)$ ). Before we address these is-sues, we will consider some very special choices for the polynomial $\overline{A}^*(\gamma) = \overline{A}(\gamma)\overline{L}(\gamma) + \overline{B}(\gamma)\overline{P}(\gamma)$ in the next section.

## 8.3    Some Special Discrete-Time Control Laws

### 8.3.1    Deadbeat Control

One special choice for the closed-loop characteristic polynomial is

$$\overline{A}^*(\gamma) = (\Delta\gamma + 1)^{2n} \tag{8.3.1}$$

This gives a control law which causes the response to settle in exactly $2n$ samples. This is commonly given the name *deadbeat* control; there is no continu-ous-time counterpart to this control law. In practice, however, a rapid exponen-tially convergent response can be achieved by a continuous-time controller and this is often just as useful as a response that exactly settles in $2n$ steps. This idea is further reinforced by the fact that an *exact deadbeat* response is only obtained if the plant model is exactly known which is rarely, if ever, true in practice.

### 8.3.2    Model Reference Control

Another special choice can be made when the plant has only stable zeros (commonly called minimum phase). In this case the discrete closed-loop transfer function can be arbitrarily assigned provided only that the relative degree of the closed loop transfer function is at least that of the original plant. To do this we need to cancel the plant numerator dynamics in the closed loop and hence we should include $\overline{B}(\gamma)$ as part of the closed-loop characteristic polynomial. Thus (8.2.9) becomes:

$$\overline{A}(\gamma)\overline{L}(\gamma) + \overline{B}(\gamma)\overline{P}(\gamma) = \overline{A}_1^*(\gamma)\overline{B}(\gamma) \tag{8.3.2}$$

where $\overline{A}_1^*(\gamma)$ has degree equal to $2n$ minus the degree of $\overline{B}(\gamma)$ .

By writing $\bar{L}(\gamma) = \bar{F}(\gamma)\bar{B}(\gamma)$ , and then cancelling $\bar{B}(\gamma)$ , (8.3.2) simplifies

to

$$\bar{A}(\gamma)\bar{F}(\gamma) + \bar{P}(\gamma) = \vec{A_1}(\gamma) \tag{8.3.3}$$

This equation is always solvable for polynomials $\bar{F}(\gamma)$ and $\bar{B}(\gamma)$ .

Finally, the control law is implemented as

$$\bar{F}(\delta)\bar{B}(\delta)u[k] = -\bar{P}(\delta)y[k] + \bar{Q}(\delta)y^*[k] \tag{8.3.4}$$

as shown in Figure 8.3.1. Note also that we only need realize $1/\overline{FB}$ once to im-

plement (8.3.4) rather than twice as Figure 8.3.1 might suggest.

The controller (8.3.4) leads to the following closed loop transfer function:

$$\vec{G_c}(\gamma) = \frac{\bar{Q}(\gamma)}{\vec{A_1}(\gamma)} \tag{8.3.5}$$

Clearly by appropriate choice of $\bar{Q}(\gamma)$ and $\vec{A_1}(\gamma)$ we can achieve any de-

sired value for the closed loop transfer function, $\vec{G_c}(\gamma)$ , linking $r$ to $y$. The trans-

fer function $\vec{G_c}(\gamma)$ is often thought of as a reference model for the desired closed

loop behaviour. Thus, this control law is commonly called a *model reference con-*

*trol law*. The end result of being able to achieve any closed-loop transfer-func-

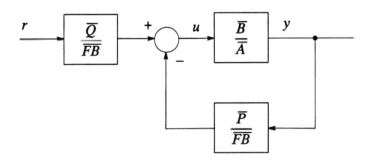

*Figure 8.3.1 : Model Reference Control.*

tion seems highly desirable. Note, however, that the design is limited to minimum phase plants since the plant numerator polynomial $\overline{B}(\gamma)$ must be included as a factor in the closed-loop characteristic polynomial – see (8.3.2).

## 8.3.3   Minimal Prototype Control

If, in model reference control, we make the very special choices $\overline{Q}(\gamma) = 1$ and $\overrightarrow{A_1}(\gamma) = (\Delta\gamma + 1)^{n'}$ then the resultant *discrete-time* closed loop system is simply a delay of $n'$ samples. In this case the control law is called a *minimal prototype* controller. This control law is the basis of other well known control laws which go by the names of *d-step ahead* control and *minimum variance* control. These controllers are frequently advocated in the literature but the reader is encouraged to re-read the cautionary note at the end of Section 8.3.2.

**Example 8.3.1**   : We will illustrate the *discrete-time* performance of the control laws described in Sections 8.3.1 and 8.3.3.

Consider the continuous-time system given by the transfer function

$$G(s) = \frac{s + 0.5}{s(s-1)}$$

We want to control this system using a discrete-time controller having a sampling period of 0.1 sec.

(a) With $\Delta = 0.1$ sec. find the discrete-time equivalent transfer function for $G(s)$, namely $\overline{G}(\gamma) = \overline{B}(\gamma)/\overline{A}(\gamma)$.

(b) Find the control law to get deadbeat control. Note that the resulting closed-loop transfer-function will have additional zeros. Can you explain the significance of these zeros?

(c) Choose $\overrightarrow{A}^*(\gamma) = (0.1\gamma +)^3\overline{B}(\gamma)$ and $\overline{Q}(\gamma) = (0.1\gamma + 1)^2$ to give a model reference controller. Find the control law to give these polynomials and find the resulting closed-loop transfer function.

(d)  Compare the unit step response with the controllers in (b) and
     (c).

## Solution

(a)  We write

$$G(s) = \frac{1.5}{s-1} - \frac{0.5}{s}$$

Using Table 4.3.1 we get

$$\overline{G}(\gamma) = \frac{1.5\alpha}{\gamma+\alpha} - \frac{0.5}{\gamma}$$

where

$$\alpha = \frac{1-e^{0.1}}{0.1} = -1.0517$$

so

$$\overline{G}(\gamma) = \frac{1.0776\gamma + 0.5259}{\gamma(\gamma - 1.0517)} \tag{8.3.6}$$

(note the resemblance between $\overline{G}(\gamma)$ and $G(s)$).

(b)  Deadbeat control is achieved if we use

$$\overline{A}^*(\gamma) = (0.1\gamma + 1)^4 = 10^{-4}\gamma^4 + 4(10)^{-3}\gamma^3 + 0.06\gamma^2 + 0.4\gamma + 1$$

and from (8.2.10) we have

$$\begin{bmatrix} 1 & 0 & 0 & 0 & 0 \\ -1.0517 & 1 & 0 & 1.0776 & 0 \\ 0 & -1.0517 & 1 & 0.5259 & 1.0776 \\ 0 & 0 & -1.0517 & 0 & 0.5259 \\ 0 & 0 & 0 & 0 & 0 \end{bmatrix} \begin{bmatrix} l_2 \\ l_1 \\ l_0 \\ p_1 \\ p_0 \end{bmatrix} = \begin{bmatrix} 10^{-4} \\ 4(10)^{-3} \\ 0.06 \\ 0.4 \\ 1 \end{bmatrix}$$

Hence

$$\begin{bmatrix} l_2 \\ l_1 \\ l_0 \\ p_1 \\ p_0 \end{bmatrix} = \begin{bmatrix} 10^{-4} \\ 0.0041 \\ 1.0914 \\ -0.9531 \\ 1.9015 \end{bmatrix}$$

and using the control law of (8.2.4) we get the control law

$$\left(10^{-4}\delta^2 + 0.0041\delta + 1.0914\right) u[k]$$
$$= \left(-0.9531\delta + 1.9015\right) \cdot \left(r[k] - y[k]\right)$$

The closed-loop transfer-function becomes

$$\overline{G}_c(\gamma) = \frac{(1.0776\gamma + 0.5259)(-0.9531\gamma + 1.9015)}{(0.1\gamma + 1)^4}$$

We see that the controller zero (which happens to be non-minimum-phase) becomes an additional zero of the closed-loop system.

(c) In this case, we have

$$\overline{A}^*(\gamma) = 0.0011\gamma^4 + 0.0329\gamma^3 + 0.3391\gamma^2 + 1.2354\gamma + 0.5259$$

Repeating the calculations of part (b) for this choice of $\overline{A}^*(\gamma)$ we obtain

$$\begin{bmatrix} l_2 \\ l_1 \\ l_0 \\ p_1 \\ p_0 \end{bmatrix} = \begin{bmatrix} 0.0011 \\ 0.034 \\ 0.0163 \\ 0.3327 \\ 1 \end{bmatrix}$$

and from (8.2.5) the controller is

$$\left(0.0011\delta^2 + 0.034\delta + 0.0163\right) u[k]$$
$$= -\left(0.3327\delta + 1\right) y[k] + \left(0.1\delta + 1\right)^2 r[k]$$

The resultant closed loop transfer function in this case is

$$\overline{G}_c(\gamma) = \frac{\overline{Q}(\gamma)\overline{B}(\gamma)}{\overline{A}^*(\gamma)} = \frac{1}{0.1\gamma + 1}$$

Notice that, the polynomial $\overline{Q}(\gamma)$ is cancelled.

(d)  In Figures 8.3.2 and 8.3.3 we show the unit step response for the given system with the two controllers. In both cases the system settles at 1. However, for the first controller the transient response is quite poor even though the closed-loop poles have been assigned to $-10$. The poor response is a result of the zeros in the closed-loop transfer-function. Clearly, these factors do not effect the solution given in (c).

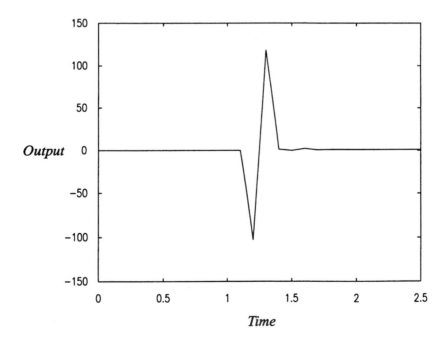

*Figure 8.3.2 :  Step response for Example 8.3.1 using deadbeat controller.*

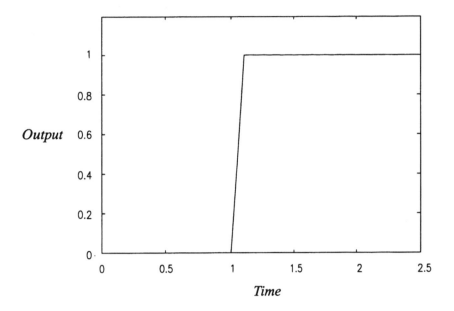

*Figure 8.3.3 : Step response for Example 8.3.1 using
model reference controller.*

## 8.4    Sensitivity and Complementary Sensitivity
Functions

We have seen above that, subject to mild restrictions, the closed-loop poles
of a system can be arbitrarily assigned. We next consider the question as to what
type of closed-loop behaviour would be desirable. A key issue in this regard is
the sensitivity of the closed-loop to various imperfections including variations
in the nominal model and the effects of noise and disturbances.

Throughout this section, we will consider the single degree-of-freedom
feedback loop shown in Figure 8.4.1.

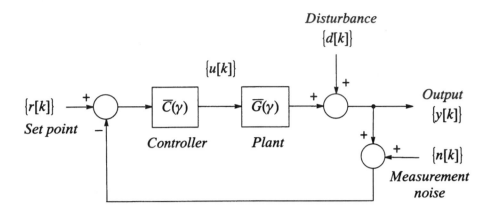

*Figure 8.4.1 : Unity feedback control loop.*

As usual, we use upper case letters to denote the discrete transform of signals. Then it is easily seen that the transform of the output response $\{y[k]\}$ in Figure 8.4.1 is given by

$$Y(\gamma) = \overline{T}(\gamma)R(\gamma) + \overline{S}(\gamma)D(\gamma) - \overline{T}(\gamma)N(\gamma) \qquad (8.4.1)$$

where $\overline{S}(\gamma)$ and $\overline{T}(\gamma)$ are the discrete-time sensitivity and complementary sensitivity functions respectively given by:

$$\overline{S}(\gamma) = \frac{1}{1 + \overline{G}(\gamma)\overline{C}(\gamma)} \qquad (8.4.2)$$

$$\overline{T}(\gamma) = \frac{\overline{G}(\gamma)\overline{C}(\gamma)}{1 + \overline{G}(\gamma)\overline{C}(\gamma)} \qquad (8.4.3)$$

We see from (8.4.1) that the closed-loop transfer function (between the reference input and the output) is equal to the complementary sensitivity function for a unity-feedback control law. Also, the transfer function between the output disturbance and the measurement noise to the output are the sensitivity function and (negative) complementary sensitivity respectively.

The sensitivity function also tells us how the closed-loop transfer function changes when the open-loop plant transfer function changes. Let $\Delta \overline{G}$ denote the change in open-loop transfer function $\overline{G}$ and let $\Delta \overline{T}$ denote the corresponding change in the closed-loop transfer function. Then by differentiation we have that

$$\Delta \overline{T} \simeq \frac{\overline{C}}{(1 + \overline{G}\overline{C})^2} \Delta \overline{G} \tag{8.4.4}$$

Hence

$$\boxed{\frac{\Delta \overline{T}}{\overline{T}} \simeq \overline{S} \frac{\Delta \overline{G}}{\overline{G}}} \tag{8.4.5}$$

We can summarize this result as follows:

**Lemma 8.4.1** For a linear time-invariant control system, the relative change in the closed loop discrete transfer function, $(\Delta \overline{T}/\overline{T})$ is given by $\Delta \overline{T}/\overline{T} \simeq \overline{S}\Delta \overline{G}/\overline{G}$ at each frequency.

$$\wedge\wedge\wedge$$

The above discussion indicates that it would be desirable to make the sensitivity function and complementary sensitive functions small at all frequencies since this would reduce the impact of disturbances, model errors and noise. Unfortunately this cannot be done as we now show. Indeed an immediate consequence of the definitions of $\overline{S}(\gamma)$ and $\overline{T}(\gamma)$ is the following fundamental constraint:

$$\boxed{\overline{S}(\gamma) + \overline{T}(\gamma) = 1} \tag{8.4.6}$$

The reader will recall that in Chapter 3 we showed that the corresponding (discrete-time) frequency response is simply obtained by replacing $\gamma$ by $\gamma_\omega = \left(e^{j\omega\Delta} - 1\right)/\Delta$.

Thus (8.4.6) has the following frequency domain interpretation

$$\overline{S}(\gamma_\omega) + \overline{T}(\gamma_\omega) = 1 \qquad\qquad (8.4.7)$$

where $\gamma_\omega = \left(e^{j\omega\Delta} - 1\right)/\Delta$ .

Equation (8.4.7) tells us that at those frequencies where $|\overline{S}|$ is small $|\overline{T}|$ must be approximately one and vice versa. In view of (8.4.1) and (8.4.7) it is clear that one can only reduce the sensitivity to disturbances by increasing the sensitivity to measurement noise and vice versa. This is a fundamental design limitation. Of course, since $\overline{S}(\gamma_\omega)$ and $\overline{T}(\gamma_\omega)$ are both complex numbers, it is quite conceivable for them to simultaneously have magnitudes that are much larger than one. (See Problem 8.1). In practice, the controller is usually designed so as to make $|\overline{T}(\gamma_\omega)|$ approach 1 at low frequencies (to give zero steady-state errors at d.c.) and to make $|\overline{T}(\gamma_\omega)|$ small at high frequencies to give insensitivity to high frequency noise. Thus, we see that it is desirable to have $|\overline{T}(\gamma_\omega)| \simeq 1$ at low frequencies and $|\overline{S}(\gamma_\omega)| \simeq 1$ at high frequencies.

## 8.5    All Stabilizing Control Laws

We have seen in the previous section that there are unavoidable trade-offs in control system design. To better understand these trade-offs it is desirable to have a way of parameterizing the controller so that closed-loop stability is preserved as different designs are considered. A remarkable way of achieving this is provided by the Youla parameterization of all stabilizing control laws which is described below.

We will first consider the case of a stable open-loop plant and later extend this to unstable open-loop plants.

### 8.5.1   Open Loop Stable Plants

The basic idea is to parameterize the controller transfer function $\overline{C}(\gamma)$ in terms of another transfer function $\overline{Q}(\gamma)$ by writing

$$\overline{C}(\gamma) = \frac{\overline{Q}(\gamma)}{1 - \hat{G}(\gamma)\overline{Q}(\gamma)} \qquad (8.5.1)$$

where $\hat{G}$ is an estimate, $\overline{G}$, of the plant transfer function. The representation given in (8.5.1) is always possible. Indeed, given any $\overline{C}(\gamma)$ we need only choose $\overline{Q}(\gamma)$ as follows:

$$\overline{Q}(\gamma) = \frac{\overline{C}(\gamma)}{1 + \hat{G}(\gamma)\overline{C}(\gamma)} \qquad (8.5.2)$$

We next consider the corresponding closed loop system as shown in Figure 8.5.1 where $\overline{G}$ is assumed to be stable.

Provided $\hat{G} = \overline{G}$, it is readily verified that the closed-loop transfer functions from $r$ to $y$, $d_o$ to $y$, $d_i$ to $y$ and $r$ to $u$ are given respectively by

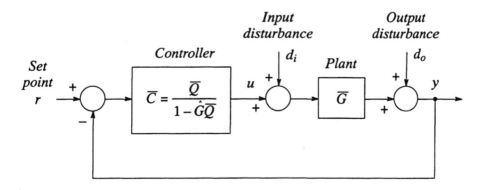

*Figure 8.5.1 : All stabilizing controllers (stable case).*

$$\frac{Y}{R} = \overline{C}\frac{\overline{G}}{1 + \overline{C}\overline{G}} = \overline{Q}\overline{G}$$

(8.5.3)

$$\frac{Y}{D_o} = \frac{1}{1 + \overline{C}\overline{G}} = 1 - \overline{Q}\overline{G}$$

(8.5.4)

$$\frac{Y}{D_i} = \frac{\overline{G}}{1 + \overline{C}\overline{G}} = (1 - \overline{Q}\overline{G})\overline{G}$$

(8.5.5)

$$\frac{U}{R} = \frac{\overline{C}}{1 + \overline{C}\overline{G}} = \overline{Q}$$

(8.5.6)

We require all of these transfer functions to be stable so that we can guarantee that the closed loop is (internally) stable.

Since $\overline{G}$ is itself stable (by assumption) then it can be seen that (8.5.3) to (8.5.6) will all be stable if and only if $\overline{Q}$ is also stable. This is a remarkable property and makes the design of $\overline{Q}$ particularly straightforward since nominal closed-loop stability is automatically guaranteed provided $\overline{Q}(\gamma)$ is restricted to stable transfer functions.

A typical design requirement is to keep $\overline{S} = 1 - \overline{Q}\overline{G}$ small up to some frequency $\omega_b$. We thus see that it is sensible to try to choose $\overline{Q}$ as some kind of stable-proper approximation to the inverse of $\overline{G}$. Thus, as mentioned in the introduction, we find model inversion appearing as a desirable part of control system design.

To obtain additional insight into the constraints on $\overline{Q}$ when the model, $\hat{G}$, and the plant, $\overline{G}$, are not identical we redraw Figure 8.5.1 as in Figure 8.5.2.

The output tracking error (i.e. $e = y - r$) can be shown to be:

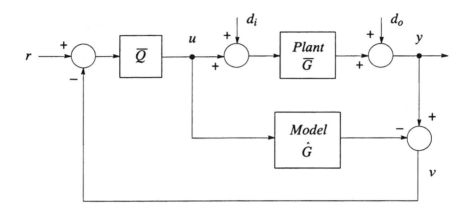

*Figure 8.5.2 : Equivalent form of all stabilizing
controllers (open-loop stable case).*

$$e = (y - r) = \left[ \frac{1 - \hat{G}\overline{Q}}{1 + \overline{G}_\Delta \overline{Q}} \right] d \qquad (8.5.7)$$

where

$$\overline{G}_\Delta = \overline{G} - \hat{G}; \quad d = \left(d_o + Gd_i\right) \qquad (8.5.8)$$

We see from (8.5.7) that, when we have model errors, in addition to keeping $\overline{Q}$ stable it is necessary to ensure that $(1 + G_\Delta Q)$ does not have zeros outside the stability boundary. A sufficient condition to ensure this is that the magnitude of $\overline{G}_\Delta \overline{Q}$ should not approach (or exceed) unity at any frequency. This is usually achieved by choosing $\overline{Q}$ so that $\overline{Q}(\gamma_\omega)$ is small at those frequencies where $\overline{G}_\Delta(\gamma_\omega)$ becomes significant.

## 8.5.2   Open-Loop Unstable Plants

The parameterization of equation (8.5.1) is valid when the open-loop poles are undesirable (unstable). However inspection of equations (8.5.3) to (8.5.6) indicates that additional constraints on $\overline{Q}$ (beyond it being stable and proper) are needed to ensure internal stability even when there is no model error, i.e. when $\hat{G} = \overline{G}$.

There are two ways that these constraints can be imposed. Firstly, we could use the simple parameterization of $\overline{C}$, with the implementation shown in Figure 8.5.1, provided that we also impose additional constraints on $\overline{Q}$ to ensure that (8.5.3) to (8.5.6) are all stable when $\overline{G}$ is unstable. These additional constraints are called *interpolation constraints*. In particular, we see that it suffices to ensure that all unstable poles of $\overline{G}$ are zeros of both $\overline{Q}$ and $1 - \overline{Q}\overline{G}$.

A second approach is to change the parameterization of $\overline{C}$. For the general open loop *unstable* case, a suitable parameterization of $\overline{C}$ turns out to be the following:

$$\overline{C} = \left[ \frac{\frac{\overline{P}}{\overline{E}_2} + \overline{Q}\frac{\hat{A}}{\hat{E}_1}}{\frac{\overline{L}}{\overline{E}_2} - \overline{Q}\frac{\hat{B}}{\hat{E}_1}} \right] \tag{8.5.9}$$

where the open-loop plant model is

$$\hat{G} = \frac{\hat{B}}{\hat{A}} \tag{8.5.10}$$

with $degree\ \{\overline{E}_1\} = degree\ \{\overline{E}_2\} = degree\ \{\hat{A}\}$, and $\overline{E}_1$ and $\overline{E}_2$ are stable polynomials:

$$\hat{A}\hat{L} + \hat{B}\hat{P} = \overline{A}^* \tag{8.5.11}$$

with $\overline{A}^*$ any stable polynomial.

Notice that in the ideal case when $\overline{G} = \hat{G}$ (no modelling error), the four transfer functions (8.5.3) to (8.5.6) become:

$$\frac{Y}{R} = \overline{C}\frac{\overline{G}}{1+\overline{CG}} = \frac{\left(\overline{PE_1 + QAE_2}\right)\overline{B}}{\overline{E_1}\overline{A}^*} = \overline{T} \qquad (8.5.12)$$

$$\frac{Y}{D_o} = \frac{1}{1+\overline{CG}} = \frac{\left(\overline{LE_1 - QBE_2}\right)\overline{A}}{\overline{E_1}\overline{A}^*} = \overline{S} \qquad (8.5.13)$$

$$\frac{Y}{D_i} = \frac{\overline{G}}{1+\overline{CG}} = \frac{\left(\overline{LE_1 - QBE_2}\right)\overline{B}}{\overline{E_1}\overline{A}^*} = \overline{SG} \qquad (8.5.14)$$

$$\frac{U}{R} = \frac{\overline{C}}{1+\overline{CG}} = \frac{\left(\overline{PE_1 + QAE_2}\right)\overline{A}}{\overline{E_1}\overline{A}^*} = \overline{SC} \qquad (8.5.15)$$

All of these transfer functions are stable if and only if $\overline{Q}$ is stable. We thus have:

**Theorem 8.5.1** Consider a system having *known* transfer function $\overline{B}/\overline{A}$. Then all (one degree-of-freedom) control laws which give internal stability can be parameterized as

$$\overline{C} = \frac{\frac{\overline{P}}{\overline{E_2}} + \overline{Q}\frac{\overline{A}}{\overline{E_1}}}{\frac{\overline{L}}{\overline{E_2}} - \overline{Q}\frac{\overline{B}}{\overline{E_1}}} \qquad (8.5.16)$$

where $\overline{E}_1$, $\overline{E}_2$ and $\overline{A}^* = \overline{AL} + \overline{BP}$ are stable but otherwise arbitrary and $\overline{Q}$ is stable.

**Proof :** Follows from (8.5.12) to (8.5.15). (See also the references and the end of the chapter which cover more general system descriptions.)

△△△

Equation (8.5.16) is known as the Youla parameterization of all stabilising controllers for the general open-loop unstable case. The controller described in (8.5.16) is shown diagramatically in Figure 8.5.3 where we have chosen to inject the reference input, $r$, as in Figure 8.4.1.

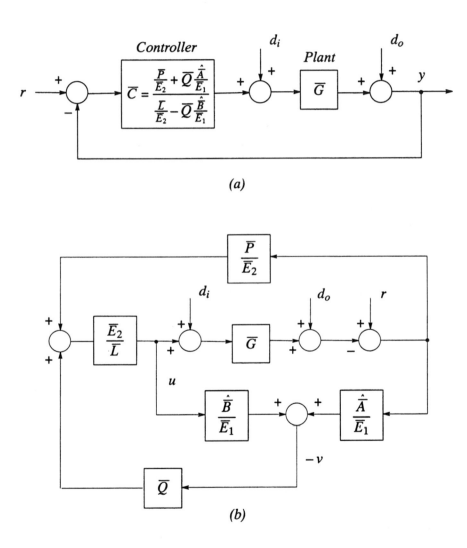

*(a)*

*(b)*

*Figure 8.5.3 : Equivalent implementations of all stabilizing controllers for open-loop unstable case.*

In cases where the model $\hat{G}$ is not necessarily equal to the plant transfer function $\overline{G}$ then additional constraints on $\overline{Q}$ are necessary to achieve closed loop stability, as described below.

The effect of model errors and disturbances on the feedback control system of Figure 8.5.3 (b) is clarified in the following result:

**Lemma 8.5.1** Consider the control system shown in Figure 8.5.3 (b) with $\overline{A}^* = \overline{E}_1 \overline{E}_2$. The closed-loop output tracking error, $e = y - r$, satisfies

$$\boxed{e = \overline{S} \overline{S}_\Delta \cdot \left[ -r + d_o + \overline{G} d_i \right]} \qquad (8.5.17)$$

where $\overline{S}$ is the nominal sensitivity function given in (8.5.13) and $\overline{S}_\Delta$ is the sensitivity due to model errors where

$$\boxed{\overline{S}_\Delta = \frac{1}{1 + \frac{\overline{G}_\Delta}{\hat{G}}(\overline{T})}} \qquad (8.5.18)$$

$\overline{T}$ is the nominal complementary sensitivity function and $\overline{G}_\Delta$ is the additive model error, i.e.

$$\overline{G} = \hat{G} + \overline{G}_\Delta; \qquad \hat{G} = \frac{\hat{B}}{\hat{A}}; \qquad \overline{T} = \frac{\left(\overline{PE}_1 + \overline{QAE}_2\right)\hat{B}}{\overline{E}_1 \overline{A}^*} \qquad (8.5.19)$$

**Proof :** See Problem 8.16.

$\triangle\triangle\triangle$

We illustrate the above discussion by a simple example.

**Example 8.5.1** : Consider the continuous-time system given by

$$G(s) = \frac{1}{(s+1)(1-s)} \qquad (8.5.20)$$

Design a digital controller with sample period $\Delta = 0.1$ sec. using the controller parameterization shown in Figure 8.5.3. Also, test the robustness to unmodelled dynamics and disturbances.

**Solution** The discrete equivalent of the given system with $\Delta = 0.1$ is

$$\overline{G}(\gamma) = \frac{-0.05\gamma - 1.0008}{\gamma^2 - 0.1001\gamma - 1.0008} = \frac{\overline{B}(\gamma)}{\overline{A}(\gamma)}$$

Let us first choose

$$\overline{A}^*(\gamma) = \overline{E}_1(\gamma)\overline{E}_2(\gamma) = (\gamma + 1)^4$$

so that $\overline{L}$ and $\overline{P}$ are found from equation (8.5.11) as follows

$$\overline{A}(\gamma)\overline{L}(\gamma) + \overline{B}(\gamma)\overline{P}(\gamma) = (\gamma + 1)^4$$

Equating coefficients gives

$$\begin{bmatrix} 1 & 0 & 0 & 0 & 0 \\ -0.1001 & 1 & 0 & 0 & 0 \\ -1.0008 & -0.1001 & 1 & 0.05 & 0 \\ 0 & -1.0008 & -0.1001 & 1.0008 & 0.05 \\ 0 & 0 & -1.0008 & 0 & 1.0008 \end{bmatrix} \begin{bmatrix} l_2 \\ l_1 \\ l_0 \\ p_1 \\ p_0 \end{bmatrix} = \begin{bmatrix} 1 \\ 4 \\ 6 \\ 4 \\ 1 \end{bmatrix}$$

Solving we obtain

$$\overline{L}(\gamma) = \gamma^2 + 4.1001\gamma + 6.9914$$

$$\overline{P}(\gamma) = -(8.397\gamma + 7.9906)$$

To complete the design of the controller we have to choose $\overline{Q}$. Since we usually desire the nominal sensitivity $\overline{S}$ to be small, we refer to (8.5.13). We then see that a logical choice for $\overline{Q}$ is some kind of stable proper approximation to $(\overline{L}\overline{E}_1)/(\overline{B}\overline{E}_2)$. We note that $\overline{B}$ is unstable and thus we choose $\overline{Q}$ as

$$\bar{Q}(\gamma) = \frac{\bar{L}(\gamma)\bar{E}_1(\gamma)}{\bar{B}(0)(\beta\gamma + 1)^2\bar{E}_2(\gamma)} = \frac{\bar{L}(\gamma)}{\bar{B}(0)(\beta\gamma + 1)^2} \qquad (8.5.21)$$

where the term $\bar{B}(0)$ in the denominator preserves the d.c. gain. Thus,

$$\bar{Q}(\gamma) = \frac{-(\gamma^2 + 4.1001\gamma + 6.9914)}{1.0008(\beta\gamma + 1)^2} \qquad (8.5.22)$$

where the term $(\beta\gamma + 1)^2$ has been introduced to keep $\bar{Q}$ proper. To give a fast response we then choose

$$\beta = 0.1$$

The above controller has been simulated in four stages. In the first we employed only the pre-stabilizing loop $\bar{L}(\delta)u[k] = -\bar{P}(\delta)[y[k] - r[k]]$. In Figure 8.5.4 we see the unit step response with the step applied at time $t = 1$. The result is typical for a system with $(\gamma + 1)^4$ in its denominator. In the next stage we employ the full controller as in Figure 8.5.3. The unit step response of this control system is presented in Figure 8.5.5.

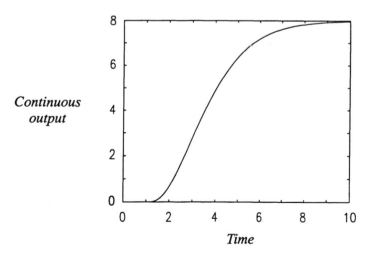

*Figure 8.5.4 : Step response with pre-stabilizing feedback alone.*

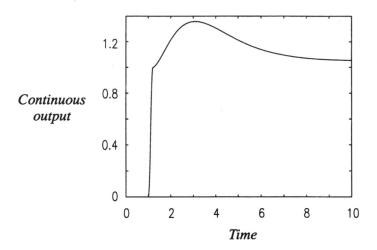

*Figure 8.5.5 : Step response with full controller* $\left(\beta = 0.1, \ \overline{A}^* = (\gamma + 1)^4\right)$.

This system can be further improved by changing the way that the reference signal is injected into the loop. An improved two degree-of-freedom design is disscussed in Problem 8.15. The resulting unit step response is as in Figure 8.5.6. We see that the (slow) dynamics of the $\overline{P}/\overline{L}$ loop have been removed from the reference response by the two degree-of-freedom design.

Next we consider the response to disturbances.

Of course, the $\overline{P}/\overline{L}$ loop still affects the disturbance response in the two degree-of-freedom design. This is illustrated in Figure 8.5.7 where a unit step input disturbance has been applied at time $t = 3$ .

If we want to improve the response to disturbances then we can either design a better pre-stabilizing loop or change the value of $\overline{Q}$ (see Problem 8.14). We choose the former of these two alternatives by changing $\overline{A}^*$ to the following (which has poles at $-5$ compared with $-1$ previously).

$$\overline{A}^*(\gamma) = \overline{E}_1(\gamma)\overline{E}_2(\gamma) = (\gamma + 5)^4$$

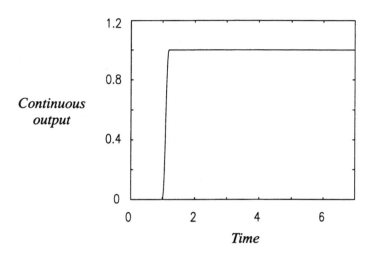

*Figure 8.5.6 : Step response with modified reference input injection* $\left(\beta = 0.1, \ \vec{A}^* = (\gamma + 1)^4\right)$.

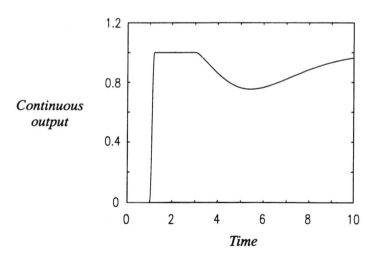

*Figure 8.5.7 : Step and disturbance response for modified controller* $\left(\beta = 0.1, \ \vec{A}^* = (\gamma + 1)^4\right)$.

Solving for $\overline{L}(\gamma), \overline{P}(\gamma)$ leads to.

$$\overline{L}(\gamma) = \gamma^2 + 20.1001\gamma + 128.2483$$

$$\overline{P}(\gamma) = -(494.8721\gamma + 752.7277)$$

We again choose $\overline{Q}(\gamma)$ as in (8.5.21) and as before we set $\beta = 0.1$. We also continue to use the improved set point injection procedure of Problem 8.15. The resulting response to a unit set point change (at 1 second) and a unit step input disturbance (at 3 seconds) is shown in Figure 8.5.8. This would seem to be a rather ideal solution.

We next consider model errors by adding an extra unmodelled pole to the system at $-2$ (this is rather severe undermodelling since the modelled system poles are at $\pm 1$). We repeat the simulation shown in Figure 8.5.8 leading to the result in Figure 8.5.9. We see that the response is extremely poor. This can be

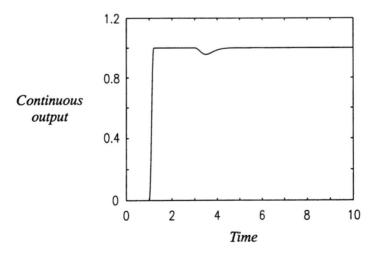

*Figure 8.5.8 : Step and disturbance response with modified*
*pre–stabilising loop $\left(\beta = 0.1, \ \overline{A}^* = (0.2\gamma + 1)^4\right)$.*

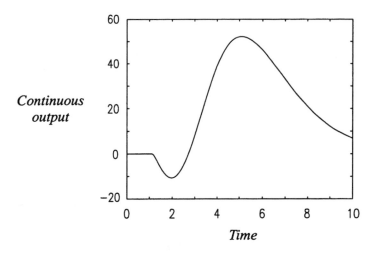

*Figure 8.5.9 : Step and disturbance response with modified*
*pre-stabilizing loop and unmodelled pole*
*at $-2$ $\left(\beta = 0.1,\ \overline{A}^* = (0.2\gamma + 1)^4\right)$.*

explained by reference to equation (8.5.18). The unmodelled dynamics have clearly made $\overline{S}_\Delta$ large.

This difficulty can be resolved by reducing the bandwidth of the loop so as to keep $\left|\overline{T}\left(\overline{G}_\Delta/\hat{G}\right)\right|$ small at all frequencies. We thus change $\beta$ in (8.5.21) to 0.2.

Figure 8.5.10 shows the response to a unit step reference and disturbance under the ideal model conditions. Comparing Figure 8.5.10 with Figure 8.5.8 we see that the performance (under ideal conditions) has slightly deteriorated. Finally, we repeat the simulation in Figure 8.5.10 but with an unmodelled pole at $-2$ again added to the system. The result is shown in Figure 8.5.11. Comparing Figure 8.5.11 with Figure 8.5.9 clearly shows the improvement resulting from the redesign of $\overline{Q}$.

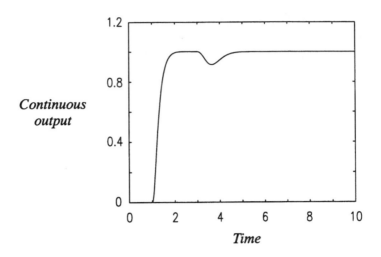

Figure 8.5.10 : *Step and disturbance response with modified pre-*
*stabilizing loop and modified* $\overline{Q}$, *no undermodelling*
$\left(\beta = 0.2, \ \overline{A}^* = (0.2\gamma + 1)^4\right)$.

## 8.6    State Estimate Feedback

We next give a state-space interpretation to the various controllers discussed above. The state-space approach is particularly useful in more difficult
problems where multi-rate or non-uniform sampling is used. These topics are
taken up later in the book.

We begin by rewriting the discrete-time system model in its equivalent discrete state-space form as

$$\delta x = A_\delta x + B_\delta u \tag{8.6.1}$$

$$y = C_\delta x \tag{8.6.2}$$

We also make the simplifying assumption that $x$ is directly measured. We
can then form the input by using state variable feedback as follows:

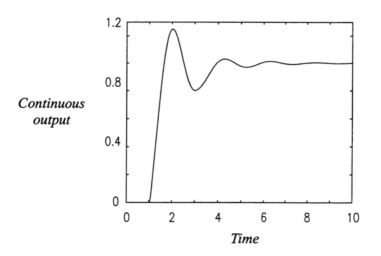

*Figure 8.5.11 : Step and disturbance response with modified pre-stabilizing loop and modified $\overline{Q}$, unmodelled pole at $-2$ $\left(\beta = 0.2, \ \overline{A}^* = (0.2\gamma + 1)^4\right)$.*

$$\boxed{u = -K_\delta x + \xi} \qquad (8.6.3)$$

where $\xi$ is some external input.

Substituting (8.6.3) into (8.6.1) gives the following closed loop system equation

$$\delta x = A_\delta' x + B_\delta \xi \qquad (8.6.4)$$

where the closed loop $A$ matrix is given by

$$\boxed{A_\delta' = A_\delta - B_\delta K_\delta} \qquad (8.6.5)$$

A key observation is that if the pair $[A_\delta, B_\delta]$ is completely reachable then $K_\delta$ can be chosen so that the eigenvalues of $A_\delta'$ are arbitrarily assigned. This fact is verified below:

## Lemma 8.6.1 *Ackermann's formula*

Let $[A_\delta, B_\delta]$ be a reachable pair with $A_\delta \in R^{n \times n}$ and $B_\delta \in R^{n \times 1}$. Let $\alpha(\lambda)$ an $n$-th order polynomial, and $Q_\delta^c$ the controllability matrix defined by

$$Q_\delta^c = \left[ B_\delta, \ A_\delta B_\delta, \ \dots, \ A_\delta^{n-1} B_\delta \right]$$

Then

$$\textit{characteristic polynomial } \left( A_\delta - B_\delta K \right) = \alpha(\lambda)$$

for

$$K_\delta = e_n^T (Q_\delta^c)^{-1} \alpha(A_\delta)$$

where $e_n^T = [0, \ \dots, \ 0, \ 1]$.

**Proof :**   Note first that the reachability of $[A_\delta, B_\delta]$ guarantees that $Q_\delta^c$ is nonsingular.

Define

$$T = \begin{bmatrix} e_n^T (Q_\delta^c)^{-1} A_\delta^{n-1} \\ e_n^T (Q_\delta^c)^{-1} A_\delta^{n-2} \\ . \\ . \\ . \\ e_n^T (Q_\delta^c)^{-1} A_\delta^{0} \end{bmatrix}$$

then

$$T B_\delta = e_1 = [1, \ 0, \ \dots, \ 0]^T$$

and

$$
TA_\delta T^{-1} = \begin{bmatrix} -\underline{a}^T \\ e_1^T \\ \cdot \\ \cdot \\ \cdot \\ e_n^T \end{bmatrix}
$$

where $\underline{a}^T = -e_n^T(Q_\delta^c)^{-1}A_\delta^n\, T^{-1}$ is a vector of the coefficients of the characteristic polynomial of $A_\delta$.

Then

$$
T(A_\delta - B_\delta K)T^{-1} = TA_\delta T^{-1} - TB_\delta e_n^T(Q_\delta^c)^{-1}\alpha(A_\delta)T^{-1}
$$

$$
= \begin{bmatrix} -\underline{a}^T \\ e_1^T \\ \cdot \\ \cdot \\ \cdot \\ e_{n-1}^T \end{bmatrix} - e_1 e_n^T(Q_\delta^c)^{-1}A_\delta^n T^{-1} - e_1 e_n^T(Q_\delta^c)^{-1}\sum_{l=1}^{n}\alpha_l A_\delta^{n-l}T^{-1}
$$

$$
= \begin{bmatrix} -\underline{a}^T \\ e_1^T \\ \cdot \\ \cdot \\ \cdot \\ e_{n-1}^T \end{bmatrix} + e_1\underline{a}^T - e_1\underline{\alpha}^T = \begin{bmatrix} -\underline{\alpha}^T \\ e_1^T \\ \cdot \\ \cdot \\ \cdot \\ e_{n-1}^T \end{bmatrix} \qquad (8.6.6)
$$

with $\underline{\alpha}^T$ the vector of the coefficients of $\alpha(\lambda)$.

Hence, $T(A_\delta - B_\delta K_\delta)T^{-1}$ which is similar to $A_\delta - B_\delta K_\delta$ has $\alpha(\lambda)$ as its characteristic polynomial.

∧∧∧

We have thus seen that it is possible to arbitrarily assign closed-loop poles using state feedback.

Next, we consider the case where only the output, $y$, is measured.

The reader who has studied Chapter 6 may suggest that since $x$ is not measured it would clearly be possible to estimate $x$ using a Kalman filter driven by $y$. Actually, the Kalman filter is a special case of an 'observer' having the following general form:

$$\delta\hat{x} = A_\delta\hat{x} + B_\delta u + J_\delta(y - C_\delta\hat{x}) \qquad (8.6.7)$$

where $J_\delta$ is a gain matrix to be chosen. (In the Kalman filter this is automatically chosen by solving a Riccati equation).

Subtracting (8.6.1) from (8.6.7) gives the following equation for the state estimation error

$$\delta\tilde{x} = A_\delta''\tilde{x} \qquad (8.6.8)$$

where

$$\tilde{x} = \hat{x} - x, \qquad A_\delta'' = A_\delta - J_\delta C_\delta \qquad (8.6.9)$$

Now, provided the pair $(C_\delta, A_\delta)$ is completely observable, it follows by the same argument as in Lemma 8.6.1 that $J_\delta$ can be chosen so that the eigenvalues of $(A_\delta - J_\delta C_\delta)$ are arbitrarily assigned. Hence, in view of (8.6.8) the state estimation error can be made to decay to zero at any desired rate by suitable choice of the vector $J$.

The logical extension of the state feedback law (8.6.3) is now to use the state estimate, $\hat{x}$, rather than the true, $x$, state when constructing the feedback law. That is, we replace (8.6.3) by

$$u = -K_\delta\hat{x} + \xi \qquad (8.6.10)$$

An interesting fact is that the set of eigenvalues of the closed-loop system resulting from the state estimate feedback law as in (8.6.10) are precisely the union of the set of eigenvalues of $A_\delta'$ and $A_\delta''$ in (8.6.9), (8.6.5). This is established below:

### Theorem 8.6.1 *Separation theorem*

The estimated state feedback control system has the following property: The resulting set of closed-loop eigenvalues is the union of the observer eigenvalues and the set of closed-loop poles which would have resulted had true state feedback been used.

**Proof :** Equation (8.6.10) can be written as

$$u = -K_\delta[x + \tilde{x}] + \xi \tag{8.6.11}$$

Combining (8.6.8), (8.6.1) and (8.6.11) gives

$$\delta \begin{bmatrix} x \\ \tilde{x} \end{bmatrix} = \begin{bmatrix} A_\delta - B_\delta K_\delta & -B_\delta K_\delta \\ 0 & A_\delta - J_\delta C_\delta \end{bmatrix} \begin{bmatrix} x \\ \tilde{x} \end{bmatrix} + \begin{bmatrix} B_\delta \\ 0 \end{bmatrix} \xi \tag{8.6.12}$$

$$y = \begin{bmatrix} C_\delta & 0 \end{bmatrix} \begin{bmatrix} x \\ \tilde{x} \end{bmatrix} \tag{8.6.13}$$

The result follows immediately from (8.6.12) since the eigenvalues of a triangular matrix are equal to the union of the eigenvalues of the diagonal blocks.

$$\wedge\wedge\wedge$$

Next we turn to the choice of the exogenous input in (8.6.10). A rather general choice is to let $\xi$ be given by some transfer function operating on the innovations process of the observer ($v = y - C\hat{x}$) plus some other transfer function operating on the reference input, $r$. Thus we might replace (8.6.10) by

$$u = -K_\delta \hat{x} + \overline{\Omega}_1(\gamma)v + \overline{\Omega}_2(\gamma)r \tag{8.6.14}$$

where $\overline{\Omega}_1(\gamma)$, $\overline{\Omega}_2(\gamma)$ are stable proper transfer functions. Since the innovation process, $v$ , is decoupled (in steady state) from the input then the term involving $v$ in (8.6.14) clearly will not affect closed loop stability.

Of course, the control law (8.6.14) can always be expressed as a member of the class of controllers given in Theorem 8.5.1 (since the latter theorem covers *all* stabilizing control laws). However, in the next section, we make this link between state estimate feedback and all stabilizing controllers explicit by showing how the latter can be implemented in state-space form.

## 8.7    Rapprochement Between State Estimate Feedback and All Stabilizing Controllers

### 8.7.1    Implicit Disturbance Modelling

In this section we show that state estimate feedback as in (8.6.14) is essentially equivalent to all stabilizing controllers as in Section 8.5. Towards this end, we present the following result:

**Lemma 8.7.1**    Consider the discrete-time plant having transfer function $\overline{G}(\gamma) = \overline{B}(\gamma)/\overline{A}(\gamma)$ and equivalent state-space model:

$$\delta x[k] = A_\delta x[k] + B_\delta u[k] \tag{8.7.1}$$

$$y[k] = C_\delta x[k] \tag{8.7.2}$$

Then, the class of all stabilizing control laws (as in Figure 8.5.3) can be implemented in either one of the following two equivalent forms:

(a)  Polynomial form

$$\frac{\overline{L}(\delta)}{\overline{E}_2(\delta)} u[k] = -\frac{\overline{P}(\delta)}{\overline{E}_2(\delta)} (y[k] - r[k]) - \overline{Q}(\delta) v[k] \tag{8.7.3}$$

where $\{v[k]\}$ is an implicit disturbance estimate given by:

$$v[k] = \frac{\overline{A}(\delta)}{\overline{E}_1(\delta)}(y[k] - r[k]) - \frac{\overline{B}(\delta)}{\overline{E}_1(\delta)}u[k] \tag{8.7.4}$$

and where $\overline{E}_1(\delta)$, $\overline{E}_2(\delta)$ and $\overline{A}(\gamma)\overline{L}(\gamma) + \overline{B}(\gamma)\overline{P}(\gamma)$ are stable polynomials.

(b) State-space form

$$u[k] = -K_\delta \hat{x}[k] - \overline{Q}(\delta)v[k] \tag{8.7.5}$$

where $\hat{x}[k]$ is the state estimate provided by an observer (with added reference input, $r[k]$ ):

$$\delta\hat{x}[k] = A_\delta\hat{x}[k] + B_\delta u[k] + J_\delta(y[k] - C_\delta\hat{x}[k] - r[k]) \tag{8.7.6}$$

and where $v[k]$ is the "innovations" sequence provided by a (possibly second) observer:

$$\delta\hat{x}'[k] = A_\delta\hat{x}'[k] + B_\delta u[k] + J_\delta'\left(y[k] - C_\delta\hat{x}'[k] - r[k]\right) \tag{8.7.7}$$

$$v[k] = y[k] - C_\delta\hat{x}'[k] - r[k] \tag{8.7.8}$$

and such that

$$\overline{E}_3(\gamma) = \det(\gamma I - A_\delta + B_\delta K_\delta) \tag{8.7.9}$$

$$\overline{E}_2(\gamma) = \det(\gamma I - A_\delta + J_\delta C_\delta) \tag{8.7.10}$$

$$\overline{E}_1(\gamma) = \det\left(\gamma I - A_\delta + J_\delta' C_\delta\right) \tag{8.7.11}$$

are all stable.

Proof :

(a) Immediate from Figure 8.5.3.

(b) Assuming $(A_\delta, B_\delta, C_\delta)$ is observable and controllable, we can design a stable observer as in equation (8.6.7) (with $y$ replaced by $y - r$ ):

$$\delta \dot{x} = (A_\delta - J_\delta C_\delta)\hat{x} + B_\delta u + J_\delta (y - r) \qquad (8.7.12)$$

such that

$$\det(\gamma I - A_\delta + J_\delta C_\delta) = \bar{E}_2(\gamma) \qquad (8.7.13)$$

We then design estimate feedback as in (8.6.10) by choosing $K_\delta$ such that $\bar{E}_3(\gamma)$ in (8.7.10) is stable leading to the feedback law:

$$u = -K_\delta \hat{x} - \bar{Q}(\delta) v \qquad (8.7.14)$$

where $v$ is some external input.

Substituting (8.7.14) into (8.7.12) gives the following equation for the control law

$$\delta \dot{x} = (A_\delta - J C_\delta)\hat{x} + B_\delta u + J_\delta (y - r) \qquad (8.7.15)$$

$$u = -K\hat{x} - \bar{Q}(\delta) v \qquad (8.7.16)$$

This can be expressed in transfer function form as

$$\frac{\bar{L}}{\bar{E}_2} u = -\frac{\bar{P}}{\bar{E}_2}(y - r) - \bar{Q}(\delta) v \qquad (8.7.17)$$

where

$$\frac{\bar{L}}{\bar{E}_2} \stackrel{\Delta}{=} I + K_\delta(\delta I - A_\delta + J_\delta C_\delta)^{-1} B_\delta \qquad (8.7.18)$$

$$\frac{\bar{P}}{\bar{E}_2} \stackrel{\Delta}{=} K_\delta(\delta I - A_\delta + J_\delta C_\delta)^{-1} J_\delta \qquad (8.7.19)$$

Further, it can be shown (see Problem 8.12) that the polynomials $\bar{L}, \bar{P}$ satisfy

$$\bar{A}\bar{L} + \bar{B}\bar{P} = \bar{E}_3 \bar{E}_2 \qquad (8.7.20)$$

where

$$\frac{\bar{B}}{\bar{A}} = C_\delta(\gamma I - A_\delta)^{-1}B_\delta \qquad (8.7.21)$$

$$\bar{E}_2 = \det(\gamma I - A_\delta + J_\delta C_\delta)$$

$$\bar{E}_3 = \det(\gamma I - A_\delta + B_\delta K_\delta) \qquad (8.7.22)$$

We see that (8.7.17) is identical to the pre-stabilizing loop in Figure 8.5.3 (b) and hence this loop can be viewed as a special case of state estimate feedback.

We will next show that $v$ can be viewed as the innovation process of another observer. We choose a (possibly second) observer gain matrix $J'_\delta$ such that

$$\det(\gamma I - A_\delta + J'_\delta C_\delta) = \bar{E}_1(\gamma) \qquad (8.7.23)$$

is a stable polynomial and write the observer as follows:

$$\delta \hat{x}' = (A_\delta - J'_\delta C_\delta)\hat{x}' + B_\delta u + J'_\delta(y - r) \qquad (8.7.24)$$

In the above observer we have again used $y - r$ instead of $y$. This does not change the poles of the observer but allows us to include the reference input $r$ in a fashion which will be compatible with Figure 8.5.3 (b). We also define the innovation process of the observer by including the reference input $r$ as follows:

$$v = y - r - C_\delta \hat{x}' \qquad (8.7.25)$$

Notice that the observer can be viewed as having two inputs ($u$ and $(r - y)$) with $-v$ as output. The respective transfer functions are (see Problem 8.13)

$$\frac{\bar{B}(\gamma)}{\bar{E}_1(\gamma)} = C_\delta\left(\gamma I - A_\delta + J'_\delta C_\delta\right)^{-1}B_\delta \qquad (8.7.26)$$

$$\frac{\overline{A}(\gamma)}{\overline{E}_1(\gamma)} = 1 - C_\delta \left( \gamma I - A_\delta + J'_\delta C_\delta \right)^{-1} J'_\delta \qquad (8.7.27)$$

Thus in transfer function form we can write

$$-v = \frac{\overline{B}}{\overline{E}_1}(u) + \frac{\overline{A}}{\overline{E}_1}(r-y) \qquad (8.7.28)$$

We thus see that the signal $v$ in Figure 8.5.3 (b) is actually the innovations process for the state observer (8.7.18).

Finally, note that if we use $J'_\delta = J_\delta$ then we need only implement one of (8.7.6) and (8.7.7).

$$\wedge\wedge$$

## 8.7.2   Explicit Disturbance Modelling

In the previous section we did not explicitly model the disturbance but instead used $v[k]$ as an implicit disturbance estimate. In some cases, it may be desirable to include an explicit estimate of the disturbance. To show how this can be achieved we generalize the discrete state-space model as in (8.6.1), (8.6.2) by including an explicit model for the disturbances. Thus, we describe the system as follows:

$$\delta x_1 = A_1 x_1 + B_1 u + Gd \qquad (8.7.29)$$

$$y = C_1 x_1 + Hd \qquad (8.7.30)$$

Note that, if the disturbance appears at the input, then we should set $G = B_1$, $H = 0$; if the disturbance appears at the output, then we should set $H = I$, $G = 0$.

We will restrict attention here to deterministic disturbances comprising a finite sum of sinusoids of different frequencies (including zero if appropriate). Note that it is also possible to include stochastic disturbances (see Chapters 5 and 6).

Thus the model for disturbances takes the following form:

$$\delta x_2 = A_2 x_2 \tag{8.7.31}$$

$$d = C_2 x_2 \tag{8.7.32}$$

For example, if $d$ is a constant, then we simply use $A_2 = 0$, $C_2 = 1$. Similarly, the discrete-time model for a sum of sinewaves is given in Problem 8.8 (see also Section 6.6.3 and Example 6.5.1).

The model (8.7.29) to (8.7.32) can be written in composite form as

$$\delta x = A_\delta x + B_\delta u \tag{8.7.33}$$

$$y = C_\delta x \tag{8.7.34}$$

where

$$x = \begin{bmatrix} x_1 \\ x_2 \end{bmatrix} ; \quad A_\delta = \begin{bmatrix} A_1 & GC_2 \\ 0 & A_2 \end{bmatrix} ; \quad B_\delta = \begin{bmatrix} B_1 \\ 0 \end{bmatrix} \tag{8.7.35}$$

$$C_\delta = \begin{bmatrix} C_1 & HC_2 \end{bmatrix} \tag{8.7.36}$$

Subject to *observability* of the model (8.7.33), (8.7.34) we can now design a stable observer for the combined state of the system and the disturbance. This takes the form:

$$\delta \hat{x} = A_\delta \hat{x} + B_\delta u + J_\delta (y - C_\delta \hat{x}) \tag{8.7.37}$$

where $[A_\delta - J_\delta C_\delta]$ is stable.

Next, we note that due to the presence of the disturbance model (8.7.31), (8.7.32), the composite system (8.7.33) is definitely *not* stabilizable. However, we need only assume that the *plant* state ($x_1$) is controllable from $u$.

We therefore choose state variable feedback of the form:

$$u = -K_\delta \dot{x}_1 + \overline{Q}_1(\delta)r - \overline{Q}_2(\delta)\hat{d} \qquad (8.7.38)$$

where $\overline{Q}_1(\delta)$ and $\overline{Q}_2(\delta)$ are stable rational proper transfer functions to be specified later and $K_\delta$ is chosen so that $(A_1 - B_1 K_\delta)$ is stable. Note that the transfer
function from $u$ to $\dot{x}_2$ is zero. Hence, in steady state, $\hat{d} = C_2 \dot{x}_2$ is a function only
of the innovation process. Thus (8.7.38) is equivalent to (8.7.5) since in the latter
equation $v$ is effectively an implicit estimate of disturbance and the reference
input.

It remains to design the reference and disturbance injection matrices
$\overline{Q}_1(\gamma)$ and $\overline{Q}_2(\gamma)$. A typical choice for $\overline{Q}_1(\gamma)$ is $\overline{Q}_1(\gamma) = 1$. However, other values of $\overline{Q}_1(\gamma)$ can be used to achieve different responses between disturbances
and output, and reference and output (commonly called two degree-of-freedom
design). A desirable choice for $\overline{Q}_2(\gamma)$ is to attempt to cancel the disturbance, as
seen at the output. One way of achieving this is outlined below.

Say we factor the numerator polynomial, $\overline{N}(\gamma)$, in the transfer function
from $u[k]$ to $y[k]$ as follows:

$$\overline{N}(\gamma) = \overline{N}_s(\gamma)\overline{N}_u(\gamma) \qquad (8.7.39)$$

where $\overline{N}_s(\gamma)$ contains the stable well-damped zeros of $\overline{N}(\gamma)$ and $\overline{N}_u(\gamma)$ contains
the remainder. We then define a desired closed loop polynomial as

$$\overline{N}'(\gamma) = \overline{N}_s(\gamma)\, \overline{N}_u(0)\, \overline{N}_e(\gamma) \qquad (8.7.40)$$

where $\overline{N}_e(\gamma)$ is a stable polynomial (with unity d.c. gain) introduced to ensure
$\overline{N}'(\gamma)$ has the same degree as the state-space model (8.7.29). We next define a
fictitious variable $z[k]$ whose transfer function to $u[k]$ has polynomial $\overline{N}'(\gamma)$
as its numerator. Note that the relationship between $z[k]$, $x_1[k]$ and $u[k]$ can
be written as

$$z[k] = C_2x_1[k] + D_2u[k] \qquad (8.7.41)$$

where $D_2 \neq 0$ by virtue of the properness (but not strict properness) of the transfer function from $u[k]$ to $z[k]$. Then, if we use (8.7.41) to set $z[k]$ equal to some desired value $z^*[k]$, it is clear that this inverts the $u[k]$ to $z[k]$ transfer function and hence gives $\overline{N}'(\gamma)$ as the new denominator polynomial.

This suggests the following feedback law:

$$u[k] = (D_2)^{-1}\left\{\xi^*[k] - C_2x[k]\right\} \qquad (8.7.42)$$

where $\xi^*[k]$ is an external input.

Since, we know from the separation theorem (Theorem 8.6.1) that state estimate feedback gives the same closed-loop poles as does state variable feedback (plus the poles of the observer), then the desired value of $K_\delta$ in (8.7.38) is clearly

$$K_\delta = (D_2)^{-1}C_2 \qquad (8.7.43)$$

Finally, we substitute (8.7.38) (with $\hat{x}_1$, $\hat{d}$ replaced by their asymptotic values $x_1$, $d$) into (8.7.29), (8.7.30), and obtain the following closed loop equation:

$$\delta x_1 = A_1x_1 + B_1\left\{-K_\delta x_1 + \overline{Q}_1(\delta)r - \overline{Q}_2(\delta)d\right\} + Gd \qquad (8.7.44)$$

$$y = C_1x_1 + Hd \qquad (8.7.45)$$

Noting that (8.7.40) gives the closed-loop denominator of (8.7.44), we see that the closed-loop output response in (8.7.45) can be written in transfer function form as

$$y[k] = \frac{1}{\overline{N}'(\gamma)}\left\{\overline{N}(\gamma)\overline{Q}_1(\gamma)r[k] - \overline{N}(\gamma)\overline{Q}_2(\gamma)d[k] + \overline{N}_2(\gamma)d[k]\right\} + Hd[k] \qquad (8.7.46)$$

where

$$\bar{N}(\gamma) = C_1\left[\mathrm{Adj}(\gamma I - A_1 + B_1 K_\delta)\right]B_1 \tag{8.7.47}$$

$$\bar{N}_2(\gamma) = C_1\left[\mathrm{Adj}(\gamma I - A_1 + B_1 K_\delta)\right]G \tag{8.7.48}$$

Thus, we now see that in order to reduce the impact of the disturbance, it is sensible to choose $\bar{Q}_2(\gamma)$ so as to make the following closed loop disturbance to output transfer function small:

$$\bar{S}'(\gamma) = \left[\frac{\bar{N}_2(\gamma)}{\bar{N}(\gamma)} - \frac{\bar{N}(\gamma)}{\bar{N}(\gamma)}Q_2(\gamma)\right] + H \tag{8.7.49}$$

Three alternative scenarios are:

1) The disturbance is an output disturbance. Then $\bar{N}_2 = 0$ and we can choose $\bar{Q}_2 = H$, leading to

$$\bar{S}'(\gamma) = H\left[1 - \frac{\bar{N}(\gamma)}{\bar{N}(\gamma)}\right] \tag{8.7.50}$$

2) The disturbance is an input disturbance. Then $H = 0$, $\bar{N}_2 = \bar{N}$ and we can choose $\bar{Q}_2 = 1$, leading to

$$S'(\gamma) = 0 \tag{8.7.51}$$

3) If the disturbance occurs mid-way between input and output, then we have the situation shown in Figure 8.7.1.

We see that $\bar{N} = \bar{N}_\alpha \bar{N}_\beta$ and $\bar{N}_2 = \bar{N}_\gamma \bar{N}_\beta$. Then, a suitable choice for $\bar{Q}_2$ is $\bar{Q}_2 = \bar{N}_\gamma/\left(\bar{N}_\alpha^s \, \bar{N}_\alpha^u(0) \, \bar{N}_\alpha^e\right)$ where $\bar{N}_\alpha^s$, $\bar{N}_\alpha^u$ are the stable and unstable parts of $\bar{N}_\alpha$, $\bar{N}_\alpha^u(0)$ denotes the d.c. gain of $\bar{N}_\alpha^u$ and $\bar{N}_\alpha^e$ is a stable polynomial (with unity d.c. gain) added to make $\bar{Q}_2$ proper.

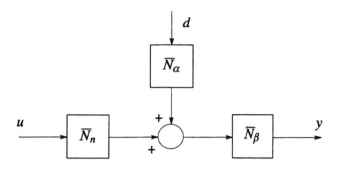

*Figure 8.7.1 : Input/Disturbance Transfer Functions*

## 8.8    Linear Quadratic Optimal Regulator

Design considerations, such as those discussed above, give valuable insights into digital control system design. However, in some more difficult problems it is helpful to be able to have a procedure that leads to an answer in one
simple step. One such procedure is based on linear quadratic regulator theory.

The basic idea in linear quadratic regulator theory is to express the various
design objectives via a quadratic cost function that includes weightings on the
states and control effort over some optimization horizon. The advantage of using
a quadratic criterion is that one obtains a one-shot analytic solution for the control law which minimizes the criterion. This is a highly desirable property. On
the other hand, a disadvantage of the method is that there is substantial *art* in the
choice of the various weightings in the cost function and thus one usually needs
to experiment with these weightings.

We will briefly outline the key elements of this theory. We will assume
throughout that the system states are directly measured. If this is not the case,
then one can combine the state-feedback developed here with an appropriate observer as in Section 8.6.

Consider a discrete-time system described in delta operator form as

$$\delta x = A_\delta x + B_\delta u \; ; \; x[0] = x_0 \qquad\qquad (8.8.1)$$

We want to find the input that minimizes a quadratic criterion of the form:

$$J = x[N]^T \Sigma_N x[N] + \sum_{k=0}^{N-1} \Delta \{ x[k]^T Q_d x[k] + u[k]^T R_d u[k] \} \qquad (8.8.2)$$

In the above expression, note that $\Sigma_N$ weights the deviation of the final state, $x[N]$, from the origin. Similarly, $Q_d$ and $R_d$ weight the states and control effort over the optimization horizon $k \in [0, 1, \; ..., \; N-1]$.

Since (8.8.2) is quadratic and (8.8.1) linear, then we might expect that we could obtain a simple analytic answer for the control sequence $\{u[k]\}$ which minimizes (8.8.2). This is indeed the case as we outline below. To develop the solution, we will use the idea of dynamic programming. To carry out this procedure, it is helpful to consider the optimal cost to go from some arbitrary time $j \in [0, k]$. Thus, let $V[j, x[j]]$ denote the *optimal* cost to go from time $j$ onwards assuming that the system is in state $x[j]$.

$$V[j, x[j]] = \min_{\{u(k)\}} \left\{ x[N]^T \Sigma_N x[N] + \sum_{k=j}^{N-1} \Delta \{ x[k]^T Q_d x[k] + u[k]^T R_d u[k] \} \right\} \quad (8.8.3)$$

where $V[N, x[N]] = x[N]^T \Sigma_N x[N]$.

An important observation is that in any sequential optimization problem, the optimal cost to go from time $j$ is the minimum of the sum of the optimal cost to go from time $j+1$ and the incremental cost at time $j$. (This is Bellman's *dynamic programming principle*). For the cost function given in (8.8.3) the incremental cost is $\Delta \{ x[j]^T Q_d x[j] + u[j]^T R_d u[j] \}$ and hence applying the dynamic programming principle we obtain:

$$V[j, x[j]] = \min_{u[j]} \{ V[j+1, x[j+1]] + \Delta x[j]^T Q_d x[j] + \Delta u[j]^T R_d u[j] \} \quad (8.8.4)$$

We then have the following result:

## Theorem 8.8.1 *The LQ Optimal Regulator*

The optimal control law is given by the following state feedback law:

$$\boxed{u[k] = -L_\delta[k+1]x[k]} \quad (8.8.5)$$

where the optimal gain $L_\delta[k+1]$ is given by:

$$\boxed{L_\delta[k] = \left( R_d + \Delta B_\delta^T \Sigma_\delta[k] B_\delta \right)^{-1} B_\delta^T \Sigma_\delta[k](I + A_\delta \Delta)} \quad (8.8.6)$$

and $\Sigma_\delta[k]$ satisfies the following reverse time Riccati equation

$$\boxed{\begin{aligned} -\delta \Sigma_\delta[k-1] &= Q_d + A_\delta^T \Sigma_\delta[k] + \Sigma_\delta[k]A_\delta + \Delta A_\delta^T \Sigma_\delta[k]A_\delta \\ &\quad - L_\delta[k]^T \left( R_d + \Delta B_\delta^T \Sigma_\delta[k]B_\delta \right) L_\delta[k] \end{aligned}} \quad (8.8.7)$$

with $\Sigma_\delta[N] = \Sigma_N$.

Proof :   We proceed by induction. We note that $V[N, x[N]]$ is a quadratic function of $x[N]$. We assume that this holds at time $j+1$ : we assume there exists a matrix $\Sigma_\delta[j+1]$ such that

$$V[j+1, x[j+1]] = x[j+1]^T \Sigma_\delta[j+1]x[j+1] \quad (8.8.8)$$

Substituting into (8.8.4) gives

$$\begin{aligned} V[j, x[j]] = \min_{u[j]} \Big\{ &x[j+1]^T \Sigma_\delta[j+1]x[j+1] \\ &+ \Delta x[j]^T Q_d x[j] + \Delta u[j]^T R_d u[j] \Big\} \end{aligned} \quad (8.8.9)$$

$$= \min_{u[j]} \left\{ [(I + \Delta A_\delta)x[j] + \Delta B_\delta u[j]]^T \Sigma_\delta[j+1][(I + \Delta A_\delta)x[j] + \Delta B_\delta u[j]] \right.$$

$$\left. + \Delta x[j]^T Q_d x[j] + \Delta u[j]^T R_d u[j] \right\} \qquad (8.8.10)$$

Differentiating with respect to $u[j]$ and setting the result to zero gives (8.8.5), (8.8.6).

Substituting into (8.8.4) gives (8.8.7).

∧∧∧

The linear quadratic regulator has some remarkable properties. For example, subject to certain mild restrictions (stabilizability of $(A_\delta, B_\delta)$ and detectability of $(C_\delta, A_\delta)$ where $C_\delta^T C_\delta = Q_d$) then if the optimization horizon $N$ is taken to infinity $L_\delta[k]$ converges to a constant vector which gives closed loop stability and $\Sigma_\delta[k]$ converges to a constant matrix $\Sigma_\delta$ which satisfies the following algebraic Riccati equation (obtained from (8.8.7) by setting $\Sigma_\delta[k-1] = \Sigma_\delta[k] = \Sigma_\delta$):

$$\boxed{Q_d + A_\delta^T \Sigma_\delta + \Sigma_\delta A_\delta + \Delta A_\delta^T \Sigma_\delta A_\delta - L_\delta^T \left( R_d + \Delta B_\delta^T \Sigma_\delta B_\delta \right) L_\delta = 0} \qquad (8.8.11)$$

where

$$\boxed{L_\delta = \left( R_d + \Delta B_\delta^T \Sigma_\delta B_\delta \right)^{-1} B_\delta^T \Sigma_\delta (I + A_\delta \Delta)} \qquad (8.8.12)$$

Additional properties of linear quadratic regulator designs are briefly explored in Problems 8.22 – 8.24. A full treatment of this approach lies beyond the scope of the present book. For further information the reader is referred to the texts listed at the end of the chapter.

## 8.9    Duality Relationships

The reader will have observed that the LQ optimal regulator discussed above and the Kalman filter discussed in Chapter 6 are closely related. For example, both depend upon Riccati equations for their solution. The two problems are dual as we show below. We will explore several aspects of this dual relationship. Firstly, we will show that the Kalman filtering problem can be directly converted to an optimal control problem. Secondly, we will show that the Riccati equations associated with the state estimation problem can be converted into the Riccati equation of the optimal control problem by simple association of variables.

## 8.9.1 Filtering as a Control Problem

We recall the form of the optimal filter given in (6.5.26) to (6.5.31). Since, the sequence $\{h[i, N]\}$ and the vector $g$ are optimal we know that they minimize the following error variance:

$$J = E\left\{(s[k] - \hat{s}[k, N])^2\right\} \tag{8.9.1}$$

We will show below that $\{h[i, N]\}$ can also be obtained as the solution of a linear optimal regulator problem. To carry out this program (and with the benefit of hindsight) we define a vector $\{\bar{x}[j]\}$ by the following *reverse time* equation:

$$-\delta \bar{x}[k-1] = A_\delta^T \bar{x}[k] - C_\delta^T h[k_1 - k, N]^T \tag{8.9.2}$$

$$\bar{x}[k_1 - 1] = f \tag{8.9.3}$$

Hence, using the summation by parts formula (3.4.16), we have from (6.5.26)

$$s[k_1] = \bar{x}[k_1 - 1]^T x[k_1]$$

$$= \bar{x}[k_0 - 1]^T x[k_0] + \sum_{k=k_0}^{k_1-1} \left\{\Delta(\delta \bar{x}[k-1])^T x[k] + \Delta \bar{x}[k]^T (\delta x[k])\right\}$$

Using (8.9.2) and (6.4.1) we have

$$s[k_1] = \bar{x}[k_0 - 1]^T x[k_0]$$

$$+ \sum_{k=k_0}^{k_1-1} \left\{ -\Delta \left( A_{\delta}^T \bar{x}[k] - C_{\delta}^T h[k_1 - k, N]^T \right)^T x[k] \right.$$

$$\left. + \Delta \bar{x}[k]^T \left( A_{\delta} x[k] + v_{\delta}[k] \right) \right\}$$

$$= \bar{x}[k_0 - 1]^T x[k_0]$$

$$+ \sum_{k=k_0}^{k_1-1} \left\{ \Delta h[k_1 - k, N] C_{\delta} x[k] + \Delta \bar{x}[k]^T v_{\delta}[k] \right\} \qquad (8.9.4)$$

Also, substituting (6.4.2) into (6.5.27) gives

$$\hat{s}[k_1, N] = \sum_{k=k_0}^{k_1-1} \Delta h[k_1 - k, N] \delta y[k] + g^T \hat{x}_0$$

$$= \sum_{k=k_0}^{k_1-1} \Delta h[k_1 - k, N] \left\{ C_{\delta} x[k] + \omega_{\delta}[k] \right\} + g^T \hat{x}_0 \qquad (8.9.5)$$

Substituting (8.9.5), (8.9.4) into (8.9.1) gives

$$J = E \left\{ \left( \bar{x}[k_0 - 1]^T x[k_0] - g^T \hat{x}_0 \right. \right.$$

$$\left. \left. + \sum_{k=k_0}^{k_1-1} \Delta \left( \bar{x}[k]^T v_{\delta}[k] + h[k_1 - k, N] \omega_{\delta}[k] \right) \right)^2 \right\} \qquad (8.9.6)$$

Taking the expected value in (8.9.6) and using (6.4.6) with $S_{\delta} = 0$ gives

$$J = \left\{ \left( [\bar{x}[k_0 - 1] - g]^T \hat{x}_0 \right)^2 + \bar{x}[k_0 - 1]^T P_0 \bar{x}[k_0 - 1] \right.$$

$$\left. + \sum_{k=k_0}^{k_1-1} \Delta \left\{ \bar{x}[k]^T \Omega_{\delta} \bar{x}[k] + h[k_1 - k, N] \Gamma_{\delta} h[k_1 - k, N]^T \right\} \right. \qquad (8.9.7)$$

The quadratic term is minimized by choosing $g = \bar{x}[k_0 - 1]$. Substituting into (8.9.7) we obtain

$$J = \bar{x}[k_0 - 1]^T P_0 \bar{x}[k_0 - 1]$$
$$+ \sum_{k=k_0}^{k_1-1} \Delta \left\{ \bar{x}[k]^T \Omega_\delta \bar{x}[k] + h[k_1 - k, N] \Gamma_\delta h[k_1 - k, N]^T \right\} \qquad (8.9.8)$$

Equation (8.9.8) together with (8.9.2) and (8.9.3) can be seen to be a *reverse-time* linear optimal control problem exactly as in Section 8.8. Note that $\{h[k_1 - k, N]\}$ is the "control signal" for the dual optimal control problem.

As in Section 8.8, the optimal control $h[\cdot, \cdot]$ can be expressed as a linear function of $\bar{x}$:

$$h[k_1 - k, N]^T = H_\delta[k]^T \bar{x}[k] \qquad (8.9.9)$$

We thus see that the optimal linear filter is closely associated with an optimal linear regulator problem.

### 8.9.2  Associations

From the above development we see that the solution of the optimal filtering problem can be converted to an optimal control problem and vice-versa. Actually a solution to either problem can be readily converted into a solution to the other problem by simply using the following table of associated variables.

## 8.10  Further Reading and Discussion

The topic of digital control is an extensive one and hence it has clearly only been possible in this chapter to only give a brief overview of some of the concepts. We have principally focused on those results that underpin further developments in later chapters.

We have limited ourselves to the single-input single-output case. However, many of the results carry over to the multivariable case if scalars are replaced by

| Kalman Filter | LQ Optimal Regulator |
|:---:|:---:|
| $k$ | $N - k$ |
| $A_\delta$ | $A_\delta^T$ |
| $C_\delta$ | $B_\delta^T$ |
| $P[k]$ | $\Sigma[k+1]$ |
| $\Omega = B_\delta B_\delta^T$ | $Q = C_\delta^T C_\delta$ |
| $\Gamma$ | $R$ |
| $H_\delta[k]$ | $L_\delta^T[k+1]$ |
| $B_\delta$ | $C_\delta^T$ |
| $P[0]$ | $\Sigma_f$ |

*Table 8.9.1 :  Associated variables in optimal filter*
*and LQ optimal regulator.*

matrices of appropriate dimension. In the multivariable case one needs to be careful about the order in which operations are performed.

In Section 8.1 we presented three simple methods for approximating continuous-time controllers by discrete-time controllers. Other methods appear in Keller and Anderson (1992). Other related work appears in Kennedy and Evans (1990). Example 8.1.1 has been drawn from Katz (1981).

We have briefly discussed quadratic optimization. There are many excellent books that deal exclusively with the topic of linear quadratic regulator design. Some possible sources are Anderson and Moore (1971), Åström (1970), Kwakernaak and Sivan (1972) and Stengel (1986).

The idea of 'loop transfer recovery' discussed in Problem 8.20 can also be given a linear quadratic regulator interpretation and is further developed in Doyle and Stein (1979).

We have not discussed other optimization methods (such as $H^\infty, L^1$). These methods are discussed in recent control literature – see for example Doyle *et. al* (1989).

An interesting issue in sampled-data control of continuous-time systems is that controllability of the continuous-time system does not necessarily imply controllability of the associated discrete-time system. It turns out that there exists certain pathological sampling intervals for which controllability can be lost - see the discussion in Kalman, Ho and Narendra (1963), Levis, Schlueter and Athens (1972) and Sivashankar and Khargonekar (1993).

The reader who wishes to take this brief introduction to digital control further should consult other books such as: Åström and Wittenmark (1984), Kuo (1992), Chen (1970), Franklin, Powell and Workman (1990) and Forsythe and Goodall (1991).

## 8.11 Problems

8.1 (a) Construct a complex plane diagram showing $T$ (the complementary sensitivity function) and $-T$ as vectors originating from the origin.

(b) Show in your diagram the vector from the point $-1$ to $-T$. Prove that this is the vector $S$.

(c) Hence show that when $T = 10e^{-j\frac{\pi}{2}}$ then $S = 1 - 10j$. Hence show that $S + T = 1$ but that $|S| + |T| = 10 + \sqrt{101}$.

(d) Show graphically that when $T = -1$, $S = 2$.

8.2 (a) Consider the following two continuous-time complementary sensitivity functions:

$$T_1(s) = \frac{1-s}{1+s}, \quad T_2(s) = \frac{1-s}{1+2s}$$

Draw $|T_1(j\omega)|$, $|S_1(j\omega)|$, $|T_2(j\omega)|$ and $|S_2(j\omega)|$. What can you conclude by comparing the two cases?

(b)   Say each system has a unit step input and no disturbances. Find the exact response in each case and compare with your conclusion in (a).

8.3   Consider the inverted pendulum system given in Figure 8.11.1.

(a)   Write the transfer function (for small $\theta$ ) from $u$ to $x$.

(b)   Write the transfer function (for small $\theta$ ) from $u$ to $\theta$ .

(c)   Compare (a) and (b) as far as the ability to control the system in each case.

8.4   Given

$$\bar{A}(\gamma) = \gamma(\gamma^3 + 2\gamma^2 - \gamma - 1) , \ \bar{B}(\gamma) = \gamma - 2$$

Find $\bar{L}(\gamma)$ and $\bar{P}(\gamma)$ so that $\bar{A}^*(\gamma) = \bar{A}(\gamma)\bar{L}(\gamma) + \bar{B}(\gamma)\bar{P}(\gamma)$ is $\bar{A}^* = (\gamma + 1)^4$. When is this a deadbeat control? When does $\bar{A}^*(\gamma)$ represent a stable system?

8.5   Draw the sensitivity functions for both controllers in Example 8.3.1. Compare the two cases. Can the poor response of the first controller be predicted from the corresponding sensitivity function?

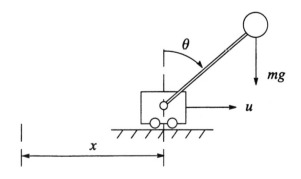

*Figure 8.11.1 : Inverted pendulum system of Problem 8.3.*

8.6  Let $\overline{G}(\gamma) = \dfrac{\gamma + 0.5}{\gamma(\gamma + 1)}$ with $\Delta = 0.1$ being the associated sampling inter-

val. Say $\hat{G}(\gamma) = \dfrac{\gamma + 0.5\alpha}{\gamma(\gamma + \alpha)}$, $0 < \alpha < 1$, is an estimate of $\overline{G}(\gamma)$ and choose

$\overline{Q}(\gamma)$ in (8.5.1) to be $\overline{Q}(\gamma) = \hat{G}(\gamma)^{-1}(\beta\gamma + 1)^{-1}$.

(a)  For at least three values of $\alpha$ find corresponding values of $\beta$ so that

the transfer function of (8.5.7) $\dfrac{1 - \hat{G}\overline{Q}}{1 + (\overline{G} - \hat{G})\overline{Q}}$ is stable.

(b)  Draw the frequency response for one of your choices of $\alpha$ and $\beta$.

8.7  Consider the special case of equation (8.5.7) where $G_\Delta = \dfrac{1}{s + 1}$ and

$Q = \dfrac{2}{s + 1}$. Show that $\dfrac{1}{1 + G_\Delta Q}$ is stable even though $|G_\Delta Q| = 2$ at d.c.

Comment on why this implies that requiring $|G_\Delta Q| < 1$ at all frequencies
is only a sufficient condition for robust stability.

8.8  (a)  Show that the following delta model generates a sinewave of frequency $\omega_0$

$$\delta x = A_\delta x; \quad y = [1\ 0]x$$

where

$$A_\delta = \begin{bmatrix} \dfrac{\cos \omega_0 \Delta - 1}{\Delta} & \dfrac{1}{\omega_0 \Delta} \sin \omega_0 \Delta \\[3mm] \dfrac{-\omega_0}{\Delta} \sin \omega_0 \Delta & \dfrac{\cos \omega_0 \Delta - 1}{\Delta} \end{bmatrix}$$

(b)  Show that a stable observer can be designed using

$$J_\delta = 2\xi\omega_0 \begin{bmatrix} \cos \omega_0 \Delta \\ \sin \omega_0 \Delta \end{bmatrix}$$

to estimate $x$ if $y$ is measured in the presence of noise or disturbances. Explain the role of $\xi$ on the performance of the observer.

(c) Compute the transfer function $\bar{F}(\gamma)$ from $y$ to $\hat{x}_1$.

(d) Show that $1 - \bar{F}(\gamma)$ for $\omega_0 = 20\pi$ and $\Delta = 0.001$ is a notch filter.

(e) Show that

$$\lim_{\Delta \to 0} \bar{F}(\gamma) = \frac{2\xi\omega_0\gamma}{\gamma^2 + 2\xi\omega_0\gamma + \omega_0^2}$$

8.9 Consider an open loop stable system. Choose $\bar{E}_1$, $\bar{E}_2$ in (8.5.12) to get the class of all stabilizing control laws for stable systems given in (8.5.1) from equation (8.5.9).

8.10 Show that the configurations in Figures 8.5.3 (a) and (b) are indeed equivalent.

8.11 Show that equation (8.7.3) is equivalent to Figure 8.5.3 (b).

8.12 Given $\bar{L}(\gamma)$ and $\bar{P}(\gamma)$ in (8.7.16), show that they satisfy (8.2.9) where

$$\bar{A}^*(\gamma) = \det(\gamma I - A_\delta + J_\delta C_\delta) \cdot \det(\gamma I - A_\delta + B_\delta K_\delta)$$

and where

$$\frac{\bar{B}(\gamma}{\bar{A}(\gamma)} = C(\gamma I - A_\delta)^{-1}B_\delta$$

Hint: Use the matrix inversion Lemma

$$\left(T \pm uv^T\right)^{-1} = T^{-1} \mp T^{-1}u\left(1 \pm v^T T^{-1}u\right)^{-1}v^T T^{-1}$$

8.13 Verify equations (8.7.26) and (8.7.27) where $\bar{A}(\gamma)$ and $\bar{B}(\gamma)$ are as in Problem 8.12.

**8.14** Say we are given a controller parameterized as in equation (8.5.9) :

$$\bar{C}_1 = \frac{\bar{N}_1}{\bar{D}_1} = \frac{\frac{\bar{P}_1}{E_2} + \bar{Q}_1 \frac{\bar{A}}{E_1}}{\frac{\bar{L}_1}{E_2} - \bar{Q}_1 \frac{B}{E_1}}$$

Say we now choose a different pre-stabilizing loop $\bar{P}_2/\bar{L}_2$ and rede-sign the transfer function $\bar{Q}$ to a new value say $\bar{Q}_2$. This leads to a new controller

$$\bar{C}_2 = \frac{\bar{N}_2}{\bar{D}_2} = \frac{\frac{\bar{P}_2}{E_2} + \bar{Q}_2 \frac{\bar{A}}{E_1}}{\frac{\bar{L}_2}{E_2} - \bar{Q}_2 \frac{B}{E_1}}$$

Show that the same controller $\bar{C}_2$ can be achieved by retaining the pre-stabilizing loop at its original value $\bar{P}_1/\bar{L}_1$ but with a new value for $\bar{Q}_1$ presribed by

$$\bar{Q}_1^1 = \frac{E_2\bar{Q}_2(\bar{L}_1\bar{A} + \bar{P}_1 B) - (\bar{P}_1\bar{L}_2 - \bar{P}_2\bar{L}_1)E_1}{E_2(\bar{L}_2\bar{A} + \bar{P}_2 B)}$$

Hence show that the choice of pre-stabilizing loop is inmaterial.

**8.15** Consider the feedback control system shown in Figure 8.5.3 (b). We wish to change the mode of injecting the reference signal $r$. Thus, instead of the configuration shown in Figure 8.5.3 (b) we define $u$ as follows:

$$u = u_p + u_q$$
$$\bar{L}u_p = -\bar{P}(y - \bar{\Gamma}r)$$
$$v = \frac{B}{E}u + \frac{\bar{A}}{E}(r - y)$$
$$u_q = \bar{Q}v$$

(a) Redraw Figure 8.5.3 (b) showing this alternative way of injecting the signal $r$.

(b) Show that (with an ideal model) the closed loop response to the reference input is now given by

$$\overrightarrow{A}^* y = \overline{BP\Gamma}r + \overline{LA}\left[\frac{\overline{QB}}{\overline{E}}\right]r$$

where $\overrightarrow{A}^* = \overline{AL} + \overline{BP}$.

(c)   Show that if we choose $\overline{\Gamma} = \dfrac{\overline{QB}}{\overline{E}}$ then the closed loop response to the reference input becomes $y = \Gamma r$.

(d)   Comment on the result in part (c) in relation to both the reference and disturbance response of the closed loop system.

8.16   Verify equation (8.5.17).

8.17   Show that any state estimate feedback controller of the form of equation (8.7.38) can be expressed as in Figure 8.5.3 (a) (with possibly a different way of injecting the reference input). Explicitly evaluate $\overline{P}$, $\overline{L}$, $\overline{E}_1$, $\overline{E}_2$, $\overrightarrow{A}^*$ and $\overline{Q}$ as a function of $A_\delta$, $B_\delta$, $C_\delta$, $J_\delta$, $K_\delta$, $\overline{\Omega}_1$ and $\overline{\Omega}_2$.

8.18   Consider the following proper (but not strictly proper) transfer function

$$\overline{G}(\gamma) = \frac{\overline{B}(\gamma)}{\overline{A}(\gamma)}$$

Let this transfer function have the following state-space realization

$$\delta x = A_\delta x + B_\delta u$$
$$y = C_\delta x + D_\delta u$$

(a)   Show that $D_\delta \neq 0$.

(b)   Prove that a state space realization for $[\overline{G}(\gamma)]^{-1}$ is

$$\delta x' = \left(A_\delta - B_\delta D_\delta^{-1}C_\delta\right)x' + B_\delta D_\delta^{-1}y$$
$$u = D_\delta^{-1}(y - C_\delta x)$$

(c)   Hence show that the following state feedback law assigns the closed loop poles to the open loop zeros (for a proper but not strictly proper) system

$$u = -K_\delta x \quad \text{where} \quad K_\delta = B_\delta D_\delta^{-1} C_\delta$$

8.19 Consider a single input single output system having state-space realization:

$$\delta x = A_\delta x + B_\delta u$$

$$y = C_\delta x$$

(a)  Show that the state observer (8.6.7) can be written in transfer function form as

$$\hat{X} = \overline{T}_1(\gamma)U + \overline{T}_2(\gamma)Y$$

where

$$\overline{T}_1(\gamma) = (\gamma I - A_\delta + J_\delta C_\delta)^{-1} B_\delta$$

$$\overline{T}_2(\gamma) = (\gamma I - A_\delta + J_\delta C_\delta)^{-1} J_\delta$$

(b)  Prove that $T_1(\gamma) + T_2(\gamma)C_\delta(\gamma I - A_\delta)^{-1}B_\delta = (\gamma I - A_\delta)^{-1}B_\delta$.

Hence show that the transfer function from $u$ to $\hat{x}$ is the same as that from $u$ to $x$. (This property is commonly known as unbiasedness).

8.20 Consider the same system as in Problem 8.19, save that the input applied to the plant is $u = u_m + \Delta u$ where $u_m$ is the measured plant input.

(a)  Show that

$$\hat{X} = \overline{T}_1(\gamma)U_m + \overline{T}_2(\gamma)C_\delta(\gamma I - A_\delta)^{-1}B_\delta(U_m + \Delta U)$$

$$X = (\gamma I - A_\delta)^{-1}B_\delta[U_m + \Delta U]$$

Hence if $\Delta x$ and $\Delta \hat{x}$ denote the error in $x$ and $\hat{x}$ respectively due to $\Delta u$, show that

$$\Delta \hat{X} = \overline{T}_2(\gamma)C_\delta(\gamma I - A_\delta)^{-1}B_\delta \Delta U$$

$$\Delta X = (\gamma I - A_\delta)^{-1}B_\delta \Delta U$$

Hence discuss the statement that "the sensitivity to input errors of a closed loop system using an observer is in general different when state estimate feedback is used in place of state feedback".

(b)    Consider the following (improper) choice for $\bar{T}_2(\gamma)$ :

$$\bar{T}_2(\gamma) = \frac{(\text{adj}(\gamma I - A_\delta))B_\delta}{\bar{B}(\gamma)}$$

where

$$C_\delta(\gamma I - A_\delta)^{-1}B_\delta = \frac{\bar{B}(\gamma)}{\bar{A}(\gamma)}$$

Show that to achieve

$$\bar{T}_1(\gamma) + \bar{T}_2(\gamma)C_\delta(\gamma I - A_\delta)^{-1}B_\delta = (\gamma I - A_\delta)^{-1}B_\delta$$

it is necessary to take $\bar{T}_1(\gamma) = 0$.

(c)    For the choice of $\bar{T}_1(\gamma)$, $\bar{T}_2(\gamma)$ as in part (b) show that $\Delta X = \Delta \hat{X}$.

Comments:

(i)    $\bar{T}_2(\gamma)$ given in part (b) can be made proper by adding some fast poles to give

$$\bar{T}_2(\gamma) = \frac{\text{adj}(\gamma I - A_\delta)B_\delta}{\bar{B}(\gamma)\bar{E}(\gamma)}$$

(ii)   Use of this observer recovers the sensitivity due to input errors achievable with state feedback. Hence the method is commonly known as 'loop transfer recovery' or LTR.

8.21   Show that the configuration in Figure 8.5.3 (b) reduces to that in Figure 8.5.2 when $\bar{G}$ is stable.

8.22   Extend Theorem 8.8.1 to the following coupled cost function:

$$J = 6x[N]^T \Sigma_n x[N] + \sum_{k=0}^{N-1} \Delta \left\{ \begin{bmatrix} x[k]^T u[k]^T \end{bmatrix} \begin{bmatrix} Q_d & S_d^T \\ S_d & R_d \end{bmatrix} \begin{bmatrix} x[k] \\ u[k] \end{bmatrix} \right\}$$

Hint : Define a new control as $\bar{u} = u + R_d^{-1}S_d x$ and optimize with respect to $\bar{u}$.

8.23 Consider a discrete time system given by $\delta x[k] = u[k]$ with $\Delta = 1$, and a cost function

$$J = 100x[k_1]^2 + \Delta \sum_{j=0}^{k_1-1}\left(x[j]^2 + 10u[j]^2\right)$$

Find the optimal regulator for $k_1 = 5,\ 10,\ 100$.

8.24 Frequency domain properties of the LQ optimal controller : Consider the LQ problem for the discrete time SISO system

$$\delta x[k] = A_\delta x[k] + B_\delta u[k]$$
$$y[k] = C_\delta x[k]$$

with the criterion

$$J = \Delta \sum_{k=0}^{\infty}\left(y^2[k] + R_d u[k]^2\right)$$

We have seen that the optimal control is of the form $u[k] = -L_\delta x[k]$ where $L_\delta = \left(R_d + \Delta B_\delta^T \Sigma_\delta B_\delta\right)^{-1} B_\delta^T \Sigma_\delta (I + \Delta A_\delta)$ and $\Sigma_\delta$ is the solution of the algebraic Riccati equation

$$0 = C_\delta^T C_\delta + A_\delta^T \Sigma_\delta + \Sigma_\delta A_\delta + \Delta A_\delta^T \Sigma_\delta A_\delta - L_\delta^T\left(R_d + \Delta B_\delta^T \Sigma_\delta B_\delta\right)L_\delta.$$

(a)  Let

$$C_\delta(\gamma I - A_\delta)^{-1}B_\delta = \frac{\bar{B}(\gamma)}{\bar{A}(\gamma)}$$

and the closed loop denominator be prescribed by

$$\det(\gamma I - A_\delta + B_\delta L_\delta) = \bar{A}_c(\gamma)$$

Show that

$$\frac{\bar{A}_c(\gamma)}{\bar{A}(\gamma)} = 1 + L_\delta(\gamma I - A_\delta)^{-1} B_\delta$$

(b)    Prove that

$$\left(R_d + \Delta B_\delta^T \Sigma_\delta B_\delta\right) \frac{\bar{A}_c(\gamma)}{\bar{A}(\gamma)} \cdot \frac{\bar{A}_d\left(-\frac{\gamma}{1+\Delta\gamma}\right)}{\bar{A}\left(-\frac{\gamma}{1+\Delta\gamma}\right)} = \frac{\bar{B}(\gamma)}{\bar{A}(\gamma)} \cdot \frac{\bar{B}\left(-\frac{\gamma}{1+\Delta\gamma}\right)}{\bar{A}\left(-\frac{\gamma}{1+\Delta\gamma}\right)} + R_d$$

(c)    Show that $\gamma$ and $\dfrac{-\gamma}{1+\Delta\gamma}$ represent reflections with respect to the

stability boundary $1 + \Delta\gamma = 0$. Hence demonstrate that the result in
(b) represents a way of finding the optimal closed loop denominator
by spectral factorization by writing

$$\alpha(\gamma) = \alpha_1(\gamma)\alpha_1\left[\frac{-\gamma}{1+\Delta\gamma}\right]$$

(d)    In light of (c) find the closed loop poles when $R_d \to 0$ and when
$R_d \to \infty$.

(e)    Use (b) to show that

$$\frac{\bar{A}_c\left(\left(e^{j\omega\Delta} - 1\right)/\Delta\right)}{\bar{A}\left(\left(e^{j\omega\Delta} - 1\right)/\Delta\right)} \geq \frac{R_d}{R_d + \Delta B_\delta^T \Sigma B_\delta} \overset{\Delta}{=} \varrho$$

Note that for $\gamma = \left(e^{j\omega\Delta} - 1\right)/\Delta$ then $\dfrac{-\gamma}{1+\Delta\gamma}$ is the complex conju-

gate, $\gamma^*$, of $\gamma$.

(f)    Express the open loop transfer function from $u$ to $L_\delta x$, namely,
$L_\delta(\gamma I - A_\delta)^{-1} B_\delta$, in terms of $\bar{A}(\gamma)$ and $\bar{A}_c(\gamma)$, then use (e) to show
that the corresponding Nyquist curve will never enter the circle in
Figure 8.11.2.

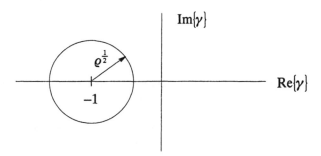

*Figure 8.11.2 : Circle criteria for Problem 8.24(f).*

(g)  Use (f) and the the Nyquist criterion to show that the system will remain stable if the plant gain increases by, up to, the factor

$$GM = \frac{1}{1 - \varrho^{\frac{1}{2}}}$$

or if the plant has additional phase lag of up to

$$PM = \cos^{-1}\left(1 - \frac{\varrho}{2}\right)$$

(In classical control, GM and PM are commonly referred to as the
gain margin and phase margin respectively.)

(h)  Discuss what happens to the above bounds on the *GM* and *PM* as
$\Delta \rightarrow 0$.

8.25  Incorporating integral action in the LQ design is a special case of using the
internal model principle with LQ design.

Consider the system given by

$$\delta x[k] = A_\delta x[k] + B_\delta u[k]$$

$$y[k] = C_\delta x[k]$$

and let the control be defined as

$$u[k] = -L_\delta \begin{bmatrix} x[k] \\ x^c[k] \end{bmatrix}$$

with

$$\delta x^c[k] = r[k] - y[k]$$

(a)   Verify that the steady state error $e[k] = r[k] - y[k]$ to a step input is zero.

(b)   To be able fo find the modified LQ optimal gain $L_\delta$ the composite state $\begin{bmatrix} x \\ x^c \end{bmatrix}$ must be reachable from $u$.

Show that sufficient conditions for the above are that $(A_\delta, B_\delta)$ is a reachable pair and that $C_\delta(\gamma I - A_\delta)^{-1}B_\delta$ does not have a zero at the origin.

8.26   Convert the optimal filter given by (6.5.26), (6.5.27) with $g = \bar{x}[k_0 - 1]$ and with $h[\cdot, \cdot]$ as in (8.9.9) back to recursive form by carrying out the following steps:

(a)   Let $\psi[k, k_0]$ be the state transition matrix defined by

$$\delta\psi[k, k_0] = (A_\delta - H_\delta[k]C_\delta)\psi[k, k_0]$$
$$\psi[k_0, k_0] = I$$

Show that $\bar{x}[k] = \psi[k_1, k + 1]^T f$.

(b)   Show that $\hat{s}[k, N] = f^T \left[ \sum_{i=k_0}^{k_1-1} \Delta\psi[k_1, i]H_\delta[i]y^0[i] + \psi[k_1, k_0]\hat{x}_0 \right]$.

(c)   Choose $\hat{x}[k_0] = \hat{x}_0$ and

$$\hat{x}[k] = \sum_{i=k_0}^{k-1} \Delta\psi[k, i+1]H_\delta[i]y^0[i] + \psi[k, k_0]\hat{x}_0 \quad k > k_0$$

and show that $\hat{x}[k]$ satisfies

$$\delta\hat{x}[k] = (A_\delta - H_\delta[k]C_\delta)\hat{x}[k] + H_\delta[k]y^0[k]$$

# Chapter 9

# Sampled Data Control

## 9.1  Introduction

The introduction to discrete-time control given in Chapter 8 followed the traditional pattern of focusing entirely on the sampled output response. Clearly this would be quite adequate if one could be certain that the response between samples would not deviate too dramatically from the response as seen at the sample points. However, there is no a-priori reason to presume that this will be true. Indeed, we shall see later in this chapter that it is quite possible for the intersample-response to be markedly different from the sampled response.

With this as background, the aim of this chapter is to develop tools for analyzing the *continuous-time* response of a system under the action of a digital control law. In doing this we shall draw on the results previously developed in Chapters 1 and 4. Using these results we shall find that it is actually relatively straightforward to analyze both sampled and continuous signals in a digital control loop within a common framework. This will allow us, in a natural way, to simultaneously calculate both the continuous and sampled output response.

The development in this chapter will be aided by some simplifying notation. As before, we use $\gamma_\omega$ to denote the discrete frequency response variable:

$$\gamma_\omega = \frac{e^{j\omega\Delta} - 1}{\Delta} \tag{9.1.1}$$

We also introduce a simplified notation for a shifted frequency variable; that is where $\omega_k = \omega - k\omega_s$ and $\omega_s$ is the sampling frequency.

We will then use the following special notation for the discrete (i.e. folded) version of the continuous transfer function $X(j\omega)$

$$[X]^s \overset{\Delta}{=} \sum_{k=-\infty}^{\infty} X(j(\omega - k\omega_s)) = \sum_{k=-\infty}^{\infty} X(j\omega_k) \qquad (9.1.2)$$

We also make extensive use of the results in Chapter 4, especially Section 4.3; the reader may find it helpful to review the earlier material.

## 9.2    Mixing Continuous and Discrete Transfer Functions

The arrangement we consider here is, again, as in Figure 8.1.1. However, whereas in Chapter 8 we focused entirely on the sampled response, here we wish to also analyze the continuous-time signal $y(t)$. We thus consider an equivalent continuous-time representation for the plant and controller as in Figure 4.3.2 of Chapter 4. We recast Figure 8.1.1 in the form shown in Figure 9.2.1 so as to capture all signals of interest.

In Figure 9.2.1 we have the following transfer functions:

$\overline{C}(\gamma)$    is the discrete controller transfer function,

$H_0(s)$    is the continuous-time transfer function of a zero order hold,

$G_p(s)$    is the continuous-time transfer function of a plant,

$F(s)$    is the continuous-time transfer function of an anti-aliasing filter.

We also use the following notation for signals and their Fourier transforms:

$u(t)$, $U(\omega)$    is the continuous-time plant input;

$z(t)$, $Z(\omega)$    is the disturbance-free plant output;

$v(t)$, $V(\omega)$    is the output disturbance;

$y(t)$, $Y(\omega)$    is the continuous-time plant output including disturbance;

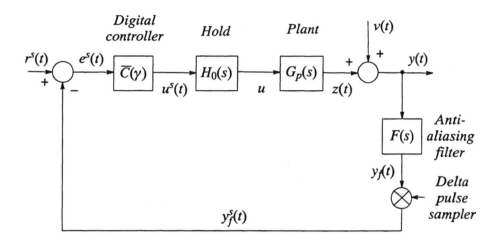

*Figure 9.2.1 : Continuous-time representation of sampled data control loop.*

$y_f(t)$, $Y_f(\omega)$   is the filtered plant output;

$y_f^s(t)$, $Y_f^s(\omega)$   is the impulse-sampled version of $y_f(t)$ ;

$r^s(t)$, $R^s(\omega)$   is the impulse-sampled reference input;

$u^s(t)$, $U^s(\omega)$   is the impulse stream at the output of the digital controller;

$e^s(t)$, $E^s(\omega)$   is the impulse-sampled tracking error.

Based on the feedback structure presented in Figure 9.2.1, we have the following key result which describes the continuous output response under the action of the digital control law:

**Theorem 9.2.1** Subject to closed-loop stability, the Fourier transform of the continuous-time output is given by:

$$Y(\omega) = P(\omega)R^s(\omega) + D(\omega)V(\omega)$$
$$- P(\omega) \sum_{\substack{k=-\infty \\ k \neq 0}}^{\infty} F(j\omega_k)V(\omega_k)$$
$$= P(\omega)R^s(\omega) + V(\omega) - P(\omega)[FV]^s \tag{9.2.1}$$

where $P(\omega)$ and $D(\omega)$ are frequency response functions given respectively by

$$P(\omega) \triangleq \overline{C}(\gamma_\omega)G_p(j\omega)H_0(j\omega)\overline{S}(\gamma_\omega) = \frac{\overline{C}(\gamma_\omega)G_p(j\omega)H_0(j\omega)}{1 + \overline{C}(\gamma_\omega)[FG_pH_0]^s} \tag{9.2.2}$$

$$D(\omega) = 1 - P(\omega)F(\omega) \tag{9.2.3}$$

where $\overline{S}(\gamma_\omega)$ is the usual discrete sensitivity function defined previously in (8.4.2):

$$\overline{S}(\gamma_\omega) \triangleq \frac{1}{1 + \overline{C}(\gamma_\omega)[FG_pH_0]^s} \tag{9.2.4}$$

and where, as in Section 9.1, the notation $[X]^s$ denotes the discrete (i.e. folded) version of the continuous frequency transfer function $X(j\omega)$ :

$$[X]^s = \sum_{k=-\infty}^{\infty} X(j(\omega - k\omega_s)) = \sum_{k=-\infty}^{\infty} X(j\omega_k) \tag{9.2.5}$$

**Proof :** Since $y_f^s(t)$ is an impulse sampled version of $y_f(t)$ we have, using the folding property (equation (1.6.12)):

$$Y_f^s(\omega) = \sum_{k=-\infty}^{\infty} Y_f(\omega - k\omega_s) = \sum_{k=-\infty}^{\infty} Y_f(\omega_k)$$

$$= \sum_{k=-\infty}^{\infty} F(j\omega_k)(z(\omega_k) + V(\omega_k))$$

$$= \sum_{k=-\infty}^{\infty} F(j\omega_k) \left[ G_p(j\omega_k)H_0(j\omega_k) \right.$$
$$\left. \bar{C}(\gamma_\omega)\left(-Y_f^s(\omega_k) + R^s(\omega_k)\right) + V(\omega_k) \right] \quad (9.2.6)$$

Now $\bar{C}_c$, $Y_f^s$ and $R^s$ are periodic functions of $\omega$ and hence

$$Y_f^s(\omega) = \bar{C}(\gamma_\omega) \sum_{k=-\infty}^{\infty} F(j\omega_k)G_p(j\omega_k)H_0(j\omega_k) \cdot \left(-Y_f^s(\omega) + R^s(\omega)\right) + V_f^s(\omega)$$

$$(9.2.7)$$

where $V_f^s(\omega)$ is the transform of the sampled filtered disturbance:

$$V_f^s(\omega) \overset{\Delta}{=} \sum_{k=-\infty}^{\infty} F(j\omega_k)V(\omega_k) \qquad (9.2.8)$$

Then using (9.2.5) in (9.2.7) we obtain

$$Y_f^s(\omega) = \bar{C}(\gamma_\omega)[FG_pH_0]^s\left(-Y_f^s(\omega) + R^s(\omega)\right) + V_f^s(\omega) \qquad (9.2.9)$$

Hence

$$Y_f^s(\omega) = \left(1 - \bar{S}(\gamma_\omega)\right)R^s(\omega) + \bar{S}(\gamma_\omega)V_f^s(\omega) \qquad (9.2.10)$$

From Figure 9.2.1

$$E^s(\omega) = S(\gamma_\omega)\left(R^s(\omega) - V_f^s(\omega)\right) \qquad (9.2.11)$$

Finally, the result follows since

$$Y(\omega) = \overline{C}(\gamma_\omega)G_p(j\omega)H_0(j\omega)E^s(\omega) + V(\omega) \qquad (9.2.12)$$

$$\wedge\wedge\wedge$$

Note that the functions $P(\omega)$ and $D(\omega)$ allow us to compute the continuous output frequency response $Y(\omega)$ using the (periodic) reference input spectrum $R^s(\omega)$ and the disturbance spectrum $V(\omega)$ respectively. We will thus refer to $P(\omega)$ and $D(\omega)$ as the *reference* and *disturbance gain functions* respectively.

In Theorem 9.2.1 we have assumed stability of the closed loop system. It can be shown that under mild restrictions this is equivalent to stability of the sampled data system. We also observe that the infinite sum in (9.2.5) is convergent provided that the composite transfer function $F(j\omega)G_p(j\omega)H_0(j\omega)$ is strictly proper.

The result in Theorem 9.2.1 holds for general reference and disturbance inputs. However it is insightful to consider the special case of sinusoidal signals. In this case, the result simplifies as follows.

Let $R^s(\omega)$ take the particular form of a delta pulse sampled cosine wave:

$$r^s(t) = \sum_{k=-\infty}^{\infty} \Delta \cos(\omega_0 t)\delta(t - k\Delta) \qquad (9.2.13)$$

In this case $R^s(\omega)$ is a periodically repeated pair of delta functions, i.e.

$$R^s(\omega) = \pi \sum_{k=-\infty}^{\infty} \left[\delta(\omega - \omega_0 - k\omega_s) + \delta(\omega + \omega_0 - k\omega_s)\right] \qquad (9.2.14)$$

Hence using (9.2.1), $Y(\omega)$ is

$$Y(\omega) = \pi \sum_{k=-\infty}^{\infty} P(\omega)\left[\delta(\omega - \omega_0 - k\omega_s) + \delta(\omega + \omega_0 - k\omega_s)\right] \qquad (9.2.15)$$

Thus the output, in this case, is *multi-frequency* with corresponding magnitudes and phases determined by the reference gain function $P(\omega)$. In particular, for a sinusoidal reference signal of amplitude $A$ at frequency $\omega_0$ where $\omega_0$ lies in the range $0 < \omega_0 < \omega_s/2$, the first two components in the output are at frequencies $\omega_0$ and $\omega_s - \omega_0$ and have amplitude $|P(\omega_0)|A$ and $|P(\omega_s - \omega_0)|A$ respectively. Similarly, for the special case of a sinusoidal disturbance of amplitude $A$ at frequency $\omega_0$ lying in the range $0 < \omega_0 < \omega_s/2$, the first two components in the output are at frequencies $\omega_0$ and $\omega_s - \omega_0$ and have amplitude $|D(\omega_0)|A$ and $|P(\omega_s - \omega_0)F(j(\omega_s - \omega_0))|A$ respectively. We thus see that the functions $P(\omega)$ and $D(\omega)$ play a central role in calculating the continuous-time response to sinusoidal reference and disturbance inputs in a similar fashion that $\overline{T}(\gamma_\omega)$ and $\overline{S}(\gamma_\omega)$ are used in discrete-time systems. The connection with $\overline{T}(\gamma_\omega)$ and $\overline{S}(\gamma_\omega)$ is even deeper as we see below.

## 9.3   Sensitivity Considerations

In Section 9.2 we found that the reference gain function $P(\omega)$ and the disturbance gain function $D(\omega)$ allow us to compute the *continuous-time* output response under closed-loop *digital* control. For convenience, we recall the following definitions:

$$P(\omega) = \overline{C}(\gamma_\omega)G_p(j\omega)H_0(j\omega)\overline{S}(\gamma_\omega) \tag{9.3.1}$$

$$D(\omega) = 1 - P(\omega)F(j\omega) \tag{9.3.2}$$

$$\overline{S}(\gamma_\omega) = \frac{1}{1 + \overline{C}(\gamma_\omega)[FG_pH_0]^s} \tag{9.3.3}$$

where $\overline{C}$, $F$, $G_p$ and $H_0$ are the transfer functions of the controller, anti-aliasing filter, plant and zero order hold respectively.

The reader will recall the discussion in Section 8.4 where we showed how $\bar{S}$ and $\bar{T}$ play a fundamental role in digital control system design at the sample points. The functions $P(\omega)$ and $D(\omega)$ play a similar role for sampled data systems with mixed discrete- and continuous-time signals.

For example, we immediately see from (9.2.2) that the following identity holds:

$$\boxed{D(\omega) + P(\omega)F(j\omega) = 1} \qquad (9.3.4)$$

This is analogous to the usual constraint that the sum of sensitivity and complementary sensitivity is unity. If the reader is concerned about the explicit presence of the anti-aliasing filter in (9.3.4), then this can be absorbed into $P(\omega)$ by simply looking at the gain between the reference input $R^s(\omega)$ and the filtered output $Y_f(\omega)$. In particular, we see that

$$Y_f(\omega) = P_f(\omega)R^s(\omega) + D(\omega)V_f(\omega) - P_f \sum_{\substack{k=-\infty \\ k \neq 0}}^{\infty} V_f(\omega_k) \qquad (9.3.5)$$

where $P_f(\omega)$ is the *filtered reference gain function* defined by

$$P_f(\omega) = P(\omega)F(j\omega) \qquad (9.3.6)$$

and $V_f(\omega)$ is the filtered noise $F(j\omega)V(\omega)$. We then have

$$\boxed{P_f(\omega) + D(\omega) = 1} \qquad (9.3.7)$$

The reason for this minor difference is the location of the anti-aliasing filter mid-way between the disturbance $v(t)$ and the reference input $r^s(t)$.

It is also interesting to note that the *open-loop continuous-time* zeros of the plant appear as zeros of $P(\omega)$ and $P_f(\omega)$. Thus, irrespective of any discrete-time considerations, one also needs to worry about the location of the *continuous-time* plant open-loop zeros since these effect the continuous-time output response.

We notice that the ratio of $|P_f(\omega)|$ to the discrete complementary sensitivity function $|\overline{T}(\gamma_\omega)|$ at frequency $\omega_0$ is

$$\left| \frac{P_f(\omega_0)}{\overline{T}(\gamma_{\omega_0})} \right| = \left| \frac{F(j\omega_0)G_p(j\omega_0)H_0(j\omega_0)}{[FG_pH_0]^s} \right| \tag{9.3.8}$$

Hence to avoid large peaks in $|P_f(\omega)|$ one needs to avoid having $|\overline{T}(\gamma_\omega)|$ near unity at any frequency where the gain of the composite continuous-time transfer function $FG_pH_0$ is significantly greater than the gain of the equivalent discrete transfer function, since this will automatically mean $|P_f(\omega)|$ is large. Further, if $|P_f(\omega)| \gg 1$ then from (9.3.7) $D(\omega)$ will also be very large. Thus large sensitivity to disturbances will follow. There are several reasons why the gain of $FG_pH_0$ might be significantly greater than that of $[FG_pH_0]^s$. Two common reasons for this are:

(i) For continuous plants having relative degree exceeding one, there is usually a discrete zero near the point $-2/\Delta$. Thus, the gain of $[FG_pH_0]^s$ typically falls near $\omega = \pi/\Delta$ (the folding frequency). Hence, it is rarely a good idea to have a discrete closed loop bandwidth that approaches the folding frequency. This will be brought out in subsequent examples.

(ii) Sometimes high frequency resonances can perturb the discrete transfer function away from the continuous-time transfer function by folding effects leading to differences between $FG_pH_0$ and $[FG_pH_0]^s$.

One needs to be careful about the effect these factors have on the difference between $P_f$ and $\bar{T}$. In particular, the bandwidth must be kept well below any frequency where folding effects reduce $[G_pH_0F]^s$ relative to the continuous plant transfer function. This will be illustrated in the examples presented later.

Finally, we look at the sensitivity of the closed-loop system to changes in the open loop plant transfer function as was done in Section 8.4.

We recall from equation (8.4.3) that

$$\bar{T}(\gamma_\omega) = \frac{\bar{C}(\gamma_\omega)[FG_pH_0]^s}{1 + \bar{C}(\gamma_\omega)[FG_pH_0]^s} \tag{9.3.9}$$

We then have the following result which extends the sensitivity result of Lemma 8.4.1 to the case of mixed continuous and discrete signals:

Lemma 9.3.1 Let $\Delta G_p(j\omega)$ denote a change in the continuous plant frequency response $G_p(j\omega)$ at frequency $\omega$. Then

(a) The relative change in the closed-loop *discrete-time* transfer function, $(\Delta\bar{T}/\bar{T})$ at frequency $\omega$ depends on the relative change in the open-loop *continuous-time* transfer function, $(\Delta G_p/G_p)$, at an *infinite* set of frequencies $\omega \pm k\omega_s$; $k = 0, 1, 2 \ldots$ via the following formula:

$$\left| \frac{\Delta\bar{T}(\gamma_\omega)}{\bar{T}(\gamma_\omega)} \right| \approx \frac{\bar{S}(\gamma_\omega)}{[FG_pH_0]^s} \sum_{k=-\infty}^{\infty} \left\{ F(j\omega_k)G_p(j\omega_k)H_0(j\omega_k)\frac{\Delta G_p(j\omega_k)}{G_p(j\omega_k)} \right\}$$

$$\tag{9.3.10}$$

where $\omega_k = \omega - k\omega_s$.

(b) Similarly, the relative change in the reference gain function $(\Delta P/P)$ satisfies:

$$\frac{\Delta P(\omega)}{P(\omega)} = D(\omega)\frac{\Delta G_p(j\omega)}{G_p(j\omega)} - \sum_{\substack{k=-\infty \\ k \neq 0}}^{\infty} P_r(\omega_k)\frac{\Delta G_p(j\omega_k)}{G_p(j\omega_k)}$$

Proof :

(a) By differentiation of (9.3.9) we see that, up to first order, we have

$$\Delta \overline{T} \simeq \frac{\overline{C}(\gamma_\omega)}{\left(1 + \overline{C}(\gamma_\omega)[FG_pH_0]^s\right)^2} \sum_{k=-\infty}^{\infty} F(j\omega_k)H_0(j\omega_k)\Delta G_p(j\omega_k) \quad (9.3.11)$$

Dividing by $\overline{T}$ leads immediately to (9.3.10).

(b) As for part (a).

⋀⋀⋀

We note that the magnitude of $(\Delta G_p/G_p)$ typically approaches unity at high frequencies. Hence (9.3.10) shows that there will be sensitivity problems if $[FG_pH_0]^s$ is small at a frequency where $FG_pH_0$ is large unless, of course, $\overline{S}$ is small at the same frequency. This further reinforces our claim that the bandwidth must be kept well below any frequencies where folding effects reduce $[FG_pH_0]^s$ relative to the continuous plant transfer function $FG_pH_0$.

## 9.4 Modified Discrete Transforms

The disturbance and reference gain functions introduced in the previous sections allow one to compute the frequency content of the *continuous-time* output of a system under the action of digital control. From these functions one can easily compute other related functions. For completeness we will show how one can compute the sampled output response when the samples are taken with an extra delay of $\epsilon$ beyond the time when the samples are taken that are used to drive the controller, where $0 \leq \epsilon < \Delta$. Our reason for making the calculation is that traditional books in digital control also evaluate this response using modified $z$-transform (Kuo [1980], page 114). Hence it is useful to show that the disturbance and reference gain functions easily cover this special case. In particular, we have the following result:

**Lemma 9.4.1** For the zero disturbance case, the Fourier transform of the $\epsilon$-delayed, sampled filtered output is given by

$$Y_\epsilon^s(\omega) = P_\epsilon^s(\omega) R^s(\omega)$$

(9.4.1)

where

$$P_\epsilon^s(\omega) = \sum_{k=-\infty}^{\infty} P_f(\omega_k) e^{-j(\omega_k)\epsilon}$$

(9.4.2)

Thus the output is multi-frequency with corresponding magnitudes and phases determined by the filtered reference gain function $P_f(\omega)$.

Proof : Applying the result in equation (1.3.9) to (9.2.1) gives the following result for the transform of the continuous filtered output delayed by $\epsilon$ seconds:

$$Y_\epsilon(\omega) = P_f(\omega) R^s(\omega) e^{-j\omega\epsilon}$$

(9.4.3)

The transform of the corresponding time-domain signal when it is sampled at period $\Delta$, is

$$Y_\epsilon^s(\omega) = \sum_{k=-\infty}^{\infty} Y_\epsilon(\omega_k)$$

(9.4.4)

The result then follows from (9.4.3), (9.4.4) on noting the periodicity of $R^s(\omega)$.

Of course, if we consider the result in Lemma 9.4.1 for the special case of $\epsilon = 0$, then we recover the normal discrete system transfer functions given in Chapter 8. Specifically from (9.4.1), (9.4.2) and (9.2.2) we have

$$P_{\epsilon}^s(\omega)\Big|_{\epsilon = 0} = \overline{C}(\gamma_\omega)\left[\sum_{k=-\infty}^{\infty} F(j\omega_k)G_p(j\omega_k)H_0(j\omega_k)\right]\overline{S}(\gamma_\omega) \qquad (9.4.5)$$

$$= \overline{T}(\gamma_\omega)$$

as expected.

## 9.5    Examples

In this section, we present some simple examples which illustrate the application of Theorem 9.2.1 in computing the intersample behaviour of continuous-time systems under the action of sampled-data control. Note that the examples are not intended to illustrate good sampled-data control system designs but have been chosen to show the utility of the functions $P(\omega)$, $P_f(\omega)$ and $D(\omega)$ in giving the correct qualitative and quantitative understanding of the continuous response in difficult situations.

### 9.5.1    Servo System with Minimal Prototype Controller

Consider the following continuous-time system

$$G_p(s) = \frac{1}{s(s + 1)} \qquad (9.5.1)$$

We assume a sampling period of 0.4 seconds with zero order hold input and no anti-aliasing filter. Figure 9.5.1 shows the frequency response of $G_p(s)H_0(s)$ and the frequency response of the sampled data transfer function $[G_pH_0]^s$ defined in (9.2.5). We note the periodic nature of $[G_pH_0]^s$ and that, over the range $(0, \pi/\Delta)$, the two frequency responses are very nearly equal with a small dis-

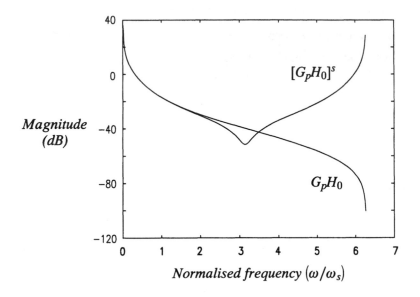

*Figure 9.5.1 :   Continuous and discrete frequency responses for
servo system with minimal prototype controller.*

crepancy near $\omega = \pi/\Delta$ due to the extra zero in the discrete transfer function
arising from sampling.

A minimal prototype sampled data controller (see Section 8.3.3) was then
designed to give a discrete complementary sensitivity function of $1/(\Delta\gamma + 1)$ .
Figure 9.5.2 compares the magnitude of the frequency response of the discrete
complementary sensitivity function $\overline{T}(\gamma_\omega)$ and the reference gain function
$P(\omega)$ . Figure 9.5.3 compares the magnitude of the frequency response of the discrete sensitivity function $\overline{S}(\gamma_\omega)$ and the disturbance gain function $D(\omega)$ .

Whilst the complementary sensitivity function may seem ideal, we notice
that the gain functions display large magnifications in the vicinity of $\pi/\Delta$
rad $\text{sec}^{-1}$ . This result is easily predictable using (9.3.8). Indeed, from Figure
9.5.1, we see that the ratio of $\left|[G_pH_0]^s\right|$ to $\left|G_p(j\omega)H_0(j\omega)\right|$ is approximately

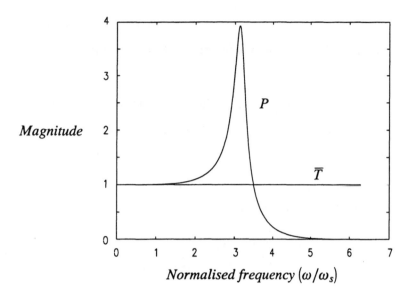

*Figure 9.5.2 : Complementary sensitivity and reference gain function
for system with minimal prototype controller.*

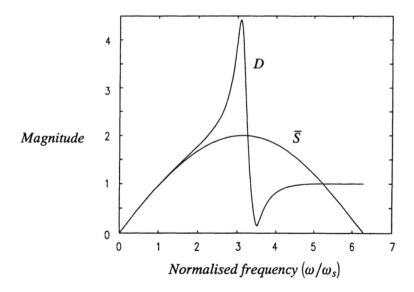

*Figure 9.5.3 : Sensitivity and disturbance gain function for
system with minimal prototype controller.*

12 dB at $\omega = \pi/\Delta$. This same ratio appears between $P$ and $\bar{T}$ in Figure 9.5.2 as predicted by (9.3.8). Also, the link between $P(\omega)$ and $D(\omega)$ in (9.3.4) explains the difference observed in Figure 9.5.3.

From these results we expect to obtain a large continuous-time output response at frequencies near $\pi/\Delta$ whenever reference or disturbance inputs are applied to the plant having significant frequency content at this frequency.

For a step reference input, it is readily shown using Theorem 9.2.1 that the Fourier transform of the continuous output response is

$$Y(\omega) = \frac{14.22(e^{j0.4\omega} - 0.6703)(e^{j0.4\omega} - 1)}{(e^{j0.4\omega} + 0.876)(j\omega)^2(j\omega + 1)} \tag{9.5.2}$$

Inversion of this transform gives the step response as shown in Figure 9.5.4.

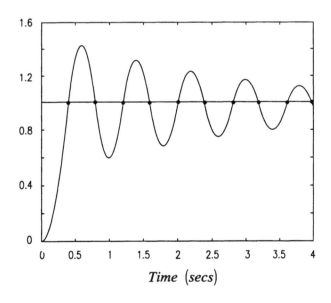

*Figure 9.5.4 : Step response for system with minimal prototype controller.*

Note that the sampled control error is zero for $k \geq 1$ as predicted by the discrete-time specification. However, we see there is a strong continuous-time response at $\pi/\Delta$ exactly as predicted by the function $P(\omega)$.

Next consider a sinusoidal disturbance input of $3/\Delta$ rad sec$^{-1}$ and unit amplitude. Equation (9.2.1) predicts a multi-frequency continuous-time plant output. In particular, the first two frequency components occur at $3/\Delta$ and $(2\pi - 3)/\Delta$ rad sec$^{-1}$. The amplitude of these two components can be determined as in Section 9.2 to have amplitude $|D(3/\Delta)|$ and $|P(2\pi - 3)/\Delta|$. From Figures 9.5.2 and 9.5.3 the amplitudes are seen to be approximately 4.0 and 2.0 respectively. The sum of these two components gives an amplitude modulated waveform having carrier frequency of $\pi/\Delta$ and envelope frequency of $\pi - 3/\Delta$. This prediction is verified in Figure 9.5.5 which shows the disturbance response at the output. Note that this is a very different response than might be

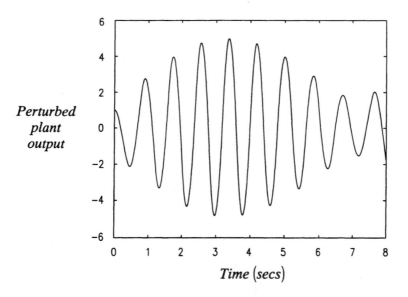

*Figure 9.5.5 : Disturbance response for system with minimal protoype controller.*

expected from the discrete sensitivity function which predicts a pure sinewave of magnitude 2 at frequency $3/\Delta$. Of course, this is precisely what is obtained *at the sample points* in Figure 9.5.5.

### 9.5.2 Servo System with Dead-Beat Control

Consider the same system as in the example of Section 9.5.1, except that now the target complementary sensitivity function is

$$\frac{\bar{B}_p(\gamma)}{\bar{B}_p(0)(1 + \Delta\gamma)^2}$$

where $\bar{B}_p(\gamma)$ is the numerator of the discrete system transfer function $\bar{G}_p(\gamma)$. Note that this controller does not attempt to compensate the apparent system gain loss for the sampled data system in the vicinity of the discrete sampling zeros at $\omega = \pi/\Delta$.

Figure 9.5.6 compares the discrete complementary sensitivity function $\bar{T}(\gamma_\omega)$ with the reference gain function $P(\omega)$. Note that these are very close over the frequency range $(0, \pi/\Delta)$. Figure 9.5.7 compares the magnitudes of the discrete sensitivity function $\bar{S}(\gamma_\omega)$ with the disturbance gain function $D(\omega)$. From these two figures we would anticipate no significant problems in the continuous-time step response. This is indeed verified in Figure 9.5.8.

As in Section 9.5.1, we expect a multi-frequency output when a sinewave disturbance is injected. For the same disturbance input as in 9.5.1, Figures 9.5.6 and 9.5.7 now predict that the first two components in the output will be at $3/\Delta$ and $(2\pi - 3)/\Delta$ rad sec$^{-1}$ as before and that the amplitudes will now be 1.17 and 0.22 respectively. The addition of these two components leads to the same modulation effect as previously although the extent of modulation is now much less due to the diminished size of the second component. This is verified in Figure 9.5.9. In this case, the 'discrete-time' sensitivity function predicts a sinewave of amplitude 1.14 which can be checked in Figure 9.5.9 at the sample points.

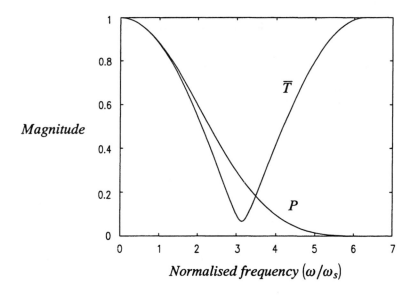

*Figure 9.5.6 : Complementary sensitivity and reference gain
functions for servo system with dead-beat control.*

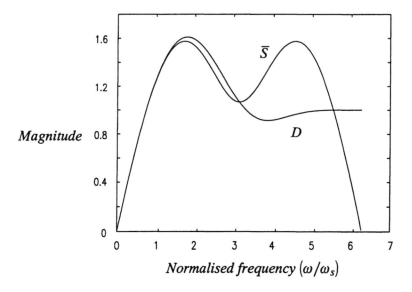

*Figure 9.5.7 : Sensitivity and disturbance gain function
for servo system with dead-beat control.*

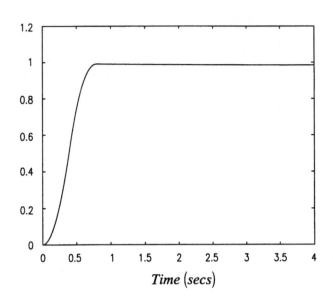

*Figure 9.5.8 :  Step response of system with dead-beat control.*

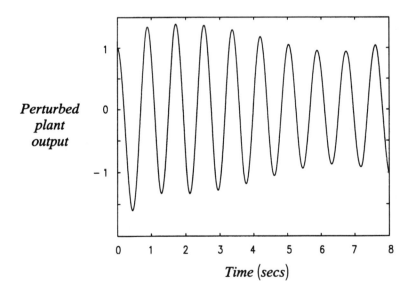

*Figure 9.5.9 :  Sinusoidal disturbance response
of system with dead-beat control.*

### 9.5.3   Resonant System #1

Consider next the following poorly damped second order plant:

$$G_p(s) = \frac{100}{s^2 + 2s + 100} \tag{9.5.3}$$

This system was sampled without anti-aliasing filter at a period of 0.5 seconds. Figure 9.5.10 compares the magnitudes of the frequency responses $[G_pH_0]^s$ and $G_p(j\omega)H_0(j\omega)$. Note that the resonant peak has been folded into the low frequency range in the discrete frequency response.

A sampled-data control system was then designed to give a discrete complementary senstivity function $\bar{T}(\gamma) = 0.5/(\Delta\gamma + 0.5)$. Figure 9.5.11 compares the complementary sensitivity function $|\bar{T}(\gamma\omega)|$ with the reference gain function $P(\omega)$. We see that $P(\omega)$ has a significant peak at $\omega = \pi/\Delta$.

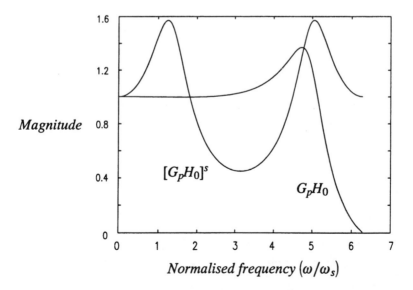

*Figure 9.5.10 : Continuous and discrete frequency responses of resonant system #1.*

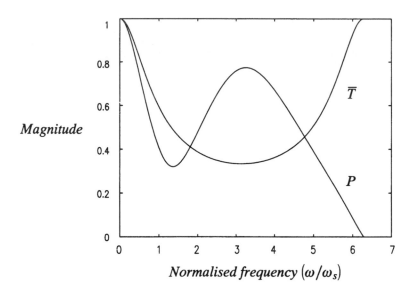

*Figure 9.5.11 : Complementary sensitivity and reference
gain function for resonant system #1.*

Figure 9.5.12 compares the magnitudes of the discrete sensitivity function $S(\gamma_\omega)$ and the disturbance gain function $D(\omega)$. Again we see that there are significant differences.

The intersample behaviour of the step response is shown in Figure 9.5.13. Note that the sampled response in Figure 9.5.13 is a simple exponential as expected from the specified value of $\overline{T}(\gamma)$.

A sinusoidal disturbance of frequency $1/\Delta$ rad sec$^{-1}$ was applied to the system. In this case, the first two components in the continuous output response are at frequencies $1/\Delta$ and $(2\pi-1)/\Delta$. From Figures 9.5.11 and 9.5.12, the amplitude of these two components are 0.86 and 0.3 respectively. These two components are clearly evident in Figure 9.5.14. However, at the sample points we see that the response is a pure sinewave of frequency $1/\Delta$ and amplitude 1.17 as is correctly predicted by $|\overline{S}(\gamma_\omega)|$ in Figure 9.5.12.

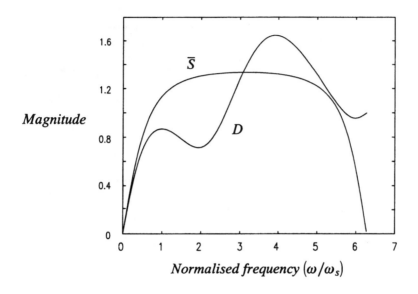

Figure 9.5.12 : Sensitivity and disturbance gain functions
for resonant system #1.

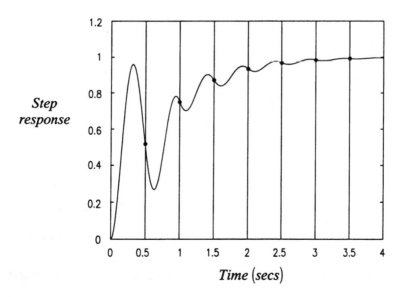

Figure 9.5.13 : Step response for resonant system #1.

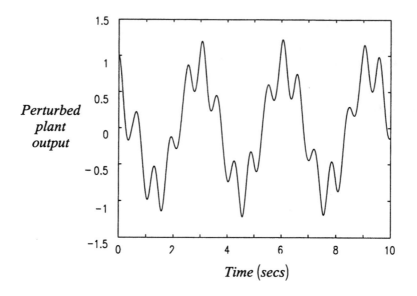

*Figure 9.5.14 : Sinusoidal disturbance response
for resonant system #1.*

## 9.5.4   Resonant System #2 with Anti-aliasing Filter

Consider the lightly damped resonant system

$$G_p(s) = \frac{100}{s^2 + 0.2s + 100} \tag{9.5.4}$$

The sampling period $\Delta$ was set at 0.5. An anti-aliasing filter in the form of a low pass second order bandwidth filter with cut-off frequency $\pi/\Delta$ was used. Figure 9.5.15 compares responses $\left| [FG_pH_0]^s \right|$ and $\left| F(j\omega)G_p(j\omega)H_0(j\omega) \right|$.

A sampled data controller was designed to achieve deadbeat control giving $\overline{T}(\gamma) = \overline{B}_p(\gamma)/\left(\overline{B}_p(0)(\Delta\gamma + 1)^4\right)$, where $\overline{B}_p(\gamma)$ is the numerator of $\overline{G}_p(\gamma)$.

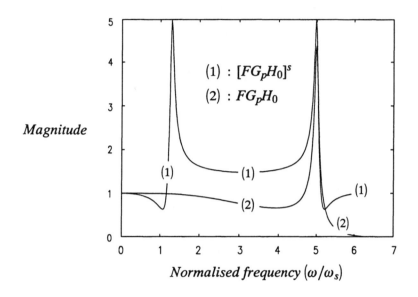

*Figure 9.5.15 : Discrete and continuous frequency responses for resonant system #2.*

Figure 9.5.16 compares the complementary sensitivity function $|\bar{T}(\gamma_\omega)|$ with the reference gain function $P(\omega)$. Figure 9.5.17 compares the sensitivity function magnitude $|\bar{S}(\gamma_\omega)|$ with the disturbance gain function $|D(\omega)|$.

There is a significant peak in $P(\omega)$ in the neighbourhood of $1/\Delta$ which manifests itself in a large component at that frequency in the continuous step response as shown in Figure 9.5.18.

A sinewave disturbance of unit amplitude and frequency $3/\Delta$ rad sec$^{-1}$ gives the response shown in Figure 9.5.19. In this case, Figures 9.5.16 and 9.5.17 predict that the first two components in the output will have amplitude 1.6 and 1.45 respectively. This is verified in Figure 9.5.19.

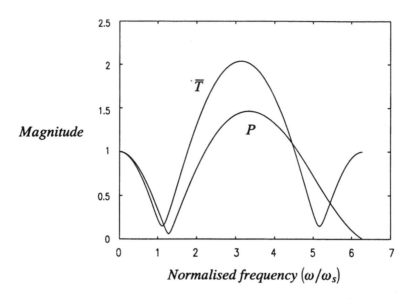

*Figure 9.5.16 : Complementary sensitivity and reference*
*gain function for resonant system #2.*

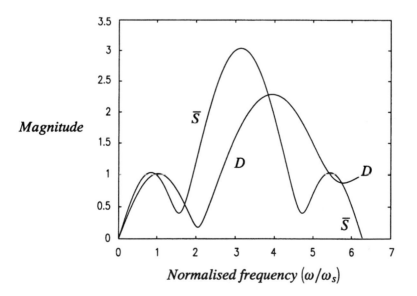

*Figure 9.5.17 : Sensitivity and disturbance gain function*
*for resonant system #2.*

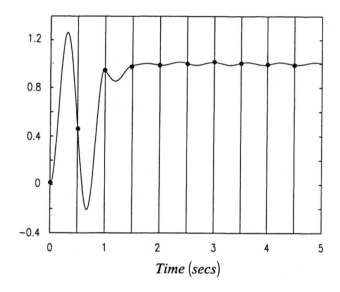

*Figure 9.5.18 : Step response (unfiltered) for resonant system #2.*

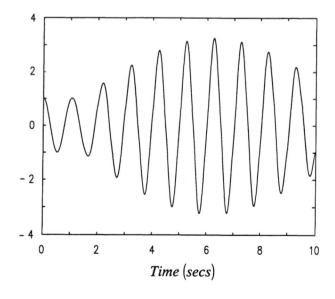

*Figure 9.5.19 : Sinusoidal disturbance response for resonant system #2.*

## 9.6    Observations and Comments from the Examples

(i) The reference and disturbance gain functions $P(\omega)$ and $D(\omega)$ are seen to give qualitative and quantitative information about the true continuous-time response resulting from either a reference or disturbance input.

(ii) In many cases, the first two components in the multi-frequency output response are sufficient to give an accurate qualitative description of the response to a sinewave disturbance input.

(iii) If we consider only the first two frequency components in the output, then for a sinewave disturbance of frequency $\omega_0 \in [0, \pi/\Delta]$, the output is an amplitude modulated waveform of carrier frequency $\pi/\Delta$ and modulating frequency of $(\pi - \omega_0)/\Delta$.

(iv) Significant resonant peaks in the system frequency response outside the range $[0, \pi/\Delta]$ can be folded back into the frequency range of interest even when anti-aliasing filters are used. These are a potential source of inter-sample problems in sampled-data design.

(v) All continuous-time systems of relative degree greater than one lead to discrete-time models having extra zeros ("sampling zeros") in the vicinity of $-\pi/\Delta$ (see Section 4.4.2). Sampled-data control systems that attempt to cancel these zeros will necessarily have high gain at frequencies near $\pi/\Delta$ and this is a potential source of poor intersample behaviour. Control laws that have this property include those based on discrete loop transfer recovery (LTR) designs. These are frequently advocated for sampled-data systems because of the recovery of the discrete-time sensitivity performance (see Problem 9.10). However, such designs have an associated difficulty of poor intersample behaviour when applied to continuous-time plants.

As might be expected, the differences between the gain functions $P_f(\omega)$, $D(\omega)$ and the traditional discrete sensitivity functions $\overline{T}(\gamma_\omega)$, $\overline{S}(\gamma_\omega)$ can be made small if one abides by the usual 'rules of thumb' for discrete control, i.e. sample ten times faster than the open or closed-loop rise-time, use anti-aliasing filters, and keep the closed-loop response time well above the sample period. However, there are situations where these rules cannot be obeyed due to hardware limitations and in these cases the gain functions described in this chapter can be used to predict the correct continuous-time output behaviour.

## 9.7   The Class of all Stabilizing Sampled-Data Controllers

As in Chapter 8, it is sometimes useful in sampled-data design to have a simple parameterization of all stabilizing sampled-data controllers. A necessary condition for boundedness of the continuous response is that the sampled response be bounded. Hence the class of all stabilizing sampled-data controllers that consider the continuous response is contained in the class of digital control laws as developed in Section 8.5.

Thus consider again the class of all stabilizing *discrete* control laws as described in Chapter 8 and as redrawn in Figure 9.7.1. Here we show explicitly the anti-aliasing filter and sampling operation. We recall that the reason for using this parameterization of the controller is that closed-loop stability of the sampled system is easily guaranteed (for the nominal system) by simply constraining $\overline{Q}$ to be stable.

Note that in Figure 9.7.1 the section inside the dashed line operates in continuous-time whilst the section outside the dashed line represents the digital control law. In Figure 9.7.1, $\overline{P}/\overline{L}$ denotes the digital pre-stabilizing controller which

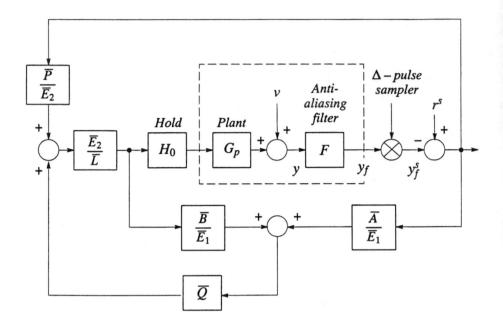

*Figure 9.7.1 : Class of all stabilizing discrete control laws.*

we take to be pre-specified and $\overline{B}/\overline{A}$ is the discrete transfer function equivalent of the (hold + plant + anti-aliasing filter) combination.

The multiple feedback loops in Figure 9.7.1 are equivalent to the following composite feedback controller transfer function:

$$\overline{C} = \frac{\frac{\overline{P}}{\overline{E}_2} + \overline{Q}\frac{\overline{A}}{\overline{E}_1}}{\frac{\overline{L}}{\overline{E}_2} - \overline{Q}\frac{\overline{B}}{\overline{E}_1}} \tag{9.7.1}$$

where the discrete plant transfer function is

$$\frac{\overline{B}}{\overline{A}} = \left[FG_pH_0\right]^s \tag{9.7.2}$$

Using the result in Theorem 9.2.1, one can immediately show that the nominal filtered output sampled response is given by

$$Y_f^s = \frac{\overline{C}[FG_pH_0]^s}{1 + \overline{C}[FG_pH_0]^s}R^s + \frac{1}{1 + \overline{C}[FG_pH_0]^s}V_f^s \qquad (9.7.3)$$

where $V_f$ is the prefiltered disturbance $(V_f = FV)$ and $V_f^s$ is its $\Delta$-pulse sampled version.

Also, the continuous filtered output response is

$$Y_f = \frac{\overline{C}FG_pH_0}{1 + \overline{C}[FG_pH_0]^s}R^s + V_f - \frac{\overline{C}FG_pH_0}{1 + \overline{C}[FG_pH_0]^s}V_f^s \qquad (9.7.4)$$

In the special case of the controller given in (9.7.1), we have in addition that

$$\frac{1}{1 + \overline{C}[FG_pH_0]^s} = \frac{\overline{AL}}{\overline{E}_1\overline{E}_2} - \overline{Q}\frac{\overline{AB}}{\overline{E}_1^2} \qquad (9.7.5)$$

Substituting (9.7.5) into (9.7.4) gives

$$Y_f = V_f + \left[\frac{\overline{AP}}{\overline{E}_1\overline{E}_2} + \overline{Q}\frac{\overline{A}^2}{\overline{E}_1^2}\right]FG_pH_0\left(R^s - V_f^s\right) \qquad (9.7.6)$$

The terms in $\overline{A}, \overline{P}, \overline{E}_1, \overline{E}_2$ and $\overline{Q}$ in (9.7.6) are evaluated at $\gamma_\omega = \left(e^{j\omega\Delta} - 1\right)/\Delta$ whereas the other terms are continuous frequency responses evaluated at $j\omega$.

Note that (9.7.6) is affine in $\overline{Q}$ as for the purely discrete-time case considered in Chapter 8. Thus equation (9.7.6) is a natural extension of these earlier results to the sampled-data problem.

## 9.8    Linear Quadratic Design of Sampled-Data Controllers

The linear quadratic optimal regulator described in Section 8.8 optimizes the at-sample response. However, in view of the arguments given in this chapter, we might expect that the inter-sample response may sometimes be rather poor if we only focus attention on the sampled response. We therefore proceed in this section to investigate how one might optimize the continuous response using a discrete-time control law.

Let us suppose that the plant (including anti-aliasing filter) has a minimal *continuous-time* state-space representation of the form

$$\frac{d}{dt}x = Ax + Bu ; \quad x(0) = x_0 \tag{9.8.1}$$

We assume that $(A, B)$ is stabilizable and that the sampling rate avoids pathological loss of stabilizability in the discrete-time system.

Now we assume that we are interested in choosing a *discrete-time* control law (using samples of $x$ to generate $u$ via a zero order hold) so as to minimize a *continuous-time* quadratic cost function of the form:

$$J_c = x(t_1)^T \Sigma_1 x(t_1) + \int_{t_0}^{t_1} \left( x(t)^T Q_c x(t) + u(t)^T R_c u(t) \right) dt \tag{9.8.2}$$

We note the analogy between (9.8.2) and (8.8.2). We see that (9.8.2) weights the continuous-time response whereas (8.8.2) focuses only on the sampled response.

Due to the zero order hold assumption, $u$ must be constant over the interval $[k\Delta, (k+1)\Delta)$. Hence we can write a composite model for $x$ and $u$ in the form

$$\frac{d}{dt}\begin{bmatrix} x \\ u \end{bmatrix} = \begin{bmatrix} A & B \\ 0 & 0 \end{bmatrix}\begin{bmatrix} x \\ u \end{bmatrix} \tag{9.8.3}$$

For $t \in [k\Delta, (k+1)\Delta)$ we can solve (9.8.3) to yield

$$\begin{bmatrix} x(t) \\ u(t) \end{bmatrix} = e^{\bar{A}(t-k\Delta)}\begin{bmatrix} x[k] \\ u[k] \end{bmatrix} \tag{9.8.4}$$

where

$$\bar{A} = \begin{bmatrix} A & B \\ 0 & 0 \end{bmatrix} \tag{9.8.5}$$

Substituting (9.8.4) into (9.8.2) gives

$$J_c = \sum_{k=(\frac{t_0}{\Delta})}^{(\frac{t_1}{\Delta})-1} \int_{k\Delta}^{(k+1)\Delta} \left( x[k]^T u[k]^T \right)\left( e^{\bar{A}(t-k\Delta)} \right)^T \begin{bmatrix} Q_c & 0 \\ 0 & R_c \end{bmatrix} \left( e^{\bar{A}(t-k\Delta)} \right)\begin{bmatrix} x[k] \\ u[k] \end{bmatrix} dt$$

$$+ x\left(\frac{t_1}{\Delta}\right)^T \Sigma_1 x\left[\frac{t_1}{\Delta}\right] \tag{9.8.6}$$

or

$$\boxed{J_c = \sum_{k=(\frac{t_0}{\Delta})}^{(\frac{t_1}{\Delta})-1} \Delta\left( x[k]^T u[k]^T \right)\begin{bmatrix} Q_d & S_d^T \\ S_d & R_d \end{bmatrix}\begin{bmatrix} x[k] \\ u[k] \end{bmatrix} + x\left[\frac{t_1}{\Delta}\right]^T \Sigma_1 x\left[\frac{t_1}{\Delta}\right]} \tag{9.8.7}$$

where

$$\begin{bmatrix} Q_d & S_d^T \\ S_d & R_d \end{bmatrix} = \frac{1}{\Delta}\int_0^\Delta \left( e^{\bar{A}\tau} \right)^T \begin{bmatrix} Q_c & 0 \\ 0 & R_c \end{bmatrix}\left( e^{\bar{A}\tau} \right)d\tau \tag{9.8.8}$$

Also, it is immediate from the model (9.8.1) that we can write

$$\boxed{\delta x[k] = A_\delta x[k] + B_\delta u[k], \quad x[0] = x_0} \tag{9.8.9}$$

We recognize (9.8.7), (9.8.9) as a standard discrete-time linear quadratic optimal control problem as in Section 8.8 with the addition of the coupling term $S_d$ which is easily treated as in Problem 8.22. We are thus led to the conclusion that a *continuous-time* linear quadratic cost function can be optimized by simply optimizing an appropriate *discrete-time* cost function provided the discrete weighting matrices are chosen as in (9.8.8). We will illustrate this remarkable property by a simple example.

**Example 9.8.1 :** Consider the continuous-time system given in state-space form by:

$$\dot{x}(t) = Ax(t) + Bu(t)$$

$$y(t) = Cx(t)$$

where

$$A = \begin{bmatrix} 0 & 0 \\ 0 & -1 \end{bmatrix}, \quad B = \begin{bmatrix} 1 \\ 1 \end{bmatrix}, \quad C = [1 \ -1], \ x(0) = \begin{bmatrix} 1 \\ -1 \end{bmatrix}$$

Design an optimal time-invariant digital regulator for this system of the form $u[k] = -L_\delta x[k]$ where $x[k] = x(k\Delta)$, $u(t) = u[k]$ for $k\Delta \le t < (k+1)\Delta$ and $\Delta = 0.1$ sec.

  (a) Find $L_\delta$ if the optimality criterion is the following discrete-time cost function

$$J_d = \sum_{k=0}^{\infty} \Delta y[k]^2 \qquad\qquad (9.8.10)$$

  (b) Calculate the corresponding continuous-time cost function

$$J_c = \int_0^{\infty} y^2(t)dt \qquad\qquad (9.8.11)$$

where

$$x[0] = \begin{bmatrix} 1 \\ -1 \end{bmatrix}$$

(c) Redesign the discrete regulator but with $J_c$ as the optimality criterion and find the resulting optimal value for $J_c$. Compare with (b).

(d) Draw the responses in (a) and (b) for $x(0) = \begin{bmatrix} 1 \\ -1 \end{bmatrix}$.

## Solution

(a) Using discrete LQ design procedures as in Section 8.8 with the cost function (9.8.10) leads to an optimal feedback gain of $L_\delta = [206.6263, -186.9588]$ and the optimal value $J_d^*$ of $J_d$ for given the initial conditions is $J_d^* = 0.4$.

Note that for this case the solution is such that the feedback gain $L_\delta$ assigns the closed loop eigenvalues so that one is at $-10$ and the other cancels the sampling zero. Then $y[k]$ goes to zero in *one* step from any initial condition.

(b) To calculate the corresponding continuous cost function $J_c$ as in (9.8.11) we repeat much of the derivation in equations (9.8.3) - (9.8.9). We get

$$\bar{A} = \begin{bmatrix} 0 & 0 & 1 \\ 0 & -1 & 1 \\ 0 & 0 & 0 \end{bmatrix}$$

in (9.8.5) and so

$$e^{\bar{A}t} = \begin{bmatrix} 1 & 0 & t \\ 0 & e^{-t} & 1-e^{-t} \\ 0 & 0 & 1 \end{bmatrix}$$

and, from (9.8.8), we get

$$\begin{bmatrix} Q_d & S_d^T \\ S_d & R_d \end{bmatrix} = \int_0^{\Delta} e^{\bar{A}^T \tau} \begin{bmatrix} 1 & -1 & 0 \\ -1 & 1 & 0 \\ 0 & 0 & 0 \end{bmatrix} e^{\bar{A} \tau} d\tau$$

$$= \int_0^{\Delta} \begin{bmatrix} 1 & -e^{-\tau} & \tau - 1 + e^{-\tau} \\ -e^{-\tau} & e^{-2\tau} & e^{-\tau} - \tau e^{-\tau} - e^{-2\tau} \\ \tau - 1 + e^{-\tau} & e^{-\tau} - \tau e^{-\tau} - e^{-2\tau} & \tau^2 - 2\tau + 1 - 2e^{-\tau} + 2\tau e^{-\tau} + e^{-2\tau} \end{bmatrix} d\tau$$

$$= \begin{bmatrix} 0.1 & -0.0952 & 0.0002 \\ -0.0952 & 0.0906 & -0.0002 \\ 0.0002 & -0.0002 & 0 \end{bmatrix}$$

so the continuous cost becomes

$$J_c = \sum_{k=0}^{\infty} \left[ x^T[k]u[k] \right] \begin{bmatrix} Q_d & S_d^T \\ S_d & R_d \end{bmatrix} \begin{bmatrix} x[k] \\ u[k] \end{bmatrix} \tag{9.8.12}$$

or, with $u[k] = -L_\delta x[k]$,

$$J_c = \sum_{k=0}^{\infty} x[k]^T \begin{bmatrix} 0.053 & -0.0519 \\ -0.0519 & 0.0508 \end{bmatrix} x[k]$$

$$= \sum_{k=0}^{\infty} x[k]^T \bar{Q} \, x[k]$$

The above sum can be evaluated using the discrete Lyapunov equation to give

$$J = x[0]^T \Omega \, x[0]$$

where

$$\Omega = \begin{bmatrix} 0.2409 & -0.2306 \\ -0.2306 & 0.2207 \end{bmatrix}$$

Finally, we see that the continuous cost becomes

$$J_c = 0.9227$$

This is considerably larger than $J_d$ of (a) which indicates that the gain $L_\delta$, though optimal for the discrete-time cost function, may lead to poor continuous-time responses.

(c) To do our redesign we use equation (9.8.7) as our criterion for designing the optimal regulator. We first obtain an equivalent coupled discrete cost function as in (9.8.7), (9.8.8). Then, using standard packages for LQ design we obtain the optimal discrete feedback gain as

$$L_\delta' = [156.7922 - 139.7698]$$

with the resulting optimal value $J_c^*$ of $J$ being $J_c^* = 0.2531$. The value of the discrete cost function $J_d$ now becomes $J_d = 0.4305$.

Note that $J_d$ is slightly larger than in (a) but $J_c$ has considerably improved (by a factor of 4:1) over the result in (b).

(d) Figure 9.8.1 (a) shows the sampled and continuous-time initial condition response of the system with the controller found in part (a). Note that the sampled output settles in one sample whereas the continuous response is highly oscillatory.

The basic reason for the failure of the controller in (a) to give a satisfactory continuous-time response is that it arises from a discrete-time LQ design with no weighting on the control effort. As shown in Problem 8.24 part (d) and Problem 9.10, this im-

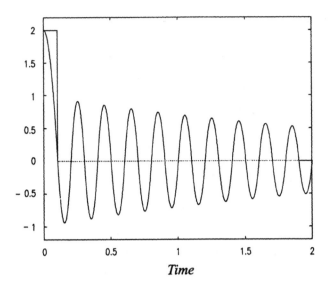

*(a) Sampled and Discrete Response for criterion (9.8.10).*

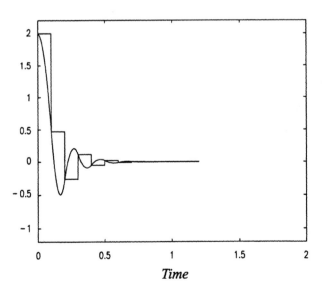

*(b) Sampled and Discrete Response for criterion (9.8.11).*

*Figure 9.8.1 : Responses for Example 9.8.1.*

plies that the closed loop poles cancel the open-loop discrete-time zeros including the so called sampling zero. However, as discussed in Section 9.6, an inevitable consequence of cancelling these zeros is a large intersample response.

Figure 9.8.1 (b) shows the sampled and continuous-time initial condition response of the system with the controller found in part (b). Note now that the continuous-time response is quite acceptable. This is a reasonable outcome since the cost function used for this design weights the continuous-time response.

# 9.9 Duality Relationships for Hybrid Optimal Controller

We have seen earlier in Section 8.9 that the *discrete-time* optimal filter is dual to the *discrete-time* optimal LQ regulator with the associations given in Table 8.9.1. We next show that two *hybrid* problems are also dual.

## 9.9.1 Review of Hybrid Optimal Controller

We recall from Section 9.8 that the hybrid LQ optimal regulator uses the following continuous-time state-space model:

$$\dot{x} = Ax + Bu \; ; \; x(0) = x_0 \tag{9.9.1}$$

and continuous-time cost function

$$J_c = x(t_1)^T \Sigma_f \, x(t_1) + \int_{t_0}^{t_1} \left( x(t)^T Q_c x(t) + u(t)^T R_c u(t) \right) dt \tag{9.9.2}$$

The integrand in (9.9.2) can be expressed as $\| \phi \|^2 + \| \psi \|^2$ where

$$\phi \overset{\Delta}{=} Q_c^{\frac{1}{2}} x \; ; \quad \psi = R_c^{\frac{1}{2}} u \tag{9.9.3}$$

The hybrid optimal control problem then aims to minimize $J_c$ subject to the constraint that the input is piecewise constant with period $\Delta$ ; i.e.

$$u(s) = u[k], \quad s \in [k\Delta, (k+1)\Delta] \tag{9.9.4}$$

It is shown in (9.8.3) that the joint process $(x, u)$ satisfies

$$\frac{d}{dt}\begin{bmatrix} x \\ u \end{bmatrix} = \begin{bmatrix} A & B \\ 0 & 0 \end{bmatrix}\begin{bmatrix} x \\ u \end{bmatrix} = \bar{A}\begin{bmatrix} x \\ u \end{bmatrix} \tag{9.9.5}$$

for $t \in [k\Delta, (k+1)\Delta)$. Using (9.9.5) the problem is transformed into the following equivalent discrete-time problem (see (9.8.9), (9.8.7)):

$$\delta x[k] = A_\delta x[k] + B_\delta u[k] \tag{9.9.6}$$

$$J_c = \sum_{k=\left(\frac{t_0}{\Delta}\right)}^{\left(\frac{t_1}{\Delta}\right)-1} \Delta \left( x[k]^T u[k]^T \right) \begin{bmatrix} Q_d & S_d^T \\ S_d & R_d \end{bmatrix}\begin{bmatrix} x[k] \\ u[k] \end{bmatrix} + x\left[\frac{t_1}{\Delta}\right]^T \Sigma_1 x\left[\frac{t_1}{\Delta}\right] \tag{9.9.7}$$

where

$$\begin{bmatrix} A_\delta & B_\delta \\ 0 & 0 \end{bmatrix} = \frac{1}{\Delta}\left[ e^{\bar{A}\Delta} - I \right] \tag{9.9.8}$$

$$A_\delta = \frac{e^{A\Delta} - I}{\Delta} \tag{9.9.9}$$

$$B_\delta = \frac{1}{\Delta}\int_0^\Delta e^{A\Delta}ds\, B \tag{9.9.10}$$

and

$$\begin{bmatrix} Q_d & S_d^T \\ S_d & R_d \end{bmatrix} = \frac{1}{\Delta}\int_0^\Delta \left(e^{\bar{A}\tau}\right)^T \begin{bmatrix} Q_c & 0 \\ 0 & R_c \end{bmatrix}\left(e^{\bar{A}\tau}\right)d\tau \tag{9.9.11}$$

Solution of this problem leads to the discrete control law $u[k] = -Lx[k]$ .

## 9.9.2   Review of Hybrid Optimal Filter

The above equations will next be compared with the results in Section 6.4 and Section 6.6.2 for the hybrid optimal filter. In the latter case, we began with a continuous-time stochastic model:

$$\begin{bmatrix} dx \\ dy \end{bmatrix} = \begin{bmatrix} A & 0 \\ C & 0 \end{bmatrix} \begin{bmatrix} x \\ y \end{bmatrix} dt + \begin{bmatrix} \Omega^{\frac{1}{2}} dv' \\ \Gamma^{\frac{1}{2}} d\omega' \end{bmatrix} \tag{9.9.12}$$

$$= \bar{A}_f \begin{bmatrix} x \\ y \end{bmatrix} dt + \begin{bmatrix} \Omega^{\frac{1}{2}} dv' \\ \Gamma^{\frac{1}{2}} d\omega' \end{bmatrix} \tag{9.9.13}$$

where $dv'$ and $d\omega'$ are zero mean, vector valued Wiener processes with unit incremental variances.

The sampling process was defined in Section 6.3 using an anti-aliasing filter leading to the sampled output

$$\delta y[k] = \frac{1}{\Delta} \int_{k\Delta}^{(k+1)\Delta} dy \tag{9.9.14}$$

It was then shown in Lemma 6.4.1 that this leads to an equivalent discrete filtering problem

$$\delta x[k] = A_\delta x[k] + v_\delta[k] \tag{9.9.15}$$

$$\delta y[k] = C_\delta x[k] + \omega_\delta[k] \tag{9.9.16}$$

where

$$\begin{bmatrix} A_\delta & 0 \\ C_\delta & 0 \end{bmatrix} = \frac{1}{\Delta} \left[ e^{\bar{A}_f\Delta} - I \right] \tag{9.9.17}$$

and where $v_\delta, \omega_\delta$ are iid sequences having covariance

$$\begin{bmatrix} \Omega_\delta & S_\delta \\ S_\delta^T & \Gamma_\delta \end{bmatrix} = \frac{1}{\Delta} \int_0^\Delta \left( e^{\bar{A}_F t} \right) \begin{bmatrix} \Omega & 0 \\ 0 & \Gamma \end{bmatrix} \left( e^{\bar{A}_F t} \right) dt \qquad (9.9.18)$$

From (6.6.4), (6.6.5) the Hybrid optimal filter then satisfies

$$\delta \dot{x}[k] = A_\delta \dot{x}[k] + H_\delta \left[ \delta y[k] - C_\delta \dot{x}[k] \right] \qquad (9.9.19)$$

$$\dot{x}(t) = e^{A(t-k\Delta)} \dot{x}[k] \qquad (9.9.20)$$

### 9.9.3   Duality

We have already seen that, with the associations in Table 8.9.1, the discrete optimal filter and discrete LQ regulator are dual to each other. In addition, the two *hybrid* problems are also dual. To explore this we recall the definition of the sampling operations that converts a continuous-time signal $s(\cdot)$ into $\Delta$ pulse samples

$$\overline{S}(s(\cdot))(t) = \sum_{k=-\infty}^{\infty} \Delta s[k] \delta(t - k\Delta) \qquad (9.9.21)$$

Similarly, the reconstruction or hold operation $\overline{H}$ is defined via its impulse response

$$h_0(t) = \begin{cases} 1/\Delta , & t \in [0, \Delta) \\ 0 & \text{otherwise} \end{cases} \qquad (9.9.22)$$

with corresponding transfer function $\left( 1 - e^{-s\Delta} \right)/s\Delta$. Also, if the input to $\overline{H}$ is a sequence of $\Delta$-pulses, $s(t) = \Sigma \omega[k] \delta(t - k\Delta)$ then

$$\overline{H}(\omega(\cdot))(t) = \omega[k] \quad \text{for } k\Delta \le t < (k + 1)\Delta \qquad (9.9.23)$$

Using this notation, the differential equation governing the hybrid optimal regulator can be obtained by substituting the optimal control law $u[k] = -Lx[k]$ into (9.9.1) leading to

$$\frac{d}{dt}x = (A - B\overline{H}L\overline{S})x \tag{9.9.24}$$

with the elements in the cost function (9.9.3) given by

$$\begin{bmatrix} \phi \\ \psi \end{bmatrix} = \begin{bmatrix} Q_c^{\frac{1}{2}} \\ -R_c^{\frac{1}{2}}\overline{H}L\overline{S} \end{bmatrix} x \tag{9.9.25}$$

Similarly, the differential equation governing the evolution of the estima-
tion error $\tilde{x} \triangleq x - \hat{x}$ of the hybrid optimal filter is from (9.9.19), (9.9.20):

$$\frac{d}{dt}\tilde{x} = [A - \overline{S}H_\delta\overline{H}C]\tilde{x} + [\Omega^{\frac{1}{2}}, -\overline{S}H_\delta\overline{H}\Gamma^{\frac{1}{2}}]\begin{bmatrix} \dot{v} \\ \dot{\omega} \end{bmatrix} \tag{9.9.26}$$

Equations (9.9.24) to (9.9.26) are shown diagramatically in Figure 9.9.1.
Note that this is quite a remarkable result since the two figures ((a) for the filter
and (b) for the regulator) can be simply obtained from each other by reversing
all arrows and changing inputs to outputs.

We illustrate the duality of the hybrid optimal filter and hybrid optimal
controller by the following example.

**Example 9.9.1 :** The following filtering problem is the dual to the hybrid
optimal control problem addressed in Example 9.8.1.

Consider the following continuous-time stochastic state-space model:

$$\dot{x} = Ax + \dot{v}$$
$$y' = Cx + \dot{\omega}$$

where

$$E\left\{\begin{bmatrix} \dot{v} \\ \dot{\omega} \end{bmatrix} [\dot{v} \ \dot{\omega}]^T\right\} = \begin{bmatrix} \Omega & 0 \\ 0 & \Gamma \end{bmatrix} \delta(t)$$

Take

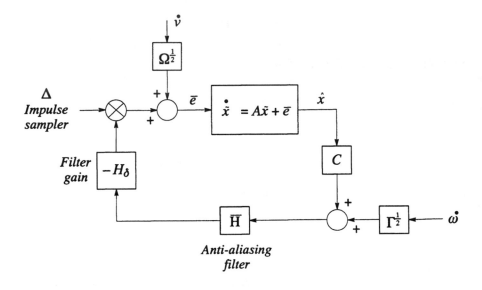

*(a)  Filter*

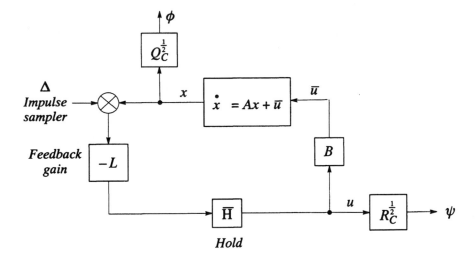

*(b)  Regulator*

*Figure 9.9.1 :  Dual hybrid filtering and control configurations.*

$$A = \begin{bmatrix} 0 & 0 \\ 0 & -1 \end{bmatrix} \; ; \; C = \begin{bmatrix} 1 & 1 \end{bmatrix}$$

$$\Omega = \begin{bmatrix} 1 & -1 \\ -1 & 1 \end{bmatrix} \; ; \; \Gamma \text{ negligible} \left( \text{say } 10^{-10} \right)$$

We assume that an anti-aliasing filter as in (6.3.1) is employed and the sampling period is $\Delta = 0.1$ sec.

(a) Calculate the optimal Kalman filter gain if we approximate the discrete covariances as $\Omega \delta_d$ and $\Gamma \delta_d$ respectively.

(b) Calculate the correct covariance matrix for the discrete case and redesign the Kalman filter.

(c) Simulate the response of the *continuous*-time state estimate as in (6.6.7) for cases (a) and (b) for an initial state error and no noise.

(d) Discuss the relationship with the hybrid regulator problem of Example 9.8.1.

Solution

(a) Using standard packages for solving the discrete algebraic Riccati equation (in shift form):

$$P = \Omega + A_s P A_s^T - A_s P C_s^T (\Gamma + C_s P C_s^T)^{-1} C_s P A_s^T$$

where $A_s$ and $C_s$ are the corresponding discrete matrices (in shift form) given by

$$A_s = \begin{bmatrix} 1 & 0 \\ 0 & 0.9084 \end{bmatrix} \text{ and } C_s = C \int_0^{\Delta} e^{A\sigma} d\sigma = \begin{bmatrix} 0.1 & 0.0952 \end{bmatrix},$$

we get $P = \begin{bmatrix} 1.0002 & -1.0002 \\ 1.0002 & 1.0002 \end{bmatrix}$ and the corresponding Kalman gain

$$H = A_s P C_s^T \left( \Gamma + C_s P C_s^T \right)^{-1}$$

$$= \begin{bmatrix} 206.6263 \\ -186.9588 \end{bmatrix}$$

(b) We calculate the exact discrete covariances as

$$\begin{bmatrix} \bar{\Omega} & \bar{S} \\ \bar{S}^T & \bar{\Gamma} \end{bmatrix} = \int_0^{\Delta} e_f^{\bar{A}\sigma} \begin{bmatrix} \Omega & 0 \\ 0 & \Gamma \end{bmatrix} e_f^{\bar{A}^T\sigma} d\sigma$$

where

$$\bar{A}_f = \begin{bmatrix} A & 0 \\ C & 0 \end{bmatrix}$$

and get

$$\begin{bmatrix} \bar{\Omega} & \bar{S} \\ \bar{S}^T & \bar{\Gamma} \end{bmatrix} = \begin{bmatrix} 0.1 & -0.0952 & 0.0002 \\ -0.0952 & 0.0906 & -0.0002 \\ 0.0002 & -0.0002 & 0.0000 \end{bmatrix}$$

Recalculating the solution for the new discrete algebraic Riccati equation (in shift form):

$$P = \bar{\Omega} + A_s \bar{P} A_s^T - \left( A_s P C_s^T + \bar{S} \right)\left( \bar{\Gamma} + C_s P C_s^T \right)^{-1}\left( C_s P A_s^T + \bar{S}^T \right)$$

we get

$$P = \begin{bmatrix} 0.0649 & -0.0633 \\ -0.0633 & 0.0616 \end{bmatrix}$$

and the corresponding gain

$$H = \left( A_s P C_s^T + \bar{S} \right)\left( \bar{\Gamma} + C_s P C_s^T \right)^{-1}\left( C_s P A_s^T + \bar{S}^T \right)$$

$$= \begin{bmatrix} 156.7922 \\ -139.7698 \end{bmatrix}$$

(c)   We simulated the continuous-time observers corresponding to the gains in (a) and (b). The results are shown in Figures 9.9.2, 9.9.3, where we have used

$$x(0) = \begin{bmatrix} 1 \\ 1 \end{bmatrix}, \quad \text{and} \quad \hat{x}(0) = \begin{bmatrix} 0 \\ 0 \end{bmatrix}$$

Clearly the design in (a) is a poor design and it takes a long time for the estimation errors to go down to zero.

(d)   Note that the gains obtained in parts (a) and (b) above are identical to the optimal feedback gain found in parts (a) and (c) of Example 9.8.1 thus verifying the dual relationship.

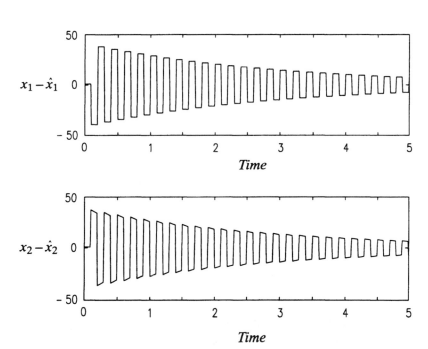

*Figure 9.9.2 :  Estimation errors from design method (a) of Example 9.9.1.*

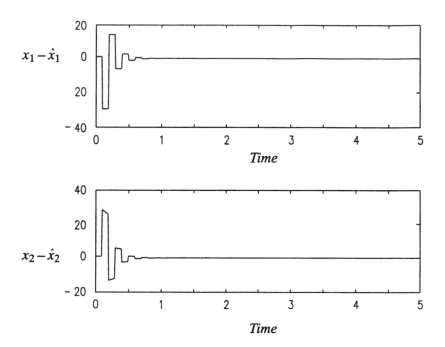

*Figure 9.9.3 :  Estimation errors from design
method (b) of Example 9.9.1.*

## 9.10   Further Reading and Discussion

The material in Section 9.2 is largely based on the paper by Goodwin and
Salgado (1993). Closely related work appears in Araki and Ito (1992 a, b, c) and
Feuer and Goodwin (1995). A comprehensive survey of recent developments in
sampled data control with many references is contained in Araki (1993). The authors wish to acknowledge the assistance of Mario Salgado in generating the examples in Section 9.5.

The modified $z$-transform (also known as the delayed $z$-transform) was developed by Tsypkin (1950), Barker (1951) and Jury (1956). The method is described in most books on digital control e.g. Kuo (1992, page 113). The disadvantage of these methods is that the response has to be calculated for a range of $\epsilon$'s

in the interval $(0, \Delta)$. The advantage of using the disturbance and reference gain functions is that they allow one to predict the complete continuous-time response in one step.

A closely related tool for analyzing sampled data systems is that of the parametric transfer function – see Rosenvasser (1994). This idea has been used to analyze multivariable sampled data systems subject to deterministic and stochastic signals (Rosenvasser (1995a), (1995b), (1995c), Rosenvasser, Polyakov and Laupe (1996).

The linear quadratic optimization results described in Section 9.8 are based on the work of Chen and Francis (1991). We have seen in Section 9.9 that this result is the dual of the standard discrete optimal filtering result of Åström (1970, p.82) and Salgado, Middleton and Goodwin (1986). Related work also appears in Bamieh and Pearson (1992b), Bernstein, Davis and Greeley (1986) and Osburn and Bernstein (1993). The result in Osburn and Bernstein (1993) explores the cost function as the sampling period is varied. In particular, it is shown that the continuous-time cost arising from an optimal sampled controller converges to the continuous-time cost for an optimal continuous-time controller as the sampling period goes to zero.

We have restricted our treatment of optimization to a quadratic criterion for reasons of simplicity and clarity. However, other criteria are also possible. For example, the following recent papers deal with design of sampled-data controllers using various optimization criteria (see references for bibliographic detail):

- o Bamieh and Pearson (1992a) - Raising techniques and $H^{\infty}$ design;

- o Hara and Kabamba (1990) - optimization of induced norm in sampled data control;

- o Toivonen (1990) - $H^{\infty}$ optimization;

- o Yamamoto (1990) - function space approach;

- o Dullerud and Francis (1992) - $L^{1}$ analysis and design;

- Sivashankar and Khargonekar (1991) - $L^\infty$ induced norms;
- Sivashankar and Khargonekar (1993) - performance analysis and robust stability;
- Linnemann (1992) - $L^\infty$ optimal performance;
- Bamieh, Dahley and Pearson (1993) - optimization of $L^\infty$ induced norm.
- Yamamoto and Khargonekar (1996)

Also, several recent books have appeared on the topic of design of sampled data control systems. Chen and Francis (1995) gives a comprehensive treatment of $H_2$ and $H_\infty$ optimal design of sampled data controllers. Dullerud (1996) gives a detailed analysis of robustness issues for uncertain sampled data systems.

## 9.11  Problems

9.1  Consider a continuous-time system given by its frequency response

$$G_p(j\omega) = \frac{1}{j\omega + 1}$$

where the anti-aliasing filter and the hold each have the same impulse response

$$f(t) = h_0(t) = \begin{cases} \dfrac{1}{\Delta} & \text{for } 0 \le t < \Delta \\ 0 & \text{otherwise} \end{cases}$$

Let the sampling period be 0.1 sec and the controller simply a gain of $\overline{C}(\gamma_\omega) = 2$.

(a)  Calculate $\overline{S}(\gamma_\omega)$, $P(\omega)$ and $D(\omega)$.

(b)  Find the first two frequency components of the output for

$$r^s(t) = \sum_{k=-\infty}^{\infty} 0.1 \cos(2\pi t)\, \delta(t - 0.1k)$$

9.2   Design the digital controller to give minimal prototype control as in Section 9.5.1.

9.3   Verify equation (9.5.2).

9.4   Show that the result in equation (9.3.10) reduces to that in (8.4.5) if one considers the relative change in the *discrete* transfer function, i.e. in $[G_pH_0F]^s$ .

9.5   Design the deadbeat controller of Section 9.5.2.

9.6   Explicitly evaluate the sensitivity of the closed loop discrete transfer function obtained in Section 9.5.1 to changes in the open loop *continuous-time* plant transfer function. Comment on the difference in sensitivity to changes in $[G_pH_0]^s$ and $G_p$ .

9.7   Design the control system described in Section 9.5.3.

9.8   Design the control system described in Section 9.5.4.

9.9   Explain why it is usually only necessary to consider the first two components of $P(\omega)$ and $D(\omega)$ . What implicit assumptions have been made so that this is possible?

9.10  Consider the LTR method described in Problems 8.18, 8.19. Show that the state estimator given by LTR involves cancelling *all* of the (stable) zeros of the discrete plant transfer function. Discuss why this is not a good idea for discrete-time systems arising from the sampling of continuous-time systems having relative degree greater than one.

   How would you suggest that the discrete LTR method be modified so as to avoid continuous-time sensitivity problems whilst retaining the essential features of the idea?

9.11  Consider the continuous-time transfer function

$$G(s) = \frac{1}{s(s+1)}$$

and the criteria

$$J = \int_0^\infty y^2(t)dt$$

(a) Find the discretized system, $\bar{G}(\gamma) = \bar{B}(\gamma)/\bar{A}(\gamma)$, with $\Delta = 0.1$ and $F(s) = 1$, and assuming zero order hold.

(b) Choose $\bar{E}_1(\gamma) = \bar{E}_2(\gamma) = (\gamma + 5)^2$. Find $\bar{L}$ and $\bar{P}$ so that $\overline{AL} + \overline{BP} = \bar{E}_1(\gamma)\bar{E}_2(\gamma)$.

(c) Assume $v = 0$ and $r^s = 1$ and

$$Q(\gamma) = \frac{\alpha_1}{\gamma + 4} + \frac{\alpha_2}{\gamma + 6}$$

Use equation (9.7.6) to derive and expression for $Y(\omega)$ in terms of $\alpha_1$ and $\alpha_2$.

(d) Compute $\alpha_1$, $\alpha_2$ which minimize $J$.

9.12 Let $(C, A, B)$ be a minimal realization of

$$G(s) = \frac{1}{s^2 - 1}$$

and consider the following continuous-time cost function:

$$J_c = \int_0^\infty \left( y^2(t) + R_c u^2(t) \right) dt$$

with $R_c$ negligible (take $R_c = 10^{-10}$).

Design a discrete-time controller with sample period $\Delta = 0.1$ using two approaches:

(a) Use the approximation

$$J \approx \sum_{k=0}^{\infty} \left( y[k]^2 + R_c u[k]^2 \right) \Delta$$

(b) Follow the methodology discussed in Section 9.8 to directly optimize the exact value of $J_c$.

(c) Discuss and compare your results in (a) and (b).

# Chapter 10

# Generalized Sample-Hold Functions

## 10.1  Introduction

As we have noted in previous chapters the combination of a continuous-time plant with a digital controller raises a number of interfacing issues. Both plant input and output are continuous-time signals while the digital controller accepts and generates sequences of numbers. Sampling – the process of transforming a continuous-time signal into a sequence of numbers – and reconstruction – the process of transforming a sequence of numbers into a continuous-time signal – have been discussed in Chapter 2. However, in that chapter, these two processes have been addressed mainly from the general signal processing point of view. The main concern was to preserve, as much as possible, the information content of the signal. Hence, the sampling rates were determined by signal bandwidth; the hold filters on the reconstruction side were typically a compromise between the ideal low pass filter and technical limitations.

For a long time control engineers accepted this approach. In sampling, they were frequently constrained by the measurement processes : the zero-order-hold (ZOH) was typically chosen for the reconstruction of the continuous-time input. In recent years the possibility of designing specialized reconstruction processes to help in the system control process intrigued many researchers. A number of published results have shown that the use of these specialized reconstruction methods leads to many benefits in the control system.

In this chapter we will introduce the specialized reconstruction processes, or, as they are commonly referred to, generalized sample-hold functions

(GSHF). We will show, in the time domain, that the system can be made to perform arbitrarily well *at its sample points* using GSHF. However, a frequency domain analysis reveals that the cost is usually poor intersample behaviour and *sensitivity* to high frequency uncertainty in the plant model.

## 10.2   Generalized Sample-Hold Function: A Time Domain Perspective

Let us consider the simple configuration of Figure 10.2.1. $G$ is a linear continuous-time, time-invariant system. The GSHF and the sampler are synchro-
nised, and each has $\Delta$ as the associated sampling interval.

To understand how $u(t)$ is generated from $r[k]$ via the GSHF, we introduce a function $h_g(t)$ which is zero outside the interval $0 \le t < \Delta$. We also de-
fine $\bar{h}_g(t)$ as the periodic extension of $h_g(t)$, i.e. $\bar{h}_g(t) = \sum_{k=-\infty}^{\infty} h_g(t - k\Delta)$. Thus
$\bar{h}_g(t)$ is a periodic function with period $\Delta$. Then $u(t)$ results from taking each
period of $\bar{h}_g(t)$ and multiplying it by the corresponding value of the sequence
$r[k]$ scaled up by $\Delta$ :

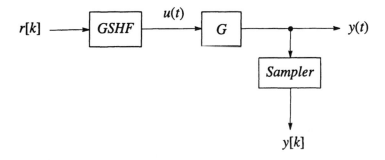

*Figure 10.2.1 : A simple generalized sample-hold
function configuration.*

$$u(t) = \Delta \bar{h}_g(t)r[k] \quad \text{for} \quad k\Delta \le t < (k+1)\Delta \qquad (10.2.1)$$

This is demonstrated in Figure 10.2.2. Note that the zero-order-hold (ZOH) can be viewed as a special case when $h_g(t) = h_0(t) = 1/\Delta$ for $0 \le t < \Delta$ and $\bar{h}_g(t) = 1/\Delta$ for all $t$.

The question now is how different choices of $h_g(t)$ affect the system output, and whether $h_g(t)$ can be chosen to have a positive impact on this output.

Let the system $G$ be described by an $n$ dimensional state-space model of the form

$$\dot{x}(t) = Ax(t) + Bu(t) \qquad (10.2.2)$$

$$y(t) = Cx(t) \qquad (10.2.3)$$

For simplicity we will assume the system to be single-input single-output (SISO), but much of the discussion can be generalized to the multiple-input multiple-output (MIMO) case.

The following result gives a discrete-time state-space model for the system when a generalized sample-hold is used.

**Lemma 10.2.1** Let $x[k] \triangleq x(k\Delta)$ and $y[k] \triangleq y(k\Delta)$. Then we have the following discrete state-space model describing the sampled system response when a generalized sample-hold function $h_g(t)$ is used:

$$\delta x[k] = A_\delta x[k] + B_g r[k]$$
$$\qquad (10.2.4)$$
$$y[k] = Cx[k]$$

where

$$\delta x[k] \triangleq \frac{1}{\Delta}(x[k+1] - x[k])$$

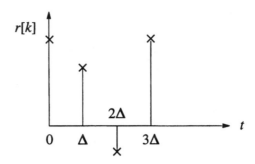

*(a) The sequence r[k].*

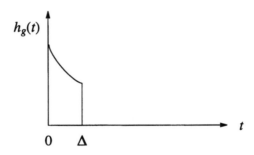

*(b) One period of the $\hat{h}_g(t)$ function.*

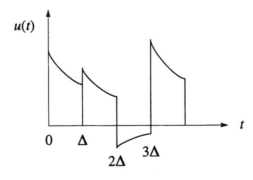

*(c) The resulting u(t).*

*Figure 10.2.2 : Generation of control signal u(t) from the sequence r[k]
using a generalized sample-hold function.*

$$A_\delta \triangleq \frac{1}{\Delta}\left(e^{A\Delta} - I\right)$$

(10.2.5)

and

$$B_g = \int_0^\Delta e^{A(\Delta - \tau)} B h_g(\tau)\ d\tau$$

(10.2.6)

Proof :  Clearly we have

$$x\big((k+1)\Delta\big) = e^{A\Delta}x(k\Delta) + \int_{k\Delta}^{(k+1)\Delta} e^{A((k+1)\Delta - \tau)}Bu(\tau)d\tau$$

Substituting (10.2.1) and using the periodicity of $\hat{h}_g(t)$ we get

$$x\big((k+1)\Delta\big) = e^{A\Delta}x(k\Delta) + \Delta\int_0^\Delta e^{A(\Delta - \tau)}Bh_g(\tau)\ d\tau \cdot r[k]$$

and (10.2.4) follows.

$$\triangle\!\triangle\!\triangle$$

Let us consider the significance of Lemma 10.2.1. Since the transfer function associated with (10.2.4) is

$$\overline{G}(\gamma) = C(\gamma I - A_\delta)^{-1}B_g$$

(10.2.7)

clearly the poles of $\overline{G}(\gamma)$ are determined by $A_\delta$, and the zeros by $C$ and $B_g$. While the choice of $h_g(t)$ has no effect on $A_\delta$ (see (10.2.5)) it does affect $B_g$ (see (10.2.6)). In other words, by different choices of $h_g(t)$ we may get different zeros for the transfer function $\overline{G}(\gamma)$. Going beyond this observation, we show in the following result that the use of a GSHF in the forward path allows us to arbitrarily assign the gain (not surprising) and zeros (very surprising) of the sampled data transfer function.

**Lemma 10.2.2** Let $(A, B)$ be a controllable pair and $(C, A_\delta)$ a detectable pair and let $\{z_i\}$; $i = 1, 2, \ldots, L \leq n - 1$ be any set of complex numbers symmetric with respect to the real axis and $K$ a real number. Then there exists a generalized hold function $h_g(t)$ such that the resulting $\overline{G}(\gamma)$ has this set as its zeros and $K$ as its gain.

**Proof :** Since $(C, A_\delta)$ is a detectable pair, it is straightforward to argue that for any given set $\{z_i\}$ and number $K$ there exist a vector $B_g$ such that

$$\overline{G}(\gamma) = C(\gamma I - A_\delta)^{-1} B_g$$

has the set $\{z_i\}$ as its zeros and $K$ as its gain.

Next we will show that since $(A, B)$ is controllable, for any real vector $B_g$ there exists a function $h_g(t)$ which satisfies (10.2.6). We do this by constructing $h_g(t)$ from $B_g$. Let $W$ be the controllability grammian, namely

$$W = \int_0^\Delta e^{A\tau} B B^T e^{A^T \tau} d\tau \tag{10.2.8}$$

It is well known that $(A, B)$ is a controllable pair if and only if $W$ is nonsingular. Define $\hat{h}_g(t)$ as

$$\tilde{h}_g(t) = B^T e^{A^T(\Delta - \text{rem}(t, \Delta))} W^{-1} B_g \tag{10.2.9}$$

where $0 \leq \text{rem}(t, \ \Delta) = t - k\Delta < \Delta$ is the remainder of $t$ divided by $\Delta$. Clearly the above $\tilde{h}_g(t)$ is periodic (see Problem 10.2) and since $\text{rem}(\tau, \ \Delta) = \tau$ for $0 \leq \tau < \Delta$ we get

$$\int_0^\Delta e^{A(\Delta-\tau)} B\tilde{h}_g(\tau) \; d\tau \; = \; \int_0^\Delta e^{A(\Delta-\tau)} BB^T e^{A^T(\Delta-\tau)} \; d\tau \cdot W^{-1}B_g$$

$$= WW^{-1}B_g$$

$$= B_g$$

as desired. This completes the proof.

$$\wedge\wedge\wedge$$

We illustrate the above result by a simple example.

**Example 10.2.1 :** Consider a linear time invariant system of the form (10.2.2) - (10.2.3) with

$$A \; = \; \begin{bmatrix} -1 & 0 \\ 0 & -2 \end{bmatrix}, \; B \; = \; \begin{bmatrix} 1 \\ 1 \end{bmatrix}, \; C \; = \; \begin{bmatrix} 2, & -3 \end{bmatrix}$$

This system is stable but has a non-minimum phase zero at $s = 1$.

(a)  Determine the resultant sampled-data transfer function using a ZOH with $\Delta = 0.1$.

(b)  Use a GSHF to move the sampled-data zero to $\gamma = -0.9516$ (to cancel the stable pole) and to give a gain of 1.

(c)  Compare the *sampled* step response for the system obtained in parts (a) and (b).

**Solution**

(a)  When sampled at $\Delta = 0.1$ with ZOH we get

$$A_\delta \; = \; \begin{bmatrix} -0.9516 & 0 \\ 0 & -1.8127 \end{bmatrix}, \; B_\delta \; = \; \begin{bmatrix} 0.9516 \\ 0.9063 \end{bmatrix}$$

with transfer function

$$C[\gamma I - A_\delta]^{-1}B_\delta \; = \; \frac{-0.8158(\gamma - 1.0573)}{(\gamma + 0.9516)(\gamma + 1.8127)}$$

(b)  Since $(C, A_\delta)$ is a detectable pair, a $B_g$ to satisfy the requirements can be computed by requiring that

$$C[\gamma I - A_\delta]^{-1}B_g \; = \; \frac{1}{\gamma + 1.8127}$$

The required value of $B_g$ is then

$$B_g \; = \; \begin{bmatrix} 0 \\ -\dfrac{1}{3} \end{bmatrix}$$

Since $(A, B)$ is controllable we can use (10.2.9) to calculate the
desired $\bar{h}_g(t)$. First

$$W \; = \; \int_0^\Delta e^{A\tau} BB^T e^{A^T\tau} \, d\tau \; = \; \begin{bmatrix} 0.0906 & 0.0864 \\ 0.0864 & 0.0824 \end{bmatrix}$$

then

$$\bar{h}_g(t) \; = \; B^T e^{A^T(\Delta - \mathrm{rem}(t,\Delta))} \, W^{-1} B_g$$

$$= \; 4206.826 e^{\mathrm{rem}(t,0.1)} - 3993.337 e^{2 \cdot \mathrm{rem}(t,0.1)}$$

(c)  In Figure 10.2.3 we show the difference in sampled response of
system to a unit step input when ZOH only was used and when
the designed GSHF was used. Note that the ZOH response un-
dershoots but the GSHF response does not.

$$\wedge\!\wedge\!\wedge$$

We have seen in Chapter 8 that zeros have a marked impact on our ability
to design a feedback control system. Since we know that the use of a GSHF al-

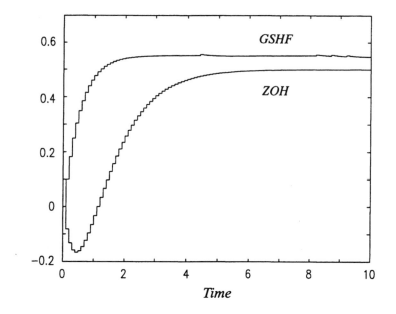

*Figure 10.2.3 : Step response for Example 10.2.1.*

lows us to arbitrarily assign the sampled-data zeros, then it would seem to be beneficial to combine the use of GSHF with feedback control. Hence consider the simple closed loop system in Figure 10.2.4.

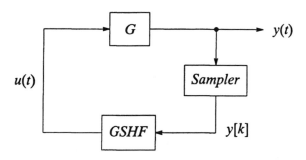

*Figure 10.2.4 : A simple closed loop GSHF configuration.*

Here we assume the GSHF to have time response $h_f(t)$. As before, we take

$\bar{h}_f(t)$ to be a periodic function with period $\Delta$ corresponding to $\sum\limits_{k=-\infty}^{\infty} h_f(t - k\Delta)$.

We also have

$$u(t) = \Delta \bar{h}_f(t) y[k] \quad \text{for } k\Delta \leq t \leq (k+1)\Delta \qquad (10.2.10)$$

Then we claim the following:

**Lemma 10.2.3** Let $(A, B)$ be a controllable pair and $(C, A_\delta)$ a detectable pair and let $\{p_i\}$ $i = 1, 2, \ldots, n$ be *any* set of complex numbers symmetric with respect to the real axis. Then there exists $h_f(t)$ so that

$$\delta x[k] = \left(A_\delta + F_g C\right) x[k]$$

with $A_\delta$ as in (10.2.5)

$$F_g = \int_0^\Delta e^{A(\Delta - \tau)} B h_f(\tau) \, d\tau \qquad (10.2.11)$$

and where $\{p_i\}$ are the eigenvalues of the closed loop matrix $(A_\delta + F_g C)$.

**Proof :** The proof of this lemma is very similar to the proofs of Lemmas 10.2.1, 10.2.2 and is given as an exercise (Problem 10.4).

$\wedge\wedge\wedge$

To summarize our discussion so far, let us combine the configurations of Figures 10.2.1 and 10.2.4 into Figure 10.2.5.

In this case

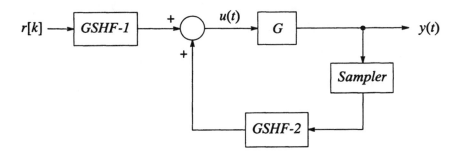

*Figure 10.2.5 : A complete GSHF feedback system.*

$$u(t) = \Delta \bar{h}_f(t) y[k] + \Delta \bar{h}_g(t) r[k]; \quad k\Delta \leq t < (k+1)\Delta \qquad (10.2.12)$$

and the result will then be:

**Theorem 10.2.1** Let the system $G$ be given by equations (10.2.2) – (10.2.3) and the control defined by (10.2.12). Then the *sampled* system will be

$$
\begin{aligned}
\delta x[k] &= (A_\delta + F_g C) x[k] + B_g r[k] \\
y[k] &= C x[k]
\end{aligned}
\qquad (10.2.13)
$$

with $B_g$ and $F_g$ given by (10.2.6) and (10.2.11) respectively.

Given that $(A, B)$ and $(C, A_\delta)$ are controllable and detectable pairs respectively the gain, zeros and poles of the transfer function

$$\bar{G}(\gamma) = C\left[\gamma I - (A_\delta + F_g C)\right]^{-1} B_g \qquad (10.2.14)$$

can be *arbitrarily* assigned via the choice of $h_g(t)$ and $h_f(t)$ in the GSHF's.

**Proof :** Immediate from Lemmas 10.2.2 and 10.2.3.

To reveal more of the properties of the GSHF, we make use of the fact that any periodic function can be written in a Fourier series form. Thus we write (see (1.2.1), (1.2.2))

$$\tilde{h}_g(t) = \frac{1}{\Delta} \sum_{p=-\infty}^{\infty} a_g[p] e^{jp\omega_s t} \qquad (10.2.15)$$

where

$$a_g[p] = \int_{\frac{-\Delta}{2}}^{\frac{\Delta}{2}} \tilde{h}_g(t) e^{-jp\omega_s t} dt \qquad (10.2.16)$$

and

$$\omega_s = \frac{2\pi}{\Delta} \qquad (10.2.17)$$

is the sampling frequency.

Substituting (10.2.15) into (10.2.6) we get

$$B_g = \frac{1}{\Delta} \sum_{p=-\infty}^{\infty} a_g[p] \int_0^{\Delta} e^{A(\Delta-\tau)} e^{jp\omega_s \tau} d\tau \cdot B$$

So, if we denote

$$v_p = \frac{1}{\Delta} \int_0^{\Delta} e^{A(\Delta-\tau)} e^{jp\omega_s \tau} d\tau \, B \qquad (10.2.18)$$

we get

$$B_g = \sum_{p=-\infty}^{\infty} a_g[p]v_p \qquad (10.2.19)$$

Note that the vectors $v_p$ depend on the continuous system and the sampling rate while the vector $B_g$ is expressed as a linear combination of these vectors. Clearly, the ability to achieve any desired $B_g \in \mathcal{C}^n$ (and hence to achieve arbitrary assignment of the zeros of the sampled-data system) depends on whether or not the set $\{v_p\}$ spans the whole $n$-dimensional vector space. To that effect we show the following:

**Lemma 10.2.4** Let $(A, B)$ be a controllable pair. If $A$ is nonsingular, then any set of $n$ vectors $\{v_p\}$ as in (10.2.18) will be linearly independent hence spans $\mathcal{C}^n$. If $A$ is singular, then any set of $n$ vectors $\{v_p\}$ will be linearly independent provided $v_0$ is included in the set.

Proof :

*Case 1 – A nonsingular*

Generically (10.2.18) can be written as follows (by generically we mean avoiding pathological sampling rates where $jp\omega_s$, $p \neq 0$ integer, is an eigenvalue of $A$).

$$v_p = \frac{1}{\Delta}[jp\omega_s I - A]^{-1}\left[I - e^{A\Delta}\right] B \quad \text{for all } p \in Z \qquad (10.2.20)$$

We will argue by contradiction. Let us assume that the lemma is false and suppose $\{v_{p_i}\}$ $i = 1, 2, \ldots, n$ are linearly dependent. Then, there exists a vector $0 \neq a \in \mathcal{C}^N$ such that

$$a^T v_{p_i} = 0 \qquad (10.2.21)$$

This implies, in view of (10.2.20), that $jp_i\omega_s$ are zeros of the transfer function

$$a^T[sI-A]^{-1}\left[I-e^{A\Delta}\right]B \qquad\qquad (10.2.22)$$

However, since $A$ is nonsingular $a^T\left[I-e^{A\Delta}\right] \neq 0$ and since $(A,B)$ is controllable $\left[I-e^{A\Delta}\right]B \neq 0$. So the transfer function in (10.2.22) has at most $(n-1)$ zeros. This is a contradiction which implies that $a = 0$.

*Case 2 – A singular*

Again, we will argue by contradiction and suppose $\{v_{p_i}\}$ $i = 1, 2, \ldots, n$ are linearly dependent (say $p_1 = 0$). Then there exists $0 \neq a \in \mathbb{C}^N$ such that (10.2.21) holds.

Considering the transfer function in (10.2.22) we first note that the controllability of $(A,B)$ implies $\left[I-e^{A\Delta}\right]B \neq 0$ and the reachability of $(e^{A\Delta}, \int_0^\Delta e^{A\sigma}d\sigma B)$. Then, using the PBH criteria for reachability, clearly, if $a^Tv_0 = a^T\int_0^\Delta e^{A\sigma}\,d\sigma \cdot B = 0$ we must have $a^T\left[I-e^{A\Delta}\right] \neq 0$. Moreover, since $A$ is singular, it can be shown (see Problem 10.7) that this transfer function has a pole-zero cancellation at the origin. So, it has at most $n-2$ zeros, which contradicts the assumption that (10.2.21) holds for $i = 2, 3, \ldots, n$. The conclusion again is that $a = 0$, which completes the proof.

ᐯᐯᐯ

From Lemma 10.2.4 it follows that in order to find an appropriate $h_g(t)$ satisfying (10.2.6), one could solve a set of linear equations (10.2.19). Since the number of unknowns is larger than the number of equations, there are infinitely many solutions for the coefficients $\{a_g[p]\}$. As the lemma implies, (10.2.19) can be replaced with a finite sum as follows:

$$B_g = \sum_{p=-P}^{P} a_g[p]v_p \qquad (10.2.23)$$

where the symmetric use of $P$ and $-P$ in the sum has been chosen to guarantee a real $\hat{h}_g(t)$. Whether to include $p = 0$ or not depends on whether $A$ is singular and the value of $n$ (see Lemma 10.2.4). Other reasons for including the term $p = 0$ will be discussed below.

**Example 10.2.2 :** Use (10.2.23) to recalculate $\bar{h}_g(t)$ for the plant considered in Example 10.2.1.

**Solution** Since $n = 2$ and $A$ is nonsingular we need only calculate $v_{-1}$ and $v_1$. Using (10.2.20) (since $A$ is nonsingular)

$$v_1 = 10 \begin{bmatrix} 1 + 20\pi j & 0 \\ 0 & 2 + 20\pi j \end{bmatrix}^{-1} \begin{bmatrix} 0.0952 & 0 \\ 0 & 0.1813 \end{bmatrix} \begin{bmatrix} 1 \\ 1 \end{bmatrix}$$

$$= \begin{bmatrix} 0.00024 - 0.01514j \\ 0.00092 - 0.02882j \end{bmatrix}$$

$$v_{-1} = \bar{v}_1 = \begin{bmatrix} 0.00024 + 0.01514j \\ 0.00092 + 0.02882j \end{bmatrix}$$

So, by (10.2.23) we have

$$B_g = \begin{bmatrix} v_{-1}, & v_1 \end{bmatrix} \begin{bmatrix} a_g[-1] \\ a_g[1] \end{bmatrix}$$

hence

$$\begin{bmatrix} a_g[-1] \\ a_g[1] \end{bmatrix} = \begin{bmatrix} -363.349 - 5.705j \\ -363.349 + 5.705j \end{bmatrix}$$

and, using (10.2.15)

$$\tilde{h}_g(t) = 10\left(a_g[-1]e^{-j20\pi t} + a_g[1]e^{j20\pi t}\right) = 7267.90\cos(20\pi t + 3.1257)$$

$\wedge\wedge\wedge$

Yet another approach to calculating a function $h_g(t)$ for GSHF is to choose a piecewise constant function as in Figure 10.2.6. Namely, $\Delta$ is divided into $n$ subintervals and then

$$\tilde{h}_g(t) = g_i \quad \text{for} \quad \frac{(i-1)\Delta}{n} \le < \frac{i\Delta}{n} \tag{10.2.24}$$

Substituting (10.2.24) into (10.2.26) we get

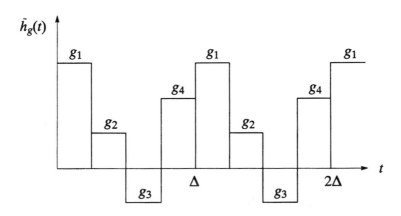

*Figure 10.2.6 : Piecewise constant GSHF.*

$$B_g = \sum_{i=1}^{n} g_i \int_{\frac{(i-1)\Delta}{n}}^{\frac{i\Delta}{n}} e^{A(\Delta-\tau)} B \, d\tau$$

$$:= \sum_{i=1}^{n} g_i \, \Gamma_i \qquad\qquad (10.2.25)$$

So

$$\begin{bmatrix} g_1 \\ \cdot \\ \cdot \\ \cdot \\ g_n \end{bmatrix} = \begin{bmatrix} \Gamma_1 \Gamma_2 \ ... \Gamma_n \end{bmatrix}^{-1} B_g \qquad\qquad (10.2.26)$$

where (generically) the controllability of $(A, B)$ guarantees the independence of the $\Gamma_i$.

**Example 10.2.3 :** Recalculate $\bar{h}_g(t)$ for Example 10.2.1 using (10.2.25).

**Solution** First we calculate the vectors $\Gamma_1, \Gamma_2$.

$$\Gamma_1 = \int_0^{\frac{\Delta}{2}} e^{A(\Delta-\tau)} B \, d\tau = \begin{bmatrix} 0.04639 \\ 0.04305 \end{bmatrix}$$

$$\Gamma_2 = \int_{\frac{\Delta}{2}}^{\Delta} e^{A(\Delta-\tau)} B \, d\tau = \begin{bmatrix} 0.00238 \\ 0.00453 \end{bmatrix}$$

Then

$$\begin{bmatrix} g_1 \\ g_2 \end{bmatrix} = \begin{bmatrix} \Gamma_1 & \Gamma_2 \end{bmatrix}^{-1} B_g$$

$$= \begin{bmatrix} 7.367 \\ -143.594 \end{bmatrix}$$

and

$$\tilde{h}_g(t) = \begin{cases} g_1 & \text{for} & 0 \le \text{rem}(t, 0.1) < 0.05 \\ g_2 & \text{for} & 0.05 \le \text{rem}(t, 0.1) < 0.1 \end{cases}$$

$\wedge\wedge\wedge$

We have thus seen that there are many alternative ways of evaluating GSHF's to achieve a specified set of zeros. We next explore other applications of GSHF's.

## 10.3   Other Applications of Generalized Sample-Hold Functions

In the previous section we saw that putting a GSHF in the forward path of a continuous-time system will assign the *sampled* system zeros. Putting the GSHF in the feedback path assigns the *sampled* systems poles. Using the combined configuration results in exact model matching for the *sampled* system. We keep emphasizing 'sampled' for an important reason which will become clear in the next section.

As is well known, the presence of non-minimum phase zeros in a system puts a constraint on the achievable gain margin (this is quite apparent via root loci considerations). Thus the ability to 'get rid of' (or hide) these zeros by GSHF's implies that one could, for example, achieve unlimited gain margins.

Furthermore, the flexibility in the choice of $\tilde{h}_g(t)$ for the GSHF as implied by (10.2.19) indicates that the same GSHF can be used for more than one purpose.

Consider $N$ linear time invariant SISO systems given via the triplets $(A_i, B_i, C_i)$ with respective state dimension $n_i$. Let $\Gamma_i$ represent the set of desired zeros for the i-th sampled system where $i$ lies in the range 1, ..., $N_1$ and let $\Lambda_i$ represent the set of desired poles for the i-th sampled system where $i$ lies in the range $N_1 + 1$, ... , $N$. Assume we wish to assign the set $\Gamma_i$ as the zeros of the sampled version of $(A_i, B_i, C_i)$, $i = 1, 2$, ..., $N_1$, by putting a GSHF in each of their forward paths. Similarly, we assign the set $\Lambda_i$ as the poles of the sampled version of $(A_i, B_i, C_i)$, $i = N_1 + 1$, ..., $N$ by putting a GSHF in each of their feedback paths.

Then we have the following result:

**Lemma 10.3.1** Let the sampling interval $\Delta$ be common to all systems. Assume all pairs $(C_i, A_\delta^i)$ are detectable. Define

$$
A = \begin{bmatrix} A_1 & & & \\ & A_2 & & \\ & & \ddots & \\ 0 & & & A_N \end{bmatrix}, \quad B = \begin{bmatrix} B_1 \\ B_2 \\ \vdots \\ B_n \end{bmatrix}
$$

Then, if $(A, B)$ is controllable, all desired assignments can be carried out with the same GSHF.

**Proof :** For $i = 1$, ..., $N_1$, let $B_g^i$ be such that $C_i(\gamma I - A_\delta^i)B_g^i$ has $\Gamma_i$ as its zeros, and for $i = N_1 + 1$, ..., $N$, let $F_g^i$ be such that $A_\delta^i + F_g^i C_i$ has $\Lambda_i$ as its eigenvalues.

As in Section 10.2

$$v_p = \frac{1}{\Delta} \int_0^{\Delta} e^{A(\Delta - \tau)} e^{jp\omega_s \tau} \, d\tau \cdot B \in \mathbb{C}^{\sum\limits_{i=1}^{N} n_i}$$

Then for $a_g[p]$ satisfying

$$\sum_{p=-\infty}^{\infty} a_g[p] v_p = \begin{bmatrix} B_g^1 \\ \cdot \\ \cdot \\ B_g^{N_1} \\ F_g^{N_1+1} \\ \cdot \\ \cdot \\ F_g^N \end{bmatrix}$$

the GSHF with

$$\tilde{h}_g(t) = \frac{1}{\Delta} \sum_{p=-\infty}^{\infty} a_g[p] e^{jp\omega_s t}$$

will satisfy all the requirements.

$\wedge\wedge\wedge$

A straightforward observation from Lemma 10.3.1 is that using two GSHF's, simultaneous sampled model matching can be achieved for any finite set of linear time invariant systems.

These results can be extended to MIMO systems.

A special generalization of model matching to MIMO system is when the matched model has a diagonal transfer matrix. Clearly, this can be done when the number of inputs equals the number of outputs and is commonly referred to in the literature as *decoupling*. The possibility of achieving decoupling for the *sampled* MIMO system is discussed in more detail in the references given at the end of the chapter.

## 10.4 Frequency Domain Analysis of GSHF

We have seen in the previous section that interesting properties can be achieved *at the sample points* using GSHF's. We will show below that the cost of achieving these properties is often poor intersample behaviour.

To simplify the frequency domain analysis of a hybrid system we use, instead of sequences, their impulse train form (as was done in Chapter 9 - see also (1.6.9)). So, the equivalent of Figure 10.2.5 is Figure 10.4.1.

From Figure 10.4.1 we note that $\tilde{h}_g(t)$ and $\tilde{h}_f(t)$ are, in fact, modulating the signals $r_1(t)$ and $y_1(t)$ to give $u(t)$, the input to the continuous-time plant. We will dwell on this point later.

We have $R^s(\omega)$, $R_1(\omega)$, $U(\omega)$, $Y(\omega)$, $Y^s(\omega)$, $Y_1(\omega)$, $\tilde{H}_f(\omega)$ and $\tilde{H}_g(\omega)$ denoting the Fourier transform of their lower case counterparts. We also have $H_0(j\omega)$ and $G(j\omega)$ denoting the frequency responses of the ZOH and the plant

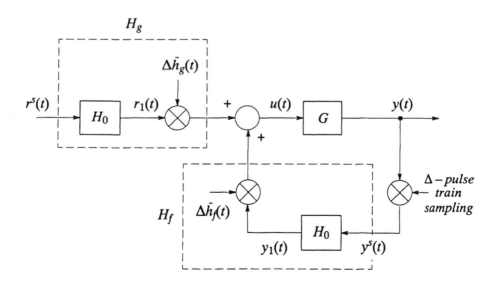

*Figure 10.4.1 : GSHF control configuration in continuous-time.*

respectively. Let $H_f(s)$ and $H_g(s)$ be the Laplace transforms of the impulse responses $h_f(t)$ and $h_g(t)$. The following continuous-time relationships are clear:

$$G(j\omega) = C(j\omega I - A)^{-1}B \qquad (10.4.1)$$

$$H_0(j\omega) = \frac{1 - e^{-j\omega\Delta}}{j\omega\Delta} \qquad (10.4.2)$$

$$Y(\omega) = G(\omega)U(\omega) \qquad (10.4.3)$$

In addition, we have the following result which describes the continuous-time response of the system in Figure 10.4.1 incorporating GSHF's. Note that this result is a generalization of Theorem 9.2.1 which applies to the case of ZOH's.

## Lemma 10.4.1

(a) The frequency response of the two generalized holds are $H_g(j\omega)$ and $H_f(j\omega)$ where $H_g(s)$ and $H_f(s)$ are the Laplace transforms of $h_g(t)$ and $h_f(t)$ respectively.

(b) The frequency response of the two generalized holds can be expressed as

$$H_g(j\omega) = \sum_{p=-\infty}^{\infty} a_g[p]H_0(j\omega_p) \qquad (10.4.4)$$

$$H_f(j\omega) = \sum_{p=-\infty}^{\infty} a_f[p]H_0(j\omega_p) \qquad (10.4.5)$$

where $\omega_p = \omega - p\omega_s$, $\{a_g[p]\}$ and $\{a_f[p]\}$ are the Fourier series coefficients of the periodic functions $\bar{h}_g(t)$ and $\bar{h}_f(t)$ respectively, and $H_0(j\omega)$ is the frequency response of the zero-order-hold.

(c) The sampled output response satisfies

$$Y^s(\omega) = H^s(\omega)R^s(\omega)$$ (10.4.6)

where

$$H^s(\omega) = \frac{[H_gG]^s}{1 - [H_fG]^s}$$ (10.4.7)

(d) The continuous output response satisfies

$$Y(\omega) = H(\omega)R^s(\omega)$$ (10.4.8)

where

$$H(\omega) = G(j\omega)H_g(j\omega) + G(j\omega)H_f(j\omega)H^s(\omega)$$ (10.4.9)

Proof :

(a) The impulse response of the zero-order-hold is $1/\Delta$ (for $0 \le t < \Delta$) and hence the impulse response of the first hold is $1/\Delta \cdot \Delta \tilde{h}_g(t)$ (for $0 \le t < \Delta$). However, this is precisely $h_g(t)$. The result follows since the frequency response is the transform of the corresponding impulse response.

(b) $\tilde{H}_g(\omega)$ and $\tilde{H}_f(\omega)$ denote the continuous Fourier transform of $\tilde{h}_g(t)$ and $\tilde{h}_f(t)$ respectively. Then using the Fourier series form (10.2.15) we have (see (1.6.6)):

$$\tilde{H}_g(\omega) = \frac{2\pi}{\Delta} \sum_{p=-\infty}^{\infty} a_g[p]\delta(\omega - p\omega_s)$$ (10.4.10)

and

$$\tilde{H}_f(\omega) = \frac{2\pi}{\Delta} \sum_{p=-\infty}^{\infty} a_f[p]\delta(\omega - p\omega_s) \tag{10.4.11}$$

Now $h_g(t)$ and $h_f(t)$ are obtained by multiplying $\Delta \tilde{h}_g(t)$ and $\Delta \tilde{h}_f(t)$ by $h_0(t)$, where $h_0(t)$ is the impulse response of the zer-o-order-hold. Equations (10.4.4), (10.4.5) follow from the con-volution property of (1.3.16) transforms.

(c) Using the convolution property (1.3.16):

$$U(\omega) = \frac{\Delta}{2\pi} \int_{-\infty}^{\infty} R_1(\eta)\tilde{H}_g(\omega - \eta)d\eta + \frac{\Delta}{2\pi} \int_{-\infty}^{\infty} Y_1(\eta)\tilde{H}_f(\omega - \eta)d\eta \tag{10.4.12}$$

$$R_1(\omega) = H_0(j\omega)R^s(\omega) \tag{10.4.13}$$

$$Y_1(\omega) = H_0(j\omega)Y^s(\omega) \tag{10.4.14}$$

Substituting (10.4.13), (10.4.14), (10.4.10), (10.4.11) in (10.4.12), and recalling that $R^s(\omega)$ and $Y^s(\omega)$ are periodic with period $\omega_s$, we get

$$U(\omega) = \sum_{p=-\infty}^{\infty} H_0(j\omega_p)\big(a_g[p]R^s(\omega) - a_f[p]Y^s(\omega)\big) \tag{10.4.15}$$

where $\omega_p = \omega - p\omega_s$, and

$$Y(\omega) = G(j\omega) \sum_{p=-\infty}^{\infty} H_0(j\omega_p)\big(a_g[p]R^s(\omega) - a_f[p]Y^s(\omega)\big) \tag{10.4.16}$$

Since, by (1.6.12)

$$Y^s(\omega) = \sum_{k=-\infty}^{\infty} Y(\omega - k\omega_s) = \sum_{k=-\infty}^{\infty} Y(\omega_k) \tag{10.4.17}$$

we substitute (10.4.16) into (10.4.17) and get

$$Y^s(\omega) = H^s(\omega)R^s(\omega) \qquad (10.4.18)$$

where

$$H^s(\omega) = \frac{\displaystyle\sum_{k=-\infty}^{\infty}\sum_{p=-\infty}^{\infty} a_g[p]H_0\big(j\omega_{p+k}\big)G(j\omega_k)}{1 - \displaystyle\sum_{k=-\infty}^{\infty}\sum_{p=-\infty}^{\infty} a_f[p]H_0\big(j\omega_{p+k}\big)G(j\omega_k)} \qquad (10.4.19)$$

Equation (10.4.6) follows using (10.4.4), (10.4.5).

(d)  Follows from (10.4.3), (10.4.15), (10.4.18).

$\wedge\!\wedge\!\wedge$

We see from the above result that $H^s(\omega)$ plays a key role in calculating the continuous response of a system incorporating GSHF's. There are several alternative ways of evaluating $H^s(\omega)$ as we next show. For example we note that

$$H^s(\omega) = \bar{S}(\gamma_\omega) \sum_{k=-\infty}^{\infty}\sum_{p=-\infty}^{\infty} a_g[p]H_0\big(j\omega_{p+k}\big)G(j\omega_k)$$

where $\bar{S}(\gamma_\omega)$ is the discrete sensitivity function discussed in Chapter 5.

Also, based on Lemmas 10.2.2 and 10.2.3 it can readily be seen that

$$H^s(\omega) = C\big(\gamma I - (A_\delta + F_g C)\big)^{-1}B_g \bigg|_{\gamma = \frac{1}{\Delta}(e^{j\omega\Delta} - 1)} \qquad (10.4.20)$$

Since

$$\big(\gamma I - (A_\delta + F_g C)\big)^{-1}$$

$$= (\gamma I - A_\delta)^{-1} + (\gamma I - A_\delta)^{-1}F_g\big(1 - C(\gamma I - A_\delta)^{-1}F_g\big)^{-1}C(\gamma I - A_\delta)^{-1}$$

we readily get

$$H^s(\omega) = \left(1 - C(\gamma I - A_\delta)^{-1}F_g\right)^{-1}C(\gamma I - A_\delta)^{-1}B_g \Bigg|_{\gamma = \frac{1}{\Delta}(e^{j\omega\Delta} - 1)}$$

$$= \frac{\Delta C\left(e^{j\omega\Delta} - e^{A\Delta}\right)^{-1}B_g}{1 - \Delta C\left(e^{j\omega\Delta} - e^{A\Delta}\right)^{-1}F_g} \tag{10.4.21}$$

Substituting (10.2.19) will result in

$$H^s(\omega) = \frac{\Delta \displaystyle\sum_{p=-\infty}^{\infty} a_g[p]C\left(e^{j\omega\Delta} - e^{A\Delta}\right)^{-1}v_p}{1 - \Delta \displaystyle\sum_{p=-\infty}^{\infty} a_f[p]C\left(e^{j\omega\Delta} - e^{A\Delta}\right)^{-1}v_p} \tag{10.4.22}$$

Comparing (10.4.19) and (10.4.22) we obtain the following alternative expression for $\{N_p\}$

$$\sum_{k=-\infty}^{\infty} H_0(j\omega_{p+k})G(j\omega_k) = \Delta C\left(e^{j\omega\Delta} - e^{A\Delta}\right)^{-1}v_p \tag{10.4.23}$$

The above analysis, and especially (10.4.4), suggests an alternative view of the GSHF as depicted in Figure 10.4.2.

Figure 10.4.2 emphasizes the modulating effect of the GSHF. We note that $z^s(t)$ (which is either $r^s(t)$ or $y^s(t)$ in the control system of Figure 10.4.1) has a periodic spectrum with period $\omega_s$. Clearly, $z_0(t)$ is the result of $z^s(t)$ passing through a ZOH, $H_0(j\omega)$. Then, $z_p(t)$; $p = \dots, -1, 0, 1, \dots,$ are in fact $z_0(t)$ modulated to the center frequencies $p\omega_s$. Each of these modulated signals is then weighted and summed. This means that the difference between using just ZOH and using GSHFs is in these *weighted modulated* components. The output

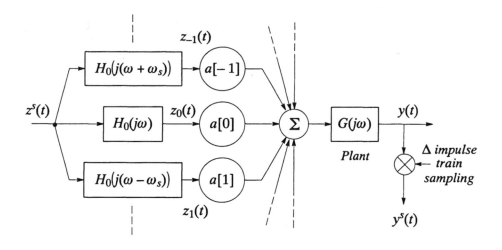

*Figure 10.4.2 : An alternative view of GSHF.*

of the GSHF is then fed through $G$. By sampling the output of $G$ we cause the modulated components to fold into the baseband.

Since, typically, $G$ will have a strictly proper transfer function and the sampling will be fast, the modulated components will go through a significant attenuation by $G$. To counteract this attenuation the values of the gains $a[p]$ will need to be large for $p \neq 0$. Thus, the signal entering $G$ will typically have high frequency components of considerably larger amplitudes than those needed in the output.

⋀⋀⋀

Let us demonstrate the above.

Example 10.4.1 :   Design a GSHF to minimize the energy in the high frequency components for the same system as in Example 9.2.2. Compare the *continuous-time* and *sampled* responses to a step input.

**Solution** Based on the arguments given above, we see that to reduce the high frequency energy it is desirable to include $p = 0$.

Repeating the calculation, $v_{-1}$ and $v_1$ remain the same and

$$v_0 = \begin{bmatrix} 0.95163 \\ 0.90635 \end{bmatrix}$$

Then equation (10.2.23) becomes

$$\begin{bmatrix} v_{-1}, & v_0, & v_1 \end{bmatrix} \begin{bmatrix} a_g[-1] \\ a_g[0] \\ a_g[1] \end{bmatrix} = \begin{bmatrix} 0 \\ 1 \\ -\dfrac{1}{3} \end{bmatrix}$$

which has infinitely many solutions. We chose the one with minimum norm which leads to

$$\begin{bmatrix} a_g[-1] \\ a_g[0] \\ a_g[1] \end{bmatrix} = \begin{bmatrix} \overline{v_{-1}}, \overline{v_0}, \overline{v_1} \end{bmatrix}^T \left\{ \begin{bmatrix} v_{-1}, v_0, v_1 \end{bmatrix} \begin{bmatrix} \overline{v_{-1}}, \overline{v_0}, \overline{v_1} \end{bmatrix}^T \right\}^{-1} \begin{bmatrix} 0 \\ 1 \\ -\dfrac{1}{3} \end{bmatrix}$$

$$= \begin{bmatrix} -0.552 + 11.548j \\ 0.368 \\ -0.552 - 11.548j \end{bmatrix}$$

So, by equation (10.2.15) we get

$$\bar{h}_g(t) = (-5.52 + 115.48j)e^{-20\pi jt} + 3.68 + (-5.52 - 115.48j)e^{20\pi jt}$$

$$= 3.68 + 231.23 \cos(20\pi t - 1.6185)$$

We now carry out two experiments in both of which we use a unit step input. In the first only ZOH is applied. Comparing the resulting continuous-time and sampled outputs as shown in Figure 10.4.3 we observe that they are quite similar. Both have a distinct undershoot typical of a system having a non-minimum phase zero.

Next we employ the GSHF designed above. In Figure 10.4.4 we show the resulting sampled and continuous-time outputs. While the sampled output be-

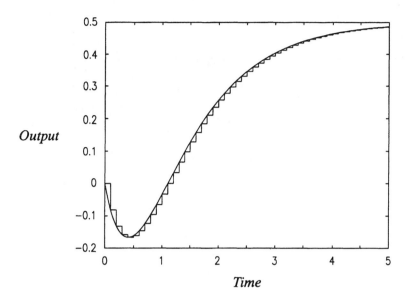

*Figure 10.4.3 : Continuous-time and sampled outputs*
*for ZOH case in Example 10.4.1.*

haves as if the effects of the non-minimum phase zero have been removed, the continuous-time output is a different story. We see clearly the presence of high frequency components : d.c. modulated to the frequency $\omega_s = 20\pi$ rad sec$^{-1}$ (at steady state) with amplitude $\sim 0.475$ .

Furthermore, we observe that the effect of the non-minimum phase zero reappears in the *continuous-time* output response (note the significant under-shoot in the continuous-time response).

We also note that the attenuation caused by $G$ at the frequency $\omega_s$ is

$$|G(j\omega_s)| = |C(j\omega_s I - A)^{-1} B| = 0.0159$$

Hence, we expect the continuous-time input to the system to have an am-plitude $0.475|G(\omega_s)|^{-1} \cong 30$ . This indeed is what we see in Figure 10.4.5.

⋀⋀⋀

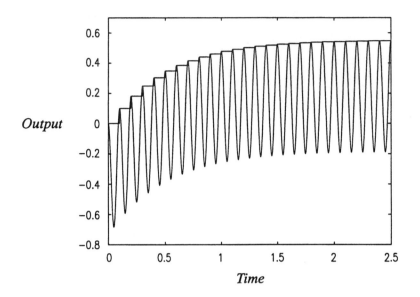

*Figure 10.4.4 : Continuous-time and sampled outputs
for GSHF case in Example 10.4.1.*

The frequency domain analysis gives insight into the mechanism of the GSHF. We see that the cost of the improved sampled output behaviour is significant high frequency components in both the continuous-time system output and input.

Moreover, since the desired frequency response of the sampled system depends on high frequency components folded into the baseband, we can imagine that the performance achieved in the system's sampled output when GSHF is used will be sensitive to system uncertainties at high frequencies.

Typically, the higher the frequency the higher the relative uncertainty. Thus, while small time delays have hardly any effect at low frequencies, they cause considerable phase shift at high frequencies. This may be an important issue for the application of GSHF's. A more in-depth treatment of this issue is carried out in the following section.

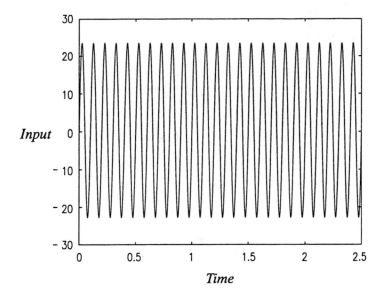

*Figure 10.4.5 : Continuous-time input for the*
*GSHF case in Example 10.4.1.*

## 10.5 Sensitivity Considerations

We have seen in Section 10.4 that the action of the generalized hold can be viewed as simply replacing the zero-order-hold transfer function by the more complex function

$$\tilde{H}_g(\omega) = \sum_{p=-\infty}^{\infty} a_g[p]H_0(j\omega_p) \qquad \omega_p = \omega - p\omega_s \qquad (10.5.1)$$

With this very simple change we can now directly use the form of analysis adopted in Section 9.2. Thus we immediately obtain from Theorem 9.2.1 that

$$Y^s(\omega) = \frac{[H_gG]^s}{1-[H_fG]^s}R^s(\omega) = H^s(\omega)R^s(\omega) \qquad (10.5.2)$$

where the notation $[\tilde{X}]^s$ was defined in Section 9.2. The corresponding continuous-time response is

$$Y(\omega) = \frac{H_g P - [H_f G]^s H_g G + H_f G [H_g G]^s}{1 - [H_f G]^s} R_s(\omega) \qquad (10.5.3)$$

As in Section 9.3 we can now find the sensitivity of the closed loop discrete transfer function to changes in $G$. By differentiation we obtain that the change, $\Delta H^s$, in $H^s$ is approximately given by:

$$\Delta H^s \simeq \frac{\sum\limits_{k=-\infty}^{\infty} H_g(j\omega_k)\Delta G(j\omega_k)}{1 - [H_f G]^s} + \frac{[H_g G]^s}{\left(1 - [H_f G]^s\right)^2} \sum\limits_{k=-\infty}^{\infty} H_f(j\omega_k)\Delta G(j\omega_k) \quad (10.5.4)$$

Hence

$$\frac{\Delta H^s}{H^s} = \frac{\sum\limits_{k=-\infty}^{\infty} H_g(j\omega_k)\Delta G(j\omega_k)}{[H_g G]^s} + \frac{1}{1 - [H_f G]^s} \sum\limits_{k=-\infty}^{\infty} H_f(j\omega_k)\Delta G(j\omega_k) \qquad (10.5.5)$$

or

$$\boxed{\begin{aligned} \frac{\Delta H^s}{H^s} &= \sum\limits_{k=-\infty}^{\infty} \frac{H_g(j\omega_k)G(j\omega_k)}{[H_g G]^s} \frac{\Delta G(j\omega_k)}{G(j\omega_k)} \\ &\quad + \bar{S} \sum\limits_{k=-\infty}^{\infty} H_f(j\omega_k)G(j\omega_k)\frac{\Delta G(j\omega_k)}{G(j\omega_k)} \end{aligned}} \qquad (10.5.6)$$

The reader will note that (10.5.6) represents a natural extension of the result in Lemma 9.3.1 for the case of a zero-order-hold.

Again noting that $|\Delta G/G|$ typically approaches unity at high frequencies, then we may conclude that high sensitivity will occur if either of the following conditions occur:

(a) $H_gG$ is large relative to $[H_gG]^s$

(b) $H_fG$ is large in a frequency range where the discrete sensitive function $\bar{S}$ is not small.

From (a) we see that if we use the generalized sample hold to shift a discrete zero to a higher frequency then $H_gG$ will typically increase before $[H_gG]^s$ does. This will necessarily lead to increased sensitivity. In particular, this implies that attempting to shift non-minimum phase zeros to more desirable locations will lead to reduced robustness margins.

Point (b) is actually very similar to point (a). Indeed, since $\bar{S} = \dfrac{1}{1-[H_fG]^s}$

then whenever $[H_fG]^s > 1$ we have $\bar{S}H_fG \simeq \dfrac{H_fG}{[H_fG]^s}$ . Thus, again, we will have

sensitivity problems if we try to use the generalized hold to reduce $[H_fG]^s$ at frequencies where $H_fG$ is large.

Again, to demonstrate the above we go back to Example 10.4.1 and introduce an extra (unmodelled) time delay of 0.02 sec. We apply the GSHF designed in the previous example and show the results in Figure 10.5.1. We can see that the GSHF now does not remove the non-minimum phase zero effect even for the sampled output.

The above mentioned sensitivity issues also apply when GSHF's are used in a feedback loop, since, in this case, the stability of the closed-loop system may be at stake.

**Example 10.5.1** : Consider again the system of Examples 10.2.1 and 10.4.1. That is, with

$$A = \begin{bmatrix} -1 & 0 \\ 0 & -2 \end{bmatrix}, \quad B = \begin{bmatrix} 1 \\ 1 \end{bmatrix}, \quad C = \begin{bmatrix} 2, & -3 \end{bmatrix}$$

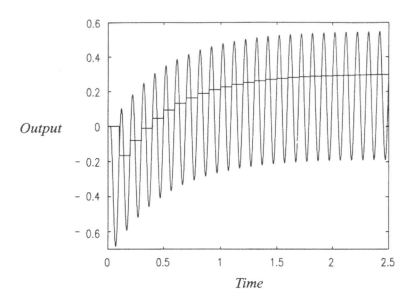

*Figure 10.5.1 : Sensitivity of GSHF to unmodelled
time delay of 0.02 sec.*

The system is sampled at $\Delta = 0.1$ with a ZOH resulting in the transfer function $\overline{G}(\gamma)$.

(a) Find the range of $K > 0$ such that a negative feedback loop around $K\overline{G}(\gamma)$ is stable – the gain margin for $\overline{G}(\gamma)$.

(b) Use the GSHF designed in Example 10.4.1 and find the subsequent improvement in the gain margin.

(c) Test the sensitivity of the result in (b) to an unmodelled time delay of 0.02 sec.

(d) Consider the sensitivity expression in (10.5.6) with $H_f = H_g$

and plot $\left| \dfrac{H_g(j\omega)KG(j\omega)}{[H_gKG]^s} \right|$.

Show that the results in (c) can be predicted from this plot.

## Solution

(a) The transfer function with ZOH is given by

$$\overline{G}(\gamma) = \frac{-0.8158(\gamma - 1.0573)}{\gamma^2 + 2.7643\gamma + 1.725}$$

So, by straightforward root-locus consideration we find that the closed loop system is stable for $0 < K < 2$.

(b) Using the GSHF we achieve the following open-loop transfer function

$$\overline{G}_{\text{GSHF}}(\gamma) = \frac{1}{\gamma + 1.8127}$$

By a simple root-locus argument we see that the closed-loop system should be stable for any $K$ in the range $0 < K < 18.2$. We chose $K = 10$ and, indeed, in Figure 10.5.2 we see that the closed -loop system is stable. We note however, that the continuous-time output displays strong non-minimum phase behaviour.

(c) Introducing an unmodelled time delay of 0.02 seconds into the system destablizes the closed-loop system with $K = 10$. This can be seen in Figure 10.5.3. The same thing happens with an unmodelled pole, say at $s = -80$.

(d) Plotting $\left| \dfrac{H_g(j\omega)KG(j\omega)}{[H_g KG]^s} \right|$ in Figure 10.5.4 we observe that

we have very high sensitivity around $\omega = 80$ rad sec$^{-1}$. At this frequency, the time delay of 0.02 seconds will result in an unmodelled phase shift of 1.6 rad ($\cong 90^o$). Thus $|\Delta G/G|$ at this frequency will be near unity. Hence, the effect on the closed-loop performance will be dramatic (as we have seen).

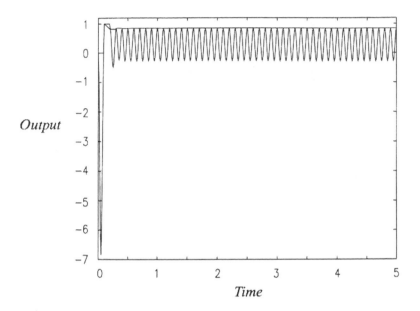

*Figure 10.5.2 : Continuous-time and sampled step*
*response for Example 10.5.1.*

## 10.6   Further Reading and Discussion

Background to the use of generalized sample holds is contained in Chammas and Leondes (1979), Kabamba (1987), Bai and Dasgupta (1989), Moore and Battacharya (1989), Yan, Anderson and Bitmead (1991). These references point to many properties which can be achieved by the use of generalized sample hold functions including decoupling, simultaneous stabilization and achieving arbitrary gain margins.

Recent work of Feuer and Goodwin (1993) and Middleton and Freudenberg (1993) has quantified the cost/benefit trade-offs associated with GSHF's. We have analyzed these issues in this chapter as well as developed new results on sensitivity to modelling errors.

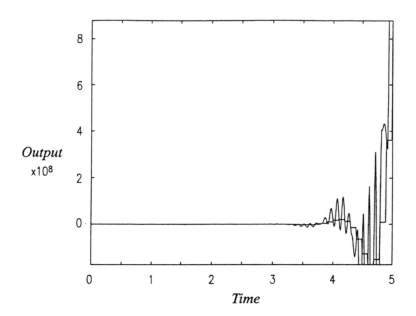

*Figure 10.5.3 : Continuous-time and sampled step response with unmodelled delay of 0.02 seconds. See Example 10.5.1.*

## 10.7 Problems

10.1 Let $r[k] = \{1, -1, 2, -1\}$ for $k = 0, 1, 2, 3$.

Draw the result of applying GSHF for the following $\tilde{h}_g(t)$ :

(a) $\quad \tilde{h}_g(t) = \dfrac{1}{\Delta}$

(b) $\quad \tilde{h}_g(t) = \dfrac{(1 + \sin \omega_s t)}{\Delta}$, where $\omega_s = \dfrac{2\pi}{\Delta}$

(c) $\quad \tilde{h}_g(t) = \dfrac{(1 + \text{rem } (t\Delta)}{\Delta}$ where $0 \le \text{rem}(t, \Delta) = t - k\Delta < \Delta$

(d) $\quad \tilde{h}_g(t) = \dfrac{(e^{\text{rem}(t,\Delta)})}{\Delta}$

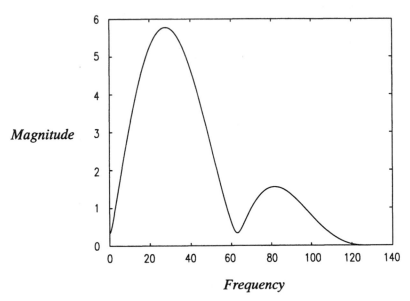

*Figure 10.5.4 : Plot of* $\left| \dfrac{\tilde{H}_g(\omega)K\tilde{G}(\omega)}{[\tilde{H}_gK\tilde{G}]^s} \right|$ *versus frequency*

*for Example 10.5.1.*

10.2  Show that $\bar{h}_g(t)$ in (10.2.9) satisfies $\bar{h}_g(t + \Delta) = \bar{h}_g(t)$ for all $t$.

10.3  Let $A = \begin{bmatrix} 0 & 1 \\ -1 & -2 \end{bmatrix}$, $\quad B = \begin{bmatrix} 0 \\ 1 \end{bmatrix}$, $\quad C = \begin{bmatrix} 1, & 1 \end{bmatrix}$.

This system is sampled with period $\Delta = 0.1$.

(a)  Calculate $A_\delta$ and verify that $(A, B)$ is controllable and $(C, A_\delta)$ detectable.

(b)  Find $B_g$ such that $C(\gamma I - A_\delta)^{-1}B_g$ will have the polynomial $5\gamma + 2$ as its numerator.

(c)  Find $\bar{h}_g(t)$ that satisfies (10.2.9) for the $B_g$ in (b).

10.4  Prove Lemma 10.2.3.

10.5  Show that out of all $\bar{h}_g(t)$ satisfying (10.2.6), the particular one defined in

(10.2.9) is of minimal energy, namely that it minimizes $\displaystyle\int_0^\Delta \bar{h}_g(t)^2 dt$ .

Find the value of the minimal energy.

10.6  Find the Fourier series expansions for Problem 10.1 (a), (b), (c), (d).

10.7  Let $(A, B)$ be a controllable pair where $A$ is singular. Show that for any $\Delta > 0$ the transfer matrix $[sI - A]^{-1}[e^{A\Delta} - I]B$ has a pole - zero cancellation at the origin.

Hint:  Assume $(A, B)$ are in the following canonical form

$$A = \begin{bmatrix} 0 & 1 & 0 & \ldots & 0 \\ 0 & 0 & 1 & \ldots & 0 \\ & \cdot & & & \\ & \cdot & & & \\ & \cdot & & & \\ 0 & 0 & 0 & \ldots & 1 \\ a_n & a_{n-1} & a_{n-2} & \cdot\cdot & a_1 \end{bmatrix} \qquad B = \begin{bmatrix} 0 \\ 0 \\ \cdot \\ \cdot \\ \cdot \\ 1 \end{bmatrix}$$

10.8  Calculate $\bar{h}_g(t)$ for Problem 10.3 using equation (10.2.23).

10.9  For the special case of a zero-order-hold, show that

$$\bar{h}_g(t) = \frac{1}{\Delta} \quad \text{for all } t$$

$$h_g(t) = \frac{1}{\Delta} \quad \text{for } 0 \le t < \Delta$$

$$a_g[p] = \begin{cases} 1 & \text{for } p = 0 \\ 0 & \text{otherwise} \end{cases}$$

Hence use (10.4.4) to confirm that $H_g(j\omega) = H_0(j\omega)$.

10.10 Show that the sequence $\{a_g[p]\}$ in (10.2.16) consists of the sample values (at spacing $2\pi/\Delta$) of the transform of $h_g(t)$. Hence, for the special case of a zero-order-hold, where $h_g(t) = 1/\Delta$ for $0 \le t < \Delta$, show that $a_g[p] = 1$ for $p = 0$ and $0$ otherwise.

# Chapter 11

# Periodic Control of Linear Time-Invariant Systems

## 11.1 Introduction

A question that has intrigued many engineers (and indeed is still the subject of ongoing research) is whether one could benefit from using time-varying linear controllers, or a more tractable sub-family of periodically time-varying digital controllers on systems which are themselves time-invariant. Similar questions have been raised with regard to continuous-time linear controllers. It has been shown that periodic controllershave interesting properties including improved gain and/or phase margins relative to linear time-invariant control. Usually this is accomplished via zero placements. We have seen in Chapter 8 that non minimum phase zeros limit the gain and phase margin. Hence, if the use of periodic control results in shifting the zeros to a harmless place, the constraints on the gain/phase margins can be mitigated. Tools for achieving this will be developed in this chapter.

However, similar to the results of Chapter 10, we will show that the advantages of periodic control are matched by other features which may be undesirable in some applications.

We will focus attention on SISO systems but most of the ideas and results carry over to the MIMO case. Also, note that for convenience we will principally use the shift operator with associated $z$-transform (see Section 3.2) to describe time-invariant discrete-time systems associated with periodic systems. We will use the notation $\bar{A}(q)$, $\bar{G}(q)$, $\bar{G}(z)$ to denote, respectively, a polynomial in $q$, a rational function of $q$, and a transfer function in $z$.

## 11.2  Periodic Control of Linear Time-Invariant Systems

Next we turn to the problem of designing periodic controllers. Here we consider their application to linear time-invariant systems. Later, in Chapter 13, we will take up their application to periodic systems.

The system we deal with in this section is presented in Figure 11.3.1. The continuous time plant, $P$, is modelled as a linear *time-invariant* system:

$$\dot{x}(t) = Ax(t) + Bu(t) \tag{11.2.1}$$

$$y(t) = Cx(t) + Du(t) \tag{11.2.2}$$

For simplicity, we assume a SISO system even though most of the results extend readily to MIMO systems. In Figure 11.3.1, $H_0$ is a ZOH and $C_d$ is a discrete-time *periodic* controller given in state-space form as in (11.2.1), (11.2.2).

Let $n$, $n_c$ be the state-space dimensions of the plant and controller respectively.

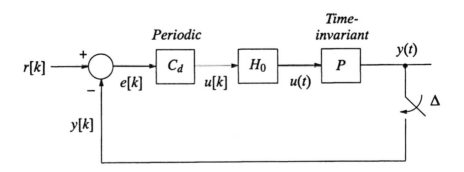

Figure 11.2.1 : Configuration for use of periodic control.

Since in this section we mostly concentrate on the relationship between $r[k]$ and $y[k]$ it will be convenient to consider the discretized version of $P$ including the ZOH and sampler. Using the results of Chapter 4 the corresponding discrete-time model is

$$x[k+1] = A_s x[k] + B_s u[k] \qquad (11.2.3)$$

$$y[k] = C_s x[k] + D_s u[k] \qquad (11.2.4)$$

where $x[k] = x(k\Delta)$, $u[k] = u(k\Delta)$, $y[k] = y(k\Delta)$ and

$$A_s = e^{A\Delta}, \quad B_s = \int_0^\Delta e^{A\sigma} d\sigma \cdot B, \quad C_s = C, \quad D_s = D \qquad (11.2.5)$$

Note that we have included the possibility of a non strictly proper system $(D \neq 0)$ since these systems, in particular, can benefit from the use of periodic control.

In Lemma 7.3.1 we have seen how a linear periodic discrete system can be converted into a linear time-invariant system. Next we make the trivial observation that any linear time-invariant system can also be viewed as periodic with arbitrary period. Considering now the discrete equations for the sampled plant model, (11.2.3) – (11.2.4), one can define $y_R[m]$ and $u_R[m]$ in a similar fashion to $e_R[m]$ in (7.3.5). Then, following the procedure described in Lemma 7.3.1 we can readily derive the relationships (see Problem 11.2):

$$x_R[m+1] = A_R x_R[m] + B_R u_R[m] \qquad (11.2.6)$$

$$y_R[m] = C_R x_R[m] + D_R u_R[m] \qquad (11.2.7)$$

where

$$x_R[m] = x[mN] \tag{11.2.8}$$

$$A_R = (A_s)^N \tag{11.2.9}$$

$$B_R = \left[ (A_s)^{N-1}B_s, \ (A_s)^{N-2}B_s, \ \dots, \ A_sB_s, \ B_s \right] \tag{11.2.10}$$

$$C_R = \begin{bmatrix} C_s \\ C_sA_s \\ \cdot \\ \cdot \\ \cdot \\ C_s(A_s)^{N-1} \end{bmatrix} \tag{11.2.11}$$

$$D_R = \begin{bmatrix} D_s & \diagdown & & 0 & 0 \\ C_sB_s & \diagdown & \diagdown & \cdot & 0 \\ \cdot & & \diagdown & \cdot & \cdot \\ \cdot & & & \diagdown & \cdot \\ \cdot & & & D_s & \cdot \\ C_s(A_s)^{N-2}B_s & & & C_sB_s & D_s \end{bmatrix} \tag{11.2.12}$$

For the case of the time-invariant plant, it is possible to develop further relationships applicable to the raised form. This is shown in the following result.

**Lemma 11.2.1** Let $(A_s, B_s, C_s, D_s)$ be a minimal realization of the discrete time-invariant plant transfer function $\bar{P}_d(z)$, and let $(A_R, B_R, C_R, D_R)$ be the corresponding $N$-raised form of this time-invariant system, as defined in (11.2.9) – (11.2.12).

Define

$$\bar{P}_1(z) = C_s(zI - A_s^N)^{-1}A_s^{N-1}B_s + D_s \tag{11.2.13}$$

and

$$\tilde{P}_i(z) = C_s A_s^{i-1}(zI - A_s^N)^{-1}A_s^{N-1}B_s + C_s A_s^{i-2}B_s \quad i = 2, \ldots, N \quad (11.2.14)$$

Then

(a)  the original transfer-function can be expressed as

$$\tilde{P}_d(z) = \tilde{P}_1(z^N) + z^{-1}\tilde{P}_2(z^N) + \ldots + z^{-N+1}\tilde{P}_N(z^N) \quad (11.2.15)$$

(b)  the raised transfer-function can be expressed as

$$\tilde{P}_R(z) = C_R(zI - A_R)^{-1}B_R + D_R$$

$$= \begin{bmatrix} \tilde{P}_1(z) & z^{-1}\tilde{P}_N(z) & - & - & - & z^{-1}\tilde{P}_2(z) \\ \tilde{P}_2(z) & \tilde{P}_1(z) & - & - & - & z^{-1}\tilde{P}_3(z) \\ \cdot & \cdot & & & & \cdot \\ \cdot & \cdot & & & & \cdot \\ \cdot & \cdot & & & & \cdot \\ \tilde{P}_N(z) & \tilde{P}_{N-1}(z) & - & - & - & \tilde{P}_1(z) \end{bmatrix} \quad (11.2.16)$$

Proof :

(a)  Substituting (11.2.13) and (11.2.14) into the right hand side of equation (11.2.15) gives

$$\tilde{P}_1(z^N) + z^{-1}\tilde{P}_2(z^N) + \cdots + z^{-N+1}\tilde{P}_N(z^N)$$

$$= D_s + C_s(z^N I - A_s^N)^{-1}A_s^{N-1}B_s$$

$$+ z^{-1}C_s A_s(z^N I - A_s^N)^{-1}A_s^{N-1}B_s + z^{-1}C_s B_s$$

$$+ \cdots + z^{-N+1}C_s A_s^{N-1}(z^N I - A_s^N)^{-1}A_s^{N-1}B_s$$

$$+ z^{-N+1}C_s A_s^{N-2}B_s$$

$$= D_s + C_s\left(I + z^{-1}A_s + \cdots + z^{-N+1}A_s^{N-1}\right)(z^N I - A_s^N)^{-1}A_s^{N-1}B_s$$

$$+ C_s z^{-1}\left(I + z^{-1}A_s^{-1} + \cdots + z^{-N+2}A_s^{N-2}\right)B_s$$

$$= D_s + C_s(I - z^{-1}A_s)^{-1}(I - z^{-N}A_s^N)(z^N I - A_s^N)^{-1}A_s^{N-1}B_s$$

$$+ z^{-1}C_s(I - z^{-1}A_s)^{-1}(I - z^{-N+1}A_s^{N-1})B_s$$

$$= D_s + C_s(I - z^{-1}A_s)^{-1}\left(z^{-N}A_s^{N-1} + z^{-1}I - z^{-N}A_s^{N-1}\right)B_s$$

$$= D_s + C_s(zI - A_s)^{-1}B_s$$

$$= \tilde{P}_d(z)$$

(b) Substituting $(A_s, B_R, C_R, D_R)$ from (11.2.9) – (11.2.12) into (11.2.16) we obtain

$$\tilde{P}_R(z) = \begin{bmatrix} C_s \\ C_s A_s \\ \cdot \\ \cdot \\ \cdot \\ C_s A_s^{N-1} \end{bmatrix} (zI - A_s^N)^{-1} \left[ A_s^{N-1}B_s, A_s^{N-2}B_s, \; \ldots, \; B_s \right]$$

$$+ \begin{bmatrix} D_s & \ddots & & & 0 & 0 \\ C_s B_s & \ddots & & & \cdot & 0 \\ \cdot & & \ddots & & \cdot & \cdot \\ \cdot & & & \ddots & \cdot & \cdot \\ \cdot & & & & D_s & \cdot \\ C_s(A_s)^{N-2}B_s & & & & C_s B_s & D_s \end{bmatrix}$$

Namely

$$\left[\tilde{P}_R(z)\right]_{i,j} = C_s A_s^{i-1}\left(zI - A_s^{N-j}\right)B_s + L_{i,j}$$

where

$$L_{i,j} = \begin{cases} 0 & \text{for } i < j \\ D_s & \text{for } i = j \\ C_s A^{i-j-1}B_s & \text{for } i > j \end{cases}$$

Then for $i > j$

$$[\tilde{P}_R(z)]_{i,j} = C_s A_s^{i-j}(zI - A_s^N)^{-1} A_s^{N-1} B_s + C_s A^{i-j-1} B_s$$

$$= \tilde{P}_{i-j+1}(z)$$

for $i = j$

$$[\tilde{P}_R(z)]_{i,j} = C_s(zI - A_s^N)^{-1} A_s^{N-1} B_s + D_s$$

$$= \tilde{P}_1(z)$$

and for $i < j$

$$[\tilde{P}_R(z)]_{i,j} = C_s(zI - A_s^N)^{-1} A_s^{N-j+1} B_s$$

$$= z^{-1} \left[ C_s A_s^{N-j+i+1}(zI - A_s^N)^{-1} A_s^{N-1} B_s + C_s A_s^{N-j+i-2} \right]$$

$$= z^{-1} \tilde{P}_{N-j+i+1}(z)$$

which establishes (b).

⋀⋀⋀

Illustrating the above ideas, the periodic controller scheme of Figure 11.3.2 (a) is changed into the linear time-invariant scheme shown in Figure 11.3.2 (b).

The procedure described so far can be a very powerful tool - it can enable the design of a periodic controller using the well studied methods of linear time-invariant controller design in the raised system. To complete this line of argument we need to convince ourselves that the linear time-invariant controller designed in the raised form is equivalent to, and hence can be implemented as, a periodic controller. This is ensured by Lemma 7.3.2.

We next show that quite remarkable properties can arise from the use of periodic control for certain types of problems for example, in the following example, we show that a single periodic controller can stabilize a plant having unknown sign for the gain.

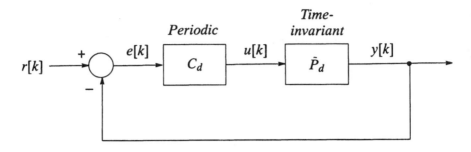

(a) Discrete linear periodic system.

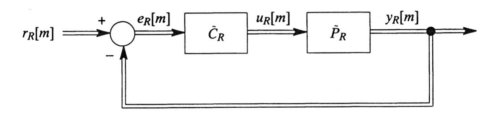

(b) Equivalent raised linear time-invariant system.

Figure 11.2.2

**Example 11.3.1 :** Consider the configuration in Figure 11.3.1 where $P(s) = K/s$, and the sampling interval is $\Delta$. The controller consists of a periodic gain only, that is $u[k] = (-1)^k e[k]$. We want to stabilize this system when $K$ is unknown.

**Solution** The sampled version of the plant is

$$\tilde{P}_d(z) = \frac{K\Delta}{z-1}$$

To obtain the raised form of $\tilde{P}_d(z)$ we note that $\tilde{P}_1(z) = \dfrac{K\Delta}{z-1}$, and

$\tilde{P}_2(z) = \dfrac{zK\Delta}{z-1}$. Hence (see equations (11.2.15), (11.2.16))

$$\tilde{P}_R(z) \;=\; \begin{bmatrix} \dfrac{K\Delta}{z-1} & \dfrac{K\Delta}{z-1} \\[2ex] \dfrac{zK\Delta}{z-1} & \dfrac{K\Delta}{z-1} \end{bmatrix}$$

The raised form of the controller is obtained noting that $F[k] = G[k]$
$= J[k] = 0$, $K[0] = 1$, $K[1] = -1$. Hence

$$\tilde{C}_R(z) \;=\; \begin{bmatrix} 1 & 0 \\ 0 & -1 \end{bmatrix}$$

Then the closed-loop raised transfer matrix is

$$\left[I + \tilde{P}_R\tilde{C}_R\right]^{-1}\tilde{P}_R\tilde{C}_R \;=\; \frac{1}{z^2 + \left[(K\Delta)^2 - 2\right]z + \left[1 - (K\Delta)^2\right]}$$

$$\cdot \begin{bmatrix} K\Delta(K\Delta + 1)(z-1) & -K\Delta(z-1) \\[2ex] z{\cdot}K\Delta(z-1) & K\Delta(K\Delta - 1)(z-1) \end{bmatrix}$$

To guarantee stability we note from the denominator that we must have

$$(K\Delta)^2 < \frac{3}{2}$$

Of particular interest is the fact that this condition does *not depend on the sign of K* and hence the one controller can stabilize the system irrespective of the sign of the high frequency gain. This is quite remarkable since it is clear that no linear time-invariant controller can stabilize the given system for both positive *and* negative K.

The system's response to a unit step input has been simulated for $|K| = 1$, $\Delta = 1$. The results are shown in Figure 11.3.3 for both $K = 1$ and $K = -1$. In both cases the system is clearly stable.

△△△

## 11.3   Time Domain Analysis

Armed with this powerful tool – raising of periodic systems – we are ready to explore further the question of what one can gain by using a periodic controller rather than a linear time-invariant one to control a linear *time-invariant* system.

We have seen in Chapter 8 that the presence of non minimum-phase zeros in a system sets some fundamental constraints on achievable performance using linear time-invariant controllers.

Hence, there is strong motivation to shift these 'nasty' zeros. The GSHF idea discussed in the previous chapter is one strategy while the use of periodic controllers, as we will see in the sequel, is another. We show below that, indeed, periodic control can shift zeros and hence improve the gain/phase margins. More

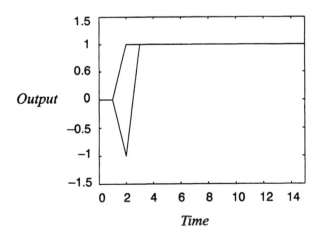

*Figure 11.2.3 : Step responses for K ± 1 in Example 10.3.1.*

specifically, in the configuration of Figure 11.4.1, say we design a controller to stabilize the nominal system $\tilde{P}_0$. The issue of gain margin is then how robust this design is to uncertainties in the gain of $\tilde{P}_0$. Thus, if we let the true system be $K\tilde{P}_0$ the question is, "For what range of values around $K = 1$ will the control system remain stable?"

For real $K$ we find that for some range $K \in (a, b)$ the system will remain stable. The ratio $b/a$ is commonly referred to as the gain margin. Similarly, if $K = e^{j\theta}$, the angle range of $\theta$ is the phase margin. The presence of non-mini-mum-phase zeros puts constraints on both margins.

**Example 11.4.1** : Consider the discrete system

$$\tilde{P}_0(z) = \frac{z-2}{z-3}$$

Suppose our control is just a proportional feedback controller with gain $-3/2$. What are the gain and phase margins for this controller?

**Solution** Say that instead of $\tilde{P}_0(z)$ we have in fact $K\tilde{P}_0(z)$. Then, the closed loop denominator will be

$$z - 3 - \frac{3}{2}K(z-2) = (1 - \frac{3}{2}K)z + 3(K-1)$$

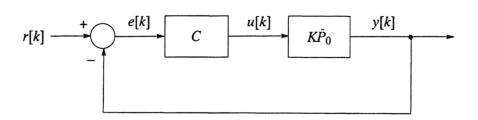

*Figure 11.3.1 : Use of complex gain K to measure gain and phase margins.*

and the question is for which values of $K$ we will have

$$-1 < \frac{3(K-1)}{1-\frac{3}{2}K} < 1$$

It can readily be shown that the above restriction holds for $K \in (8/9, 4/3)$ so that the gain margin is $3/2$. In fact, it is independent of our choice of stabilizing feedback gain for the nominal system.

It has actually been shown that the gain margin for *any* linear time-invariant controller used on the above system is less than 2.25 (see e.g. Khargonekar *et. al* (1985)).

It is quite clear that the Nyquist plot of $\frac{3}{2} \cdot \frac{2-z}{z-3}$ is a circle centered at $-15/16$ with radius $3/16$. It encircles the $-1$ point as it should. The phase margin is symmetric and is found to be $\theta \in (-10.5°, 10.5°)$.

The small margins as calculated above can be attributed to the non-minimum phase zero and unstable pole in this system.

∿∿

We next show how the use of a periodic controller can affect the discrete zeros when viewed with a sufficiently large sampling period. Hence, consider the configuration in Figure 11.4.2 where $\tilde{P}_d$ is the discretized system, $\tilde{F}_d$ is a linear time-invariant prefilter to be designed and $\tilde{F}$ is a filler operation as described in Section 1.7.1. The output, $y[k]$, is sampled so that

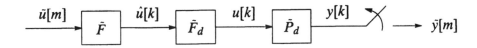

*Figure 11.3.2 : Zero placement configuration.*

$$\tilde{y}[m] = y[mN] \tag{11.3.1}$$

and (using the filler operations):

$$\hat{u}[k] = \begin{cases} \bar{u} \cdot \dfrac{k}{N} & \text{when } N \text{ divides } k \\ 0 & \text{otherwise} \end{cases} \tag{11.3.2}$$

and $N$ is an integer to be chosen.

We will proceed in a number of steps:

*Step 1: Choice of N*

We will assume in the sequel that $A_s$ (the discrete '$A$' matrix linking $u[k]$ to $y[k]$) is nonsingular. This assumption simplifies the discussion and will be true if $A_s$ results from sampling a continuous system.

With this assumption we choose $N \geq n$ (recall that $n$ is the order of $\tilde{P}_d(z)$) so that the following condition is satisfied: if $\lambda$ is an eigenvalue of $A_s$ the $N-1$ values $\lambda e^{j2\pi\frac{i}{N}}$, $i = 1, 2, \ldots, N-1$ are not eigenvalues of $A_s$. A value of $N$ having this property like this *always* exists.

*Step 2: Design of $\tilde{F}_d$*

The design of $\tilde{F}_d$ depends on our desired choice for the zeros of the transfer function from $\bar{u}[m]$ to $\tilde{y}[m]$. We establish the following result concerning zero assignment:

Lemma 11.3.1 Given the $n$-th order linear time-invariant system $\tilde{P}_d$ with no poles at the origin ($A_s$ nonsingular), an integer $N$ chosen according to Step 1, and a set $\{z_i\}$ of $n$ desired zeros when $\tilde{P}_d$ is proper (or $n-1$ zeros when $\tilde{P}_d$ is strictly proper), then there exists a prefilter $\tilde{F}_d(z)$ so that the transfer function from $\bar{u}[m]$ to $\tilde{y}[m]$ has the set $\{z_i\}$ as its zeros.

Proof :  We prove the Lemma by constructing the desired $\tilde{F}_d$. We have shown in Lemma 11.2.1 that $\tilde{P}_d(z)$ can be (uniquely) written as

$$\tilde{P}_d(z) = \tilde{P}_1(z^N) + z^{-1}\tilde{P}_2(z^N) + \ \ldots \ + z^{1-N}\tilde{P}_n(z^N) \tag{11.3.3}$$

with $(A_s, B_s, C_s, D_s)$ the minimal realization of $\tilde{P}_d(z)$ and

$$\tilde{P}_1(z) = C_s(zI - A_s^N)A_s^{N-1}B_s + D_s \tag{11.3.4}$$

$$\tilde{P}_i(z) = C_s A_s^{i-1}(zI - A_s^N)A_s^{N-1}B_s + C_s A_s^{i-2}B_s \quad i = 2, \ 3, \ \ldots, \ N \tag{11.3.5}$$

in which case the raised transfer matrix is (see Lemma 11.2.1)

$$\tilde{P}_R(z) \ = \ \begin{bmatrix} \tilde{P}_1(z) & z^{-1}\tilde{P}_N(z) & - & - & - & z^{-1}\tilde{P}_2(z) \\ \tilde{P}_2(z) & \tilde{P}_1(z) & - & - & - & z^{-1}\tilde{P}_3(z) \\ \cdot & \cdot & & & & \cdot \\ \cdot & \cdot & & & & \cdot \\ \cdot & \cdot & & & & \cdot \\ \tilde{P}_N(z) & \tilde{P}_{N-1}(z) & - & - & - & \tilde{P}_1(z) \end{bmatrix} \tag{11.3.6}$$

Similarly, for the prefilter $\tilde{F}_d(z)$ yet to be designed we can also write the raised form

$$\tilde{F}_R(z) \ = \ \begin{bmatrix} \tilde{F}_1(z) & z^{-1}\tilde{F}_N(z) & - & - & - & z^{-1}\tilde{F}_2(z) \\ \tilde{F}_2(z) & \tilde{F}_1(z) & - & - & - & z^{-1}\tilde{F}_3(z) \\ \cdot & \cdot & & & & \cdot \\ \cdot & \cdot & & & & \cdot \\ \cdot & \cdot & & & & \cdot \\ \tilde{F}_N(z) & \tilde{F}_{N-1}(z) & - & - & - & \tilde{F}_1(z) \end{bmatrix} \tag{11.3.7}$$

Clearly, the raised form of $\tilde{P}_d(z)\tilde{F}_d(z)$ is $\tilde{P}_R(z)\tilde{F}_R(z)$ which is the transfer matrix from the vector $\left[ \hat{u}[mN], \ \ldots, \ \hat{u}[(m+1)N-1] \right]^T$ to the vector $\left[ y[mN], \right.$

$\ldots, \ \left. y[(m+1)N-1] \right]^T$. Hence, from equations (11.3.1) – (11.3.2), the transfer function $\tilde{G}(z)$ from $\tilde{u}[m]$ to $\tilde{y}[m]$ is simply the (1,1) th entry of $\tilde{P}_R(z)\tilde{F}_R(z)$. That is,

$$\tilde{G}(z) = \tilde{P}_1(z)\tilde{F}_1(z) + z^{-1} \sum_{i=2}^{N} \tilde{P}_{N+2-i}(z)\tilde{F}_i(z) \tag{11.3.8}$$

Substituting (11.3.8) – (11.3.9) we get

$$\tilde{G}(z) = \tilde{F}_1(z)\left(C_s(zI - A_s^N)^{-1}A_s^{N-1}B_s + D_s\right)$$

$$+ z^{-1} \sum_{i=2}^{N} \tilde{F}_i(z)\left(C_s A_s^{N+1-i}(zI - A_s^N)^{-1}A_s^{N-1}B_s + C_s A_s^{N-i}B_s\right) \tag{11.3.9}$$

$$= \tilde{F}_i(z)\left(C_s(zI - A_s^{N-1})^{-1}A_s^{N-1}B_s + D_s\right)$$

$$+ z^{-1} \sum_{i=2}^{N} \tilde{F}_i(z)C_s A^{N-i}(zI - A_s^N)^{-1}\left(A_s^N + zI - A_s^N\right)B_s \tag{11.3.10}$$

$$= \tilde{F}_1(z)D_s + \left[\sum_{i=1}^{N} \tilde{F}_i(z)C_s A_s^{N-i}\right](zI - A_s^N)^{-1}B \tag{11.3.11}$$

Next we will argue that since the pairs $(A_s, B_s)$ and $(C_s, A_s)$ are reachable and detectable respectively, so are the corresponding pairs $(A_s^N, B_s)$ and $(C_s, A_s^N)$.

First note that by our choice of $N$ in Step 1, if $\lambda$ is an eigenvalue of $A_s$, then $\lambda e^{j2\pi \frac{i}{N}}$, $i = 1, 2, \ldots, N-1$ are not. The matrices, $\left[\lambda e^{j2\pi \frac{i}{N}}I - A_s\right]$ for $i = 1, 2, \ldots, N-1$, are all nonsingular. Also, if $\{\lambda_l\}$ are the eigenvalues of $A_s$, then $\{\lambda_l^N\}$ are the eigenvalues of $A_s^N$. So, if $(A_s^N, B_s)$ is not reachable the PBH rule for reachability implies that there exists a vector $v \neq 0$ such that

$$v^T(\lambda^N I - A_s^N) = 0 \quad \text{and} \quad v^T B_s = 0$$

Then

$$v^T(\lambda I - A_s) \prod_{i=1}^{N-1} \left(\lambda e^{j2\pi\frac{i}{N}} I - A_s\right) = 0$$

or, by the nonsingularity of the product,

$$v^T(\lambda I - A_s) = 0 \tag{11.3.12}$$

which together with $v^T B = 0$ and the PBH rule means that $(A_s, B_s)$ is not reachable either – a contradiction. The detectability of $(C_s, A_s^N)$ is similarly established.

It follows now from linear system theory that *constants* $F_i$ ($i = 1, \ldots, N \geq n$) can be chosen so that

$$\left[ \sum_{i=1}^{N} F_i C_s A_s^{N-i} \right] \left(zI - A_s^N\right)^{-1} B$$

has any desired $(n-1)$th order numerator. If $D_s = 0$ ($\tilde{P}_d$ strictly proper) we simply choose the numerator to be $\prod_{k=1}^{n-1}(z - z_i)$ with any desired gain. If $D_s \neq 0$ we choose it to be

$$F_1 D_s \left[ \prod_{i=1}^{n}(z - z_i) - \det\left(zI - A_s^N\right) \right]$$

In either case $\tilde{G}(z)$ in (11.3.4) will have the desired zeros. With the $N$ constants $\{F_i\}$ determined, $\tilde{F}_d(z)$ is calculated as in (11.2.15)

$$\tilde{F}_d(z) = \sum_{i=1}^{N} z^{1-i} \tilde{F}_i \tag{11.3.13}$$

and turns out to be a finite impulse response (FIR) prefilter.

This completes the proof of the Lemma.

$\triangle\!\triangle\!\triangle$

**Example 11.4.2 :** Consider again the system of Example 11.4.1:

$$\tilde{P}_d(z) = \frac{z-2}{z-3}$$

with $N = 2$. Say we wish to achieve a desired zero of $z_1 = 0.5$. Find $\tilde{F}_d(z)$.

**Solution** Since for $N = 2$

$$\tilde{P}_d(z) = \tilde{P}_1(z^2) + z^{-1}\tilde{P}_2(z^2)$$

where

$$\tilde{P}_1(z) = \frac{z-6}{z-9}, \quad \tilde{P}_2(z) = \frac{z}{z-9}$$

the transfer function from $\bar{e}[m]$ to $\bar{y}[m]$ will be

$$\frac{\tilde{F}_1(z)(z-6) + \tilde{F}_2(z)}{z-9} = \frac{z-0.5}{z-9}$$

so that

$$\tilde{F}_1(z) = 1, \quad \tilde{F}_2(z) = 5.5$$

Hence

$$\tilde{F}_d(z) = \tilde{F}_1(z^2) + z^{-1}\tilde{F}_2(z^2)$$

$$= 1 + 5.5z^{-1}$$

$\triangle\!\triangle\!\triangle$

Lemma 11.3.1 provides a way of placing the zeros of a system at arbitrary locations. It is important to note that the zeros in question are only as *perceived* at the slow sampling rate - a point we will get back to later. However, once any nasty zeros are out of the way one can design a controller $\tilde{C}_s$ to stabilize the system with the newly placed zeros. The resulting configuration will then be as in Figure 11.4.3.

Clearly $\tilde{C}_s$ is designed for the slow rate and is linear time-invariant for that rate. However, we can readily prove this guarantees the stability of the fast sampled part as shown by the following.

Let $(\hat{A},\hat{B},\hat{C},\hat{D})$ and $(\tilde{A},\tilde{B},\tilde{C},\tilde{D})$ be the minimal realizations of $\hat{P}_d(z)$, $\tilde{F}_d(z)$ and $\tilde{C}_s(z)$ respectively so that $\hat{D}\hat{D} \neq -1$ (this is to avoid infinite gain in the loop). We have

$$\hat{x}[k+1] = \hat{A}\hat{x}[k] + \hat{B}\hat{u}[k] \tag{11.3.14}$$

$$y[k] = \hat{C}\hat{x}[k] + \hat{D}\hat{u}[k] \tag{11.3.15}$$

and

$$\tilde{x}_s[m+1] = \tilde{A}\tilde{x}_s[m] + \tilde{B}\tilde{e}[m] \tag{11.3.16}$$

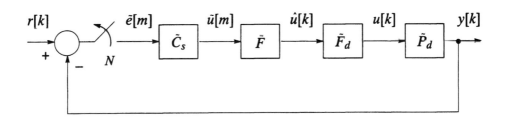

*Figure 11.3.3 : Final controller for Example 11.4.2 achieving zero placement and desired properties at the slow sampling period.*

$$\tilde{u}[m] = \tilde{C}\tilde{x}_s[m] + \tilde{D}\tilde{e}[m] \tag{11.3.17}$$

The following result then establishes the stability of the fast sampled system:

**Lemma 11.3.2** Consider the system in Figure 11.4.3 and (11.3.14) - (11.3.17). Then, if $\lim\limits_{m \to \infty} \hat{x}[mN] = 0$ and $\lim\limits_{m \to \infty} \tilde{x}_s[m] = 0$ we also have $\lim\limits_{k \to \infty} x[k] = 0$.

**Proof :** From (11.3.1) – (11.3.2) and (11.3.14) – (11.3.17) we have

$$\hat{x}[k] = \hat{A}^{(k-mN)}\hat{x}[mN] + \hat{A}^{(k-mN-1)}\hat{B}\tilde{u}[m] \quad \text{for} \quad mN < k \le (m+1)N$$

and

$$\tilde{u}[m] = \frac{1}{1 + \tilde{D}\hat{D}}\left[\tilde{C}\tilde{x}_s[m] - \tilde{D}\hat{C}\hat{x}[mN]\right]$$

So that if $\lim\limits_{m \to \infty} \hat{x}[mN] = 0$ and $\lim\limits_{m \to \infty} \tilde{x}_s[m] = 0$ it follows that $\lim\limits_{m \to \infty} \tilde{u}[m] = 0$ and $\lim\limits_{k \to \infty} \hat{x}[k] = 0$.

$\triangle\triangle\triangle$

We redraw Figure 11.4.3 as Figure 11.4.4 to highlight the effective controller.

The controller we see in Figure 11.4.4 consists of the sampler, the (slow) controller $\tilde{C}_s$, the filler $\tilde{F}$ and the fast prefilter $\tilde{F}_d$. Let us examine the equations involved in this controller's action. As a first step let us include in equations (11.3.16) – (11.3.17) the sampler and the filler. It can readily be seen that with the above we have the following state-space model linking $e[k]$ and $\hat{u}[k]$ in Figure 11.4.4:

$$x_s[k+1] = A_s[k]x_s[k] + B_s[k]e[k] \tag{11.3.18}$$

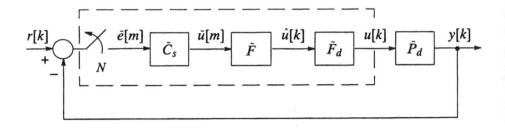

*Figure 11.3.4 : Complete controller design configuration.*

$$\hat{u}[k] = C_s[k]x_s[k] + D_s[k]e[k] \tag{11.3.19}$$

where

$$\left(A_s[k],\ B_s[k],\ C_s[k],\ D_s[k]\right) = \begin{cases} \left(\tilde{A},\ \tilde{B},\ \tilde{C},\ \tilde{D}\right) & \text{for } k = mN \\ \left(I,\ 0,\ 0,\ 0\right) & \text{otherwise} \end{cases} \tag{11.3.20}$$

So if we let

$$x_f[k+1] = A_f x_f[k] + B_f \hat{u}[k]$$

$$u[k] = C_f x_f[k] + D_f \hat{u}[k]$$

be a minimal realization of $\tilde{F}_d$, the controller equations have the form of (7.3.1) – (7.3.2), namely,

$$x_c[k+1] = F[k]x_c[k] + G[k]e[k]$$
$$u[k] = J[k]x_c[k] + K[k]e[k] \tag{11.3.21}$$

where

$$F[k] = \begin{bmatrix} A_s[k] & 0 \\ B_f C_s[k] & A_f \end{bmatrix} \qquad G[k] = \begin{bmatrix} B_s[k] \\ B_f D_s[k] \end{bmatrix} \tag{11.3.22}$$

$$J[k] = \left[D_f C_s[k],\ C_f\right] \qquad K[k] = D_f D_s[k] \tag{11.3.23}$$

Clearly, the controller we have ended up with has periodicity $N$.

**Example 11.4.3 :** Continue Example 11.4.2 to design a periodic controller with improved gain margin.

**Solution** For the (slow) sampled system with prefilter we have so far achieved the transfer function (see the solution of Example 11.4.2)

$$\frac{z - 0.5}{z - 9}$$

So if we take

$$\tilde{C}_s(z) = 40$$

the closed loop system will be stable.

Now, since (see Example 11.4.2)

$$\tilde{F}_d(z) = 1 + 5.5z^{-1}$$

we have

$$\left(A_f,\ B_f,\ C_f,\ D_f\right) = \left(0,\ 5.5,\ 1,\ 1\right)$$

and for the above

$$\left(\tilde{A},\ \tilde{B},\ \tilde{C},\ D_s\right) = \left(0,\ 0,\ 0,\ 40\right)$$

so, from (11.3.22) – (11.3.23)

$$F[k] \;=\; \begin{cases} \begin{bmatrix} 0 & 0 \\ 0 & 0 \end{bmatrix} & \text{for } k \text{ even} \\[4mm] \begin{bmatrix} 1 & 0 \\ 0 & 0 \end{bmatrix} & \text{for } k \text{ odd} \end{cases}$$

$$
G[k] = \begin{cases} \begin{bmatrix} 0 \\ 220 \end{bmatrix} & \text{for } k \text{ even} \\ \\ \begin{bmatrix} 0 \\ 0 \end{bmatrix} & \text{for } k \text{ odd} \end{cases}
$$

$$
J[k] = [0, 1], \quad K[k] = \begin{cases} 40 & \text{for } k \text{ even} \\ 0 & \text{for } k \text{ odd} \end{cases}
$$

Note that this controller has an *infinite* gain margin compared to the finite (and small) gain margins we achieved for this system with linear time-invariant controllers in Example 11.4.1.

⋀⋀⋀

**Remark 11.3.1** The periodic control design method for linear time-invariant systems discussed above is based on ideas from Francis *et.al* (1988). It was presented here as a mechanism for assigning the system's perceived zeros. However, it may also be viewed as a design method for a stabilizing controller for the *raised system* which satisfies the constraint in Lemma 7.3.2 (see also Problem 11.6). Other methods appear in the literature and the interested reader is referred to the references at the end of this chapter. The key point is that remarkable properties are achievable at the *slow* sampling rate by the use of periodic controllers. This may be a very useful property in certain applications.

## 11.4  Frequency Domain Analysis

So far we have analyzed periodic control of linear time-invariant systems in the time domain. The main concern was the question of stability. However, simulation of systems with periodic controllers indicates that the response within the periods can be quite different from that at the sample points.

The concept of raised systems and control design based on the raised system offers a powerful tool. Of course, the end result is achieved via a sequential implementation. Thus whilst each entry of the output vector in the raised system may behave 'nicely', these entries are interlaced when implemented sequentially and the result may actually be oscillatory. this issue is similar to that studied in Section 9.8. Specifically, the response at the raised (large) sample period can be very different from the response at the underlying (small) period. We will analyze this issue in this chapter. In this regard, the periodicity of the controller suggests the frequency domain as a useful tool to analyze the behaviour of the periodic controller applied to a linear time-invariant system.

It is interesting to point out that the observations resulting from this analysis are quite similar to those for the GSHF in Chapter 10. In particular, the price of forcing the raised system to behave differently from the behaviour that could be achieved at the underlying fast sampling rate by a linear time-invariant controller is usually significant intra-period oscillations and sensitivity to high frequency modelling errors.

We consider again the configuration in Figure 11.3.2 (a). To simplify our discussion we make use of our Remark 10.2.2 and assume that the state equation of the controller is time-invariant but has a periodic output map. We rewrite (7.3.2) as

$$x_c[k + 1] = Fx_c[k] + Ge[k] \tag{11.4.1}$$

$$u[k] = J[k]x_c[k] + K[k]e[k] \tag{11.4.2}$$

Our notation here is similar to the one used in Section 10.3. We apply the Fourier transform to the configuration in Figure 11.3.2 (a) and get the following relationships:

$$E^s(\omega) = R^s(\omega) - Y^s(\omega) \tag{11.4.3}$$

$$Y^s(\omega) = P_s(\omega)U^s(\omega) \tag{11.4.4}$$

(from equations (11.4.1) – (11.4.2))

$$e^{j\omega\Delta}X_c^s(\omega) = FX_c^s(\omega) + GE^s(\omega) \tag{11.4.5}$$

$$U^s(\omega) = \frac{1}{2\pi} \int_{-\frac{\pi}{\Delta}}^{\frac{\pi}{\Delta}} J_s(\eta)X_c^s(\omega - \eta)d\eta + \frac{1}{2\pi} \int_{-\frac{\pi}{\Delta}}^{\frac{\pi}{\Delta}} K_s(\eta)E^s(\omega - \eta)d\eta \tag{11.4.6}$$

where $J_s(\omega)$ and $K_s(\omega)$ are the Fourier transforms of the sequences $J[k]$ and $K[k]$ respectively.

As a side comment, it is interesting to note the resemblance between equations (11.4.2) and (10.2.12) for the GSHF. In both cases the signals involved are modulated by periodic functions (sequences). Hence, it is not surprising that the analysis here proceeds along similar lines as for the GSHF.

To calculate $J_s(\omega)$ and $K_s(\omega)$ we use the fact that $J[k]$ and $K[k]$ are periodic in $N$. Hence, these sequences have a Fourier series representation

$$\bar{J}[k] = \frac{1}{N} \sum_{i=0}^{N-1} J_i e^{j\omega_N \Delta ki}$$

$$\bar{K}[k] = \frac{1}{N} \sum_{i=0}^{N-1} K_i e^{j\omega_N \Delta ki} \tag{11.4.7}$$

where

$$J_i = \sum_{k=0}^{N-1} \bar{J}[k] e^{-j\omega_N \Delta ki}$$

$$K_i = \sum_{k=0}^{N-1} \bar{K}[k] e^{-j\omega_N k\Delta i} \tag{11.4.8}$$

and

$$\omega_s = \frac{2\pi}{\Delta}, \quad \omega_N = \frac{\omega_s}{N} \qquad (11.4.9)$$

Hence

$$J_s(\omega) = \frac{2\pi}{N} \sum_{i=-\infty}^{\infty} J_i \delta(\omega - i\omega_N)$$

and

$$K_s(\omega) = \frac{2\pi}{N} \sum_{i=-\infty}^{\infty} K_i \delta(\omega - i\omega_N)$$

Substituting in (11.4.6) we get

$$U^s(\omega) = \frac{1}{N} \sum_{i=0}^{N-1} \left[ J_i X_c^s(\omega - i\omega_N) + K_i E^s(\omega - i\omega_N) \right] \qquad (11.4.10)$$

Since (11.4.10) involves shifted versions of $X_c^s(\omega)$ and $E^s(\omega)$ it will be convenient to define the vectors

$$\tilde{X}_c^s(\omega) = \begin{bmatrix} X_c^s(\omega) \\ X_c^s(\omega - \omega_N) \\ \cdot \\ \cdot \\ \cdot \\ X_c^s(\omega - (N-1)\omega_N) \end{bmatrix} \qquad (11.4.11)$$

$$\tilde{E}^s(\omega) = \begin{bmatrix} E^s(\omega) \\ E^s(\omega - \omega_N) \\ \cdot \\ \cdot \\ \cdot \\ E^s(\omega - (N-1)\omega_N) \end{bmatrix} \qquad (11.4.12)$$

$\tilde{U}^s(\omega)$, $\tilde{R}^s(\omega)$ and $\tilde{Y}^s(\omega)$ are similarly defined. Then, from (11.4.5) and (11.4.10) we readily get

$$\boxed{\tilde{U}^s(\omega) = \tilde{C}_s(\omega)\tilde{E}^s(\omega)} \qquad (11.4.13)$$

where

$$\boxed{\tilde{C}_s(\omega) = \tilde{J}\left(e^{j\omega\Delta}\tilde{W} - \tilde{F}\right)^{-1}\tilde{G} + \tilde{K}} \qquad (11.4.14)$$

$$\tilde{J} = \frac{1}{N}\begin{bmatrix} J_0 & J_1 & & J_{N-1} \\ J_{N-1} & J_0 & & J_{N-2} \\ J_{N-2} & J_{N-1} & & J_{N-3} \\ \cdot & \cdot & \multicolumn{1}{c}{- - - -} & \cdot \\ \cdot & \cdot & & \cdot \\ J_1 & J_2 & & J_0 \end{bmatrix} \qquad (11.4.15)$$

$$\tilde{K} = \frac{1}{N}\begin{bmatrix} K_0 & K_1 & & K_{N-1} \\ K_{N-1} & K_0 & & K_{N-2} \\ K_{N-2} & K_{N-1} & & K_{N-3} \\ \cdot & \cdot & \multicolumn{1}{c}{- - - -} & \cdot \\ \cdot & \cdot & & \cdot \\ K_1 & K_2 & & K_0 \end{bmatrix} \qquad (11.4.16)$$

$$\tilde{F} = \begin{bmatrix} F & 0 & & 0 \\ 0 & F & & 0 \\ 0 & 0 & & 0 \\ \cdot & \cdot & \multicolumn{1}{c}{- - - -} & \cdot \\ \cdot & \cdot & & \cdot \\ 0 & 0 & & F \end{bmatrix} \qquad (11.4.17)$$

$$\tilde{G} = \begin{bmatrix} G & 0 & & & 0 \\ 0 & G & & & 0 \\ 0 & 0 & & & 0 \\ \cdot & \cdot & & - - - - & \cdot \\ \cdot & \cdot & & & \cdot \\ 0 & 0 & & & G \end{bmatrix} \qquad (11.4.18)$$

$$\tilde{W} = \begin{bmatrix} I & 0 & & & 0 \\ 0 & e^{-j\omega_N \Delta} \, I & & & 0 \\ 0 & 0 & & & 0 \\ \cdot & \cdot & & - - - - & \cdot \\ \cdot & \cdot & & & \cdot \\ 0 & 0 & & & e^{-j(N-1)\omega_N \Delta} \, I \end{bmatrix} \qquad (11.4.19)$$

While catching our breath for a minute we note that the matrix $\tilde{C}_s(\omega)$ in (11.4.14) must be related to the raised transfer matrix for the controller, $\tilde{C}_R(z)$. This relation is worked out in Problem 11.8.

Using (11.4.3), (11.4.4) and (11.4.13) we get

$$\boxed{\tilde{Y}^s(\omega) = \tilde{H}(\omega)\tilde{R}^s(\omega)} \qquad (11.4.20)$$

where

$$\boxed{\tilde{H}(\omega) = \left[I + \tilde{P}_s(\omega)\tilde{C}_s(\omega)\right]^{-1}\tilde{P}_s(\omega)\tilde{C}_s(\omega)} \qquad (11.4.21)$$

$$\tilde{P}_s(\omega) = \begin{bmatrix} \tilde{P}_d \cdot e^{j\omega \Delta} & 0 & & & 0 \\ 0 & \tilde{P}_d \cdot e^{j(\omega-\omega_N)\Delta} & & & 0 \\ 0 & 0 & & & 0 \\ \cdot & \cdot & & - - - - & \cdot \\ \cdot & \cdot & & & \cdot \\ 0 & 0 \cdot & & & \tilde{P}_d \cdot e^{j(\omega-(N-1)\omega_N)\Delta} \end{bmatrix} \qquad (11.4.22)$$

Specifically, we are interested in $Y^s(\omega)$ for which we have from (11.4.20)

$$Y^s(\omega) = \sum_{i=1}^{N} \tilde{H}_{1,i}(\omega) R^s(\omega - (i-1)\omega_N) \qquad (11.4.23)$$

Equation (11.4.23) immediately reveals that $Y^s(\omega)$ *contains high frequency components* resulting from $R^s(\omega)$ being shifted to the higher range of the interval $[0, \omega_s/2)$ and multiplied by the corresponding $\tilde{H}_{1,i}(\omega)$.

To demonstrate this, say $r[k] = 1$ (a constant) and let $N = 2$. Then the output $y[k]$ will be

$$y[k] = a_0 + a_1 \cos(\pi k + \phi)$$

where $a_0 = \tilde{H}_{1,1}(0)$ and $\tilde{H}_{1,2}(\pi/\Delta) = a_1 e^{j\phi}$.

To actually show that the presence of these high frequency components is an inherent part of using a periodic controller we show the following:

**Lemma 11.4.1** Consider the periodic control system in Figure 11.3.2 (a) and let $r[k]$ be a pure sinewave having frequency in the interval $[0, \omega_N/2)$. Then, the sampled output will *not* contain higher frequency components if and only if the controller is linear time-invariant.

**Proof :** The 'if' part is trivial since for a linear time-invariant system the output has the same frequency as the input.

For the 'only if' we observe from equation (11.4.23) that $Y^s(\omega)$ will have the same frequencies as $R^s(\omega)$ only if

$$\tilde{H}_{1,i}(\omega) = 0, \qquad i = 2, \ 3, \ \ldots, \ N-1$$

for all $\omega$. However, from the definition of $\tilde{Y}^s(\omega)$ and $\tilde{R}^s(\omega)$ this implies that $\tilde{H}^s(\omega)$ *must be diagonal*.

Then, from equations (11.4.21) and (11.4.22)

$$
\tilde{C}_s(\omega) = \tilde{J}
\begin{bmatrix}
\left(e^{j\omega\Delta}I - F\right)^{-1}G & 0 & 0 \\
0 & \left(e^{j(\omega-\omega_N)\Delta}I - F\right)^{-1}G & 0 \\
0 & 0 & 0 \\
\cdot & \cdot & \cdot \\
\cdot & \cdot & \cdot \\
0 & 0 & \left(e^{j(\omega-(N-1)\omega_N)\Delta}I - F\right)^{-1}G
\end{bmatrix}
$$

$$(11.4.24)$$

Hence, the diagonality of $\tilde{C}^s(\omega)$ implies the diagonality of $\tilde{K}$ and $\tilde{J}$ which in turn means the controller may be realized in a linear time-invariant form.

△△△

The conclusion from Lemma 11.4.1 is twofold : (i) the high frequency components in the system output are an inherent part of using periodic control for linear time-invariant systems, and (ii) the ability of the periodic controller to achieve anything a linear time-invariant controller cannot is a consequence of these components. Hence:

**Remark 11.4.1** In a strictly proper system the need for high frequency components in the output will typically force the input to the system to have these frequencies with larger amplitude.

**Remark 11.4.2** With the dependence of the periodic control performance on the high frequency components in the system output it is apparent that its performance may be sensitive to high frequency uncertainties in the system.

To demonstrate all these points we return to our example.

**Example 11.5.1 :** (Example 11.4.3 continued)

Simulate the controller designed for the system

$$\tilde{P}_d(z) = \frac{z-2}{z-3}$$

in Example 11.4.3 and examine the response at the fast sampling rate and associated sensitivity issues.

**Solution** The configuration is as in Figure 11.3.2 (a) and the controller equations are

$$x_c[k+1] = G[k]e[k]$$

$$u[k] = x_c[k] + K[k]e[k]$$

where

$$G[k] = 110 \cdot (1 + \cos \pi k)$$

$$K[k] = 20 \cdot (1 + \cos \pi k)$$

In Figure 11.5.1 we see the unit step response of this control system. Note
that the control system is stable. In Figure 11.5.1 (a) we show $y[k]$, while in Figure 11.5.1 (b) we show $\bar{y}[m]$. It is clear that the nice behaviour occurs only for
the slow sampled response, namely $\bar{y}[m]$, while in the fast sampled response,
$y[k]$, we observe the predicted presence of high frequency components. Furthermore, even though the 'nasty' zero effect does not show in $\bar{y}[m]$ its presence is
apparent in $y[k]$.

To test the gain margin we replace $\tilde{P}_d(z)$ with $100 \cdot \tilde{P}_d(z)$ and the results
are shown in Figure 11.5.2. We see that the system is stable as predicted by the
large gain margin achieved by the design.

To demonstrate the sensitivity to high frequency uncertainty in the system
we add an unmodelled pole at the origin which corresponds to a one sample unmodelled delay. The implication of this can be calculated (via the raising mechanism) and the result is instability (see Problem 11.11). This is also demonstrated

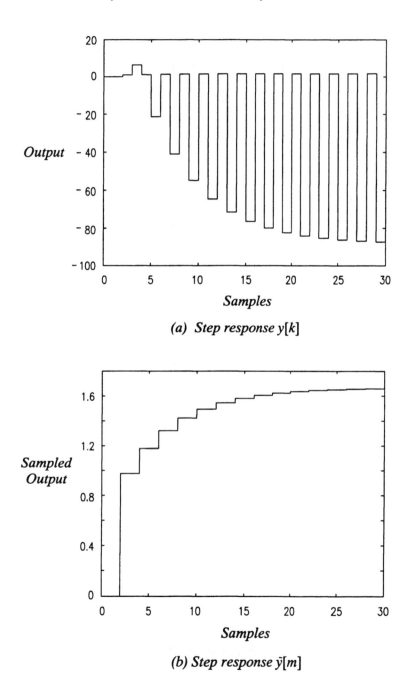

(a) *Step response y[k]*

(b) *Step response ȳ[m]*

*Figure 11.4.1 : Step responses for Example 11.5.1.*

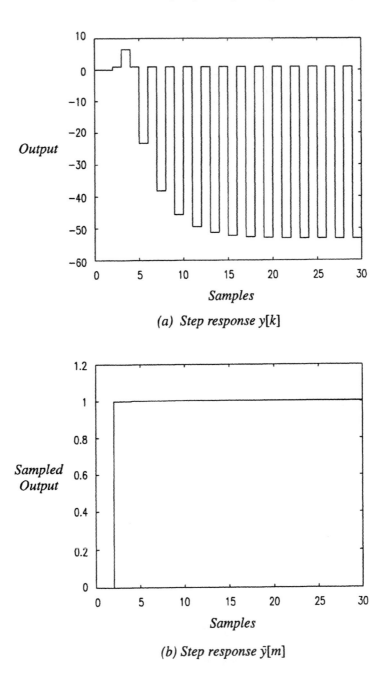

(a) Step response y[k]

(b) Step response ȳ[m]

Figure 11.4.2 :  Step responses for improved system in Example 11.5.1.

in Figure 11.5.3. Figure 11.5.4 shows the magnitude and phase of $\tilde{H}_{1,1}(\omega)$ = $\tilde{H}_{1,2}(\omega)$.

ΛΛΛ

Finally we present an example where high frequency signals do not seem, at first sight, to arise in periodic control. However, we show that this is due to rather special features of the problem.

**Example 11.5.2 :** Consider again Example 11.3.1. Use the frequency domain tools developed above to explain the lack of high frequency components in the simulated step responses shown in Figure 11.3.3.

**Solution** We calculate the matrix $\tilde{H}(\omega)$ for this example and get

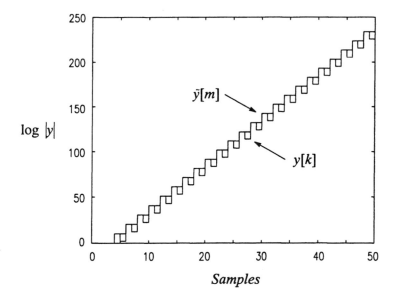

*Figure 11.4.3 : Unstable response occuring in Example 11.5.1.*

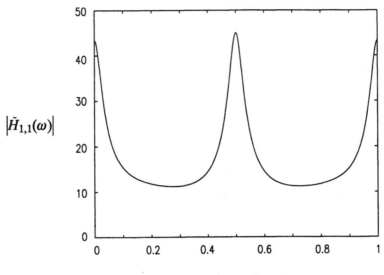

*Frequency/Sampling frequency*

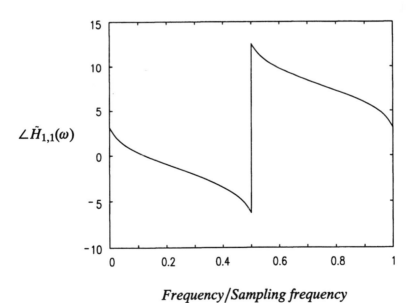

*Frequency/Sampling frequency*

*Figure 11.4.4 :  Magnitude and phase of $\tilde{H}_{1,1}(\omega) = \tilde{H}_{1,2}(\omega)$*
*from Example 11.5.1.*

$$\tilde{H}_{1,1}(\omega) = \frac{(K\Delta)^2}{e^{2j\omega\Delta} - 1 + (K\Delta)^2}$$

$$\tilde{H}_{1,2}(\omega) = \frac{K\Delta(e^{j\omega\Delta} + 1)}{e^{2j\omega\Delta} - 1 + (K\Delta)^2}$$

and we see that $\tilde{H}_{1,2}(\omega)$ has a zero at $\omega = \omega_N = \pi/\Delta$. Hence, a d.c. input will not cause any high frequency components in the output. In Figure 11.5.5 we see $\left|\tilde{H}_{1,1}(\omega)\right|$ and $\left|\tilde{H}_{1,2}(\omega)\right|$ for $\Delta = K = 1$, that is $\tilde{H}_{1,1}(\omega) = e^{-2j\omega}$,

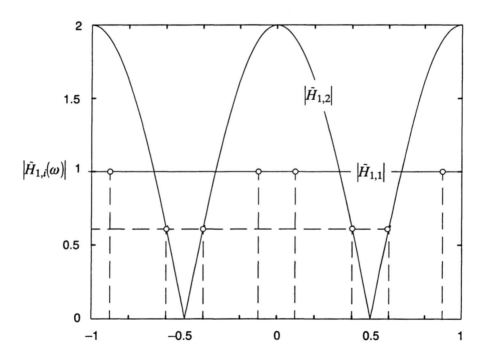

*Frequency/Sampling frequency* $(\omega\Delta/2\pi)$

*Figure 11.4.5 : Responses of $\left|\tilde{H}_{1,1}(\omega)\right|$ and $\left|\tilde{H}_{1,2}(\omega)\right|$ for $\dfrac{\omega\Delta}{2\pi} = 0.1$ from Example 11.5.2.*

$\tilde{H}_{1,2}(\omega) = e^{-j\omega}\left(1 + e^{-j\omega}\right)$. Note that $\left|\tilde{H}_{1,2}\right|$ is non-zero for the case illustrated of

$(\omega\Delta)/2\pi = 0.1$. So if we enter $r[k] = \cos(\omega_0 k)$ the output will be (see equation

(11.4.23):

$$y[k] = \cos\left(\omega_0 k - \angle\tilde{H}_{1,1}(\omega_0)\right) + \left|\tilde{H}_{1,2}(\omega_0 + \pi)\right|\cos\left((\omega_0 + \pi)k - \angle\tilde{H}_{1,2}(\omega_0 + \pi)\right)$$

For $\omega_0 = 0$ we have $y[k]$ constant. Similarly, for $\omega_0 = \pi$ we get

$$y[k] = \cos(\pi k) + 2$$

which means that in this case the 'high frequency' results in a bias since the shifted high frequency is actually at $2\pi$.

◇◇◇

## 11.5   Further Reading and Discussion

Periodic controllers and their properties are discussed in Dahleh, Voulgaris and Valavani (1992), Hyslop, Schättler and Tarn (1992), Davis (1972), Khargonekar *et. al* (1985), Lee *et. al* (1987), Francis and Georgiou (1988), Das and Rajagopalan (1992).

In Shamma and Dahleh (1991) it is claimed that in the control of certain feedback objectives, periodic compensation offers no advantage over time-invariant compensation for linear time-invariant systems.

The costs associated with some of the perceived benefits in using periodic control for linear time-invariant systems have been discussed in Goodwin and Feuer (1992) and Feuer (1993).

Example 11.3.1 was suggested to us by Andrzej Olbrot of Wayne State University.

## 11.6   Problems

11.1   Given the system $y[k + 1] = \cos(k\pi)y[k] + \sin(2\pi/3)u[k]$

(a) find this system's periodicity,

(b) write down its raised version,

(c) calculate the transfer matrix for the raised system.

11.2 Verify (11.2.6) – (11.2.12).

11.3 (a) Describe the system in Figure 11.3.2(a) in a combined state-space equation with $r[k]$ as an input and $y[k]$ as an output. (Note that this system is $N$ periodic).

(b) Describe the system in Figure 11.3.2(b) in a combined state-space equation with $r_R[m]$ as an input and $y_R[m]$ as an output – the system is linear time-invariant.

(c) Show, by *directly* raising the system you got in (a), that (b) is indeed the result.

11.4 Use the method outlined in the proof of Lemma 7.3.2 and implement the transfer matrix you found in Problem 11.1(c) sequentially. (Note that you get a realization with time-invariant state equation).

11.5 Let $\check{C}_R(z) = \begin{bmatrix} \dfrac{z}{z-0.1} & z^{-1} \\ \dfrac{z-1}{z-0.1} & 1 \end{bmatrix}$

(a) Find the corresponding sequential implementation.

(b) Is the implementation you found minimal (is the resulting state space model both reachable and detectable)?

(c) Is the implementation in Example 11.2.2 minimal?

11.6 Let $\check{P}_d(z) = \dfrac{(z+2)^2}{z(z-3)}$, $\check{N}(z) = z^2$.

(a) Find $\check{F}_d(z)$ so that $\check{N}(z)$ is the numerator of the transfer function from $\bar{e}[m]$ to $\bar{y}[m]$ in Figure 11.4.2 (choose $N = 2$ ).

(b) Find a controller to stabilize $\dfrac{\check{N}(z)}{z(z-3)}$.

(c)    Find the resulting periodic controller for the original system.

11.7  Find the raised form of the controller in Figure 11.4.3, $\check{C}_R(z)$, and show that it is indeed a stabilizing controller for the raised version of $\check{P}_d$ which satisfies the constraint that $\check{C}_R(\infty)$ is lower triangular.

11.8  Let $E^s(\omega)$, $U^s(\omega)$ be the Fourier transforms of the sequences $e[k]$, $u[k]$ respectively.

(a)    Show that $\Lambda_s(\omega)W_s\check{E}^s(\omega)$ is the Fourier transform of the vector

$$\left[ e[mN], \ e[mN+1], \ \ldots, \ e[(m+1)N-1] \right]^T, \text{where}$$

$$\Lambda_s(\omega) \ = \ \begin{bmatrix} 1 & 0 & & & 0 \\ 0 & e^{j\omega\Delta} & & & 0 \\ \cdot & \cdot & \cdot & \cdot & \cdot \\ \cdot & \cdot & \cdot & \cdot & \cdot \\ 0 & 0 & & & e^{j(N-1)\omega\Delta} \end{bmatrix}$$

and $W_s$ is the familiar DFT matrix

$$W_s \ = \ \begin{bmatrix} 1 & 1 & 1 & \cdots & 1 \\ 1 & W & W^2 & \cdots & W^{N-1} \\ \cdot & & & & \cdot \\ \cdot & & & & \cdot \\ \cdot & & & & \cdot \\ 1 & W^{N-1} & W^{2(N-1)} & \cdots & W^{(N-1)^2} \end{bmatrix} \qquad W = e^{-j\omega_N\Delta}$$

(b)    Show that $\check{C}_R(z)$, the raised transfer matrix of the controller, is related to $\check{C}_s(\omega)$ of equation (11.4.14) as follows:

$$\check{C}_s(\omega) \ = \ W_s^{-1}\Lambda_s^{-1}(\omega)\check{C}_R(e^{j\omega N\Delta})\Lambda_s(\omega)W_s.$$

11.9  Take $\check{P}_d(z) = \dfrac{z-2}{z(z-3)}$ and the same controller as in Example 11.5.1. Show that the resulting system is unstable.

11.10 Find the raised form of the closed loop system in Example 11.5.2 when $\Delta = 1$ and $P(s) = e^{-s}/s$ and show that with this unmodelled time delay the control system is unstable.

11.11 Verify the claim of instability with the addition of the unmodelled pole at the origin in Example 11.5.1.

11.12 Show that any time delay of the order of the sampling interval will lead to instability in Example 11.5.2.

# Chapter 12

# Multirate Control

## 12.1 Introduction

In this chapter we turn to another form of control which is closely related to periodic control, namely *multirate* control.

The term 'multirate' control refers to problems in which more than one sampling rate applies. We take it for granted that if one can sample all outputs rapidly and change all inputs equally rapidly, then it cannot make things better to artificially constrain the problem by demanding that one of these rates be fixed at a slower value. However, practical considerations may mean that it is not feasible to sample all outputs at the same rapid rate. In other cases, it may not be feasible to change all inputs at the same rate.

For example, in chemical process control it frequently happens that some variables (e.g. temperatures) can be measured essentially continuously whereas other variables (e.g. concentrations) may require chemical analysis to be carried out which can lead to significant times between samples. Equally, it may be possible to change some inputs continuously (e.g. by opening or closing a value) whereas some other inputs may require substantial periods between adjustments (e.g. when a human operator needs to manually go to a remote part of a plant to make an adjustment).

Another reason for using multirate control is where the process control computer does not have the speed to update all variables at the same rate. Of course, it might be argued that this is unlikely to occur with modern technology, but one might be forced to use old technology. It may not be desirable to make all inputs vary equally rapidly if it is known that there is a wide spread in time constants. Indeed, running a control loop excessively fast can lead to numerical difficulties.

We thus see that there are many practical reasons why multirate controllers might be used.

In the case of a single-input single-output plants, there are really only two possibilities: (i) when the input can change faster than the output sampling rate, and (ii) when the input cannot change as fast at the output sampling rate. For a multiple-input multiple-output system there are, of course, many possible combinations.

This chapter describes methods applicable to the design of multirate controllers. We will begin with a unifying approach which gives a conceptual solution to all multirate control problems. We will then survey some of the methods that have appeared in the literature and contrast them with the unifying approach.

## 12.2  A Unifying Approach

In a sense, the problem of multirate control is quite straightforward from an information viewpoint. In particular, we should use *all* of the available measured data to obtain the best possible estimate of the state, and then we should change the input (or inputs) as quickly as possible so as to quickly counteract the effect of the estimated disturbances and have the output(s) track the desired reference trajectories.

Say we have a slow output sampling rate but we can use fast input adjustments. Then, the best strategy would seem to be to use a state estimator to interpolate the plant state and disturbances between output samples. Conversely, say the output can be sampled more rapidly than the input can be adjusted, then clearly one should use the extra output measurements to improve the state estimates. Indeed, if output measurement noise is low, then it is possible to completely re-estimate the state and disturbances between each input adjustment.

Obviously a key consideration in the design of such systems is the model that one has for the various disturbances. For simplicity, we assume here that the disturbances are constant (this can readily be generalized - see Chapter 6 which deals with optimal filters for stochastic signals). Also, we assume that the disturbances enter at some internal (but known) point in the process. Typical points for disturbances to arise are at the input or output.

Then a suitable continuous-time model might be

$$\dot{x} = Ax + B_1u + B_2d \qquad (12.2.1)$$

$$\dot{d} = 0$$

$$y = C_1x + C_2d \qquad (12.2.2)$$

For example, if the disturbance is an input disturbance then we can set $B_2 = B_1$, $C_2 = 0$. Alternatively, if it is an output disturbance then we can set $B_2 = 0$, $C_2 = 1$.

Now let us assume that the output $y$ is measured at some period $\Delta_0$. We can then design an observer for $x$ and $d$ as follows

$$\frac{d}{dt}\begin{bmatrix} \hat{x} \\ \hat{d} \end{bmatrix} = \begin{bmatrix} A & B_2 \\ 0 & 0 \end{bmatrix}\begin{bmatrix} \hat{x} \\ \hat{d} \end{bmatrix} + J\left[y^s - C_1\hat{x}^s - C_2\hat{d}^s\right] + \begin{bmatrix} B_1 \\ 0 \end{bmatrix}u \qquad (12.2.3)$$

where $y^s$, $\hat{x}^s$ and $\hat{d}^s$ are the impulse sampled versions of $y$, $\hat{x}$ and $\hat{d}$ respectively. Note that the effect of the impulse sampled signals in (12.2.3) is to give a quantum step change to the state estimates at the sample times. As we show later, when (12.2.3) is implemented via fast sampling then the impulse sampled signals will be replaced by filled sequences (as defined in Section 1.7.1). $J$ is designed to ensure that the state errors decay to zero (at the sample points). To see how this can be done, note that the discrete version of (12.2.3) is given by

$$\delta\begin{bmatrix} \hat{x} \\ \hat{d} \end{bmatrix} = \bar{A}_\delta\begin{bmatrix} \hat{x} \\ \hat{d} \end{bmatrix} + \bar{J}_\delta[y - C_1\hat{x} - C_2\hat{d}] + \bar{B}_\delta u \qquad (12.2.4)$$

where all signals are now sequences and

$$\bar{A}_\delta = \frac{1}{\Delta_0}\left[e^{\bar{A}\Delta_0} - I\right] \tag{12.2.5}$$

$$\bar{B}_\delta = \int_0^{\Delta_0} e^{\bar{A}\sigma}\, d\sigma \begin{bmatrix} B_1 \\ 0 \end{bmatrix} \tag{12.2.6}$$

$$\bar{J}_\delta = e^{\bar{A}\Delta_0} J \tag{12.2.7}$$

$$\bar{A} = \begin{bmatrix} A & B_2 \\ 0 & 0 \end{bmatrix} \tag{12.2.8}$$

Assuming $\left([C_1, C_2],\ \bar{A}\right)$ is detectable we choose $\bar{J}_\delta$ so that $\left(\bar{A}_\delta - \bar{J}_\delta[C_1, C_2]\right)$ has desirable eigenvalues and then let

$$J = e^{-\bar{A}\Delta_0}\, \bar{J}_\delta \tag{12.2.9}$$

Clearly with the choice of $J_\delta$ as above the state estimation errors can be forced to converge to zero at the sample points, hence, so does the continuous state estimate error (see Problem 12.1).

Next we want to design the feedback to achieve three purposes : (i) to stabilize the system, (ii) to cancel the disturbance effects as quickly as possible, and (iii) to track the reference input.

To illustrate the principles, we will restrict attention to the case of an input disturbance. The extension to other disturbances is straightforward although the design must naturally reflect the disturbance source. Whether the constraint is fast sampling at the output and slow changes at the input, or vice versa, the methodology is similar. However, since the details differ, we treat each case separately. This is done in the next two sections. Note that the control system design methodology that we follow is essentially as outlined in Chapter 8 save that we

clock the observer and state estimate feedback at rates set by their respective constraints.

## 12.3   Slow Output Sampling, Fast Input Sampling

In this case, since the rate at which the control can be applied is greater than
the rate at which measurements can be made, the control action will necessarily
be open loop between samples of the output, even though the control values
might change during this time. The idea is to base that open loop control on *fast*
state *estimates*.

We thus run an observer that is clocked at the fast rate but which has error
adjustments at the slow rate as shown in the configuration of Figure 12.3.1.

Note in Figure 12.3.1 that $u[k]$, $r[k]$, $\hat{x}[k]$ and $\hat{d}[k]$ are clocked every period $\Delta$, while $y[m]$ is available only every large period $\Delta_0 = N\Delta$.

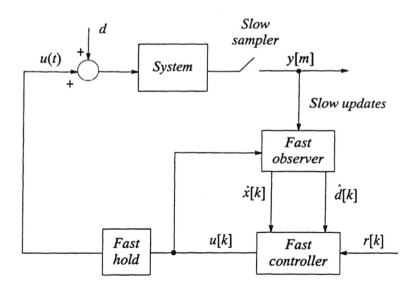

*Figure 12.3.1 :   Multirate controller with slow output and fast
input sampling.*

Equations (12.2.1) – (12.2.2), when rewritten under the simplifying assumption that we have only an input disturbance, become

$$\dot{x} = Ax + B(u + d) \tag{12.3.1}$$

$$\dot{d} = 0$$

$$y = Cx \tag{12.3.2}$$

Let us begin by converting the slow observer (clocked at period $\Delta_0 = N\Delta$ ) into one which gives estimates at period $\Delta$ . To do that we use the continuous observer equation in (12.2.3) and a filled output measurement $y_{fil}[k]$ to yield:

$$\delta_k \begin{bmatrix} \hat{x}[k] \\ \hat{d}[k] \end{bmatrix} = \bar{A}_\delta \begin{bmatrix} \hat{x}[k] \\ \hat{d}[k] \end{bmatrix} + \bar{J}_\delta \left[ y_{fil}[k] - C\hat{x}_{fil}[k] \right] + \bar{B}_\delta u[k] \tag{12.3.3}$$

where for clarity we have adopted the notation $\delta_k$ and $\delta_m$ to represent the delta operators corresponding to the fast and slow sampling rates respectively. Thus

$$\delta_k x[k] = \frac{1}{\Delta}(x[k+1] - x[k]) \quad \text{with} \quad x[k] = x(k\Delta)$$

and

$$\delta_m y[m] = \frac{1}{N\Delta}(y[m+1] - y[m]) \quad \text{with} \quad y[m] = y(mN\Delta)$$

Also

$$\bar{A}_\delta = \frac{1}{\Delta}\left[ e^{\bar{A}\Delta} - I \right] \tag{12.3.4}$$

$$\bar{B}_\delta = \frac{1}{\Delta} \int_0^\Delta e^{\bar{A}(\Delta - \sigma)} \, d\sigma \begin{bmatrix} B \\ 0 \end{bmatrix} \tag{12.3.5}$$

$$J_\delta = Ne^{\bar{A}\Delta}J \tag{12.3.6}$$

and

$$\bar{A} = \begin{bmatrix} A & B \\ 0 & 0 \end{bmatrix} \tag{12.3.7}$$

The notation $y_{fil}[k]$ in equation (12.3.3) is as explained in Section 1.7.1.

The design of $\bar{J}_\delta$ has been given by equation (12.2.9); to recap we do the following:

(i) Choose $\bar{J}_\delta$ so that the (slow) discrete observer matrix $(\bar{A}_\delta - \bar{J}_\delta[C, 0])$ has desired eigenvalues assuming $([C, 0], \bar{A}_\delta)$ is a detectable pair with

$$\bar{A}_\delta = \frac{1}{N\Delta}\left[e^{\bar{A}N\Delta} - I\right]$$

(ii) Calculate $\bar{J}_\delta$ from equations (12.2.9) and (12.3.6) as

$$\bar{J}_\delta = Ne^{\bar{A}\Delta(1-N)}\bar{J}_\delta \tag{12.3.8}$$

As argued for the continuous observer (see Problem 12.1) we can readily conclude that, even though the observer is driven by slow observations, we have at the fast rate:

$$\lim_{k \to \infty}\left[\begin{bmatrix} x(k\Delta) \\ d(k\Delta) \end{bmatrix} - \begin{bmatrix} \hat{x}[k] \\ \hat{d}[k] \end{bmatrix}\right] = 0$$

Now that we have a fast rate state estimate available, we consider the control law. To be specific, we follow the approach described in Section 8.7.2 (see also Example 8.5.1).

Let $y[k]$ be the system output sampled at the fast rate (note, again, that we have only every $N$-th value of this sequence but have an estimate, $\hat{y}[k] = C\hat{x}[k]$, of the remaining values). Denote the discrete transfer function from $u[k]$ to $y[k]$ by

$$\bar{G}_p(\gamma) = \frac{\bar{B}(\gamma)}{\bar{A}(\gamma)} \tag{12.3.9}$$

and let

$$\bar{B}(\gamma) = \bar{B}_s(\gamma)\bar{B}_u(\gamma) \tag{12.3.10}$$

where $\bar{B}_s(\gamma)$ is monic and contains the stable well damped zeros, and $\bar{B}_u(\gamma)$ contains the remainder.

Define

$$\bar{B}^1(\gamma) \overset{\Delta}{=} \bar{B}_s(\gamma)\bar{B}_e(\gamma) \tag{12.3.11}$$

with $\bar{B}_e(\gamma)$ comprising a set of fast zeros so that $\bar{B}^1(\gamma)$ is of the same degree as $\bar{A}(\gamma)$.

The controller we use is simply a state estimate feedback control, and we use $\hat{d}[k]$ to reduce the input disturbance effect. Let $u[k]$ be defined by

$$\boxed{u[k] = Gr[k] - K_\delta \hat{x}[k] - \hat{d}[k]} \tag{12.3.12}$$

where $K_\delta$ is chosen so that

$$C(\gamma I - A_\delta + B_\delta K_\delta)^{-1} B_\delta = \frac{\bar{B}(\gamma)}{\bar{B}_s(\gamma) \cdot \bar{B}_e(\gamma)} \tag{12.3.13}$$

with $\left(C, A_\delta, B_\delta\right)$ being the minimal realization of $\bar{G}_p(\gamma)$, and with $G$ in (12.3.12) chosen as $\left[\bar{B}_u(0)\right]^{-1}\bar{B}_e(0)$.

Hence, the complete controller is defined by equations (12.3.3) and (12.3.12).

To see the result of employing this controller we first make use of the separation principle to claim that, at steady state, $\hat{x}[k] = x[k]$ and $\hat{d}[k] = d[k]$. Hence, the discretized *steady state* closed loop equations can be written as

$$x[k+1] = (A_\delta - B_\delta K_\delta)x[k] + [\bar{B}_u(0)]^{-1}r[k] \qquad (12.3.14)$$

$$y[k] = Cx[k] \qquad (12.3.15)$$

which means that the transfer function from $r[k]$ to $y[k]$ is

$$[\bar{B}_u(0)]^{-1}\bar{B}_e(0)\frac{\bar{B}_u(\gamma)}{\bar{B}_e(\gamma)}$$

In the next example we demonstrate the above multirate control design procedure and compare its performance to the more conservative approach of using a *slow* controller (i.e. where we accept the fact that the output is sampled slowly and decide to similarly restrict the input).

**Example 12.3.1** : Consider the continuous-time system given by

$$\dot{x} = Ax + B(u+d)$$

$$y = Cx$$

where

$$A = \begin{bmatrix} -3 & -2 \\ 1 & 0 \end{bmatrix} \qquad B = \begin{bmatrix} 1 \\ 0 \end{bmatrix} \qquad C = \begin{bmatrix} 0 & 2 \end{bmatrix}$$

and $d$ is a unit step disturbance.

The system output is sampled at period $\Delta_0 = 0.5$ seconds but the input can be adjusted every $\Delta = 0.05$ seconds.

Design the multirate controller following the methodology presented in this section and compare its performance with a slow rate controller clocked at the slow output sampling rate.

## Solution

(a) The fast rate discrete equivalent of this system is given by

$$\delta_k x[k] = A_\delta x[k] + B_\delta(u[k] + d)$$

$$y[k] = Cx[k]$$

where

$$A_\delta = \begin{bmatrix} -2.8311 & -1.8557 \\ 0.9278 & -0.0476 \end{bmatrix} \qquad B_\delta = \begin{bmatrix} 0.9278 \\ 0.0238 \end{bmatrix}$$

with transfer function

$$\overline{G}_p(\gamma) = \frac{0.0476(\gamma - 39.0246)}{\gamma^2 + 2.8787\gamma + 1.8565}$$

Let us start with the observer design. For that we need to compute $\overline{A} = \begin{bmatrix} A & B \\ 0 & 0 \end{bmatrix}$ and

$$\overline{A}_\delta = \frac{1}{N\Delta}\left[e^{\overline{A}N\Delta} - I\right] = \begin{bmatrix} -1.7415 & -0.9546 & 0.4773 \\ 0.4773 & -0.3096 & 0.1548 \\ 0 & 0 & 0 \end{bmatrix}$$

where $N = 10$.

We use Ackermann's formula (Lemma 8.6.1) to find $\overline{J}_\delta$ so that $\left(\overline{A}_\delta - \overline{J}_\delta[C \ 0]\right)$ has all its eigenvalues at –2 (deadbeat for the slow rate). This leads to

$$\overline{J}_\delta = \begin{bmatrix} 1.7161 \\ 1.9744 \\ 8.0412 \end{bmatrix}$$

From equation (12.3.8) we get

$$\bar{J}_\delta = N e^{\bar{A}\Delta(1-N)} \bar{J}_\delta = \begin{bmatrix} 21.0315 \\ 11.057 \\ 80.4117 \end{bmatrix}$$

The fast observer will then be as in equation (12.3.3)

$$\delta_k \begin{bmatrix} \hat{x}[k] \\ \hat{d}[k] \end{bmatrix} = \tilde{A}_\delta \begin{bmatrix} \hat{x}[k] \\ \hat{d}[k] \end{bmatrix} + \tilde{J}_\delta \big[ y_{fil}[k] - C \hat{x}_{fil}[k] \big] + \tilde{B}_\delta u[k] \quad (12.3.16)$$

where

$$\tilde{A}_\delta = \frac{1}{\Delta} \big[ e^{\bar{A}\Delta} - I \big] = \begin{bmatrix} -2.8311 & -1.8557 & 0.9278 \\ 0.9278 & -0.0476 & 0.0238 \\ 0 & 0. & 0 \end{bmatrix}$$

$$\tilde{B}_\delta = \begin{bmatrix} 0.9278 \\ 0.0238 \\ 0 \end{bmatrix}$$

(Note that here $\tilde{A}_\delta = \begin{bmatrix} A_\delta & B_\delta \\ 0 & 0 \end{bmatrix}$ and $\tilde{B}_\delta = \begin{bmatrix} B_\delta \\ 0 \end{bmatrix}$, but that is

not generally true for other models of disturbances.)

Recall that

$$y_{fil}[k] = \begin{cases} y[mN] & \text{for} \quad k = mN \\ 0 & \text{otherwise} \end{cases}$$

and $\hat{x}_{fil}[k]$ is similarly generated.

To complete the design we choose

$$\bar{B}_e(\gamma) = (\gamma + 10)^2$$

and again use Ackermann's formula to find the corresponding $K_\delta$ :

$$K_\delta = \begin{bmatrix} 15.7424 & 105.7323 \end{bmatrix}$$

So, the control law is given by

$$u[k] = Gr[k] - K_\delta \hat{x}[k] - \hat{d}[k]$$

where

$$G = [\overline{B}_u(0)]^{-1}[\overline{B}_e(0)] = 53.8661$$

In Figure 12.3.2 we show the unit step response of this control system. We have also introduced a unit step disturbance at the input at $t = 3$.

(b) The conservative alternative to the multirate controller is the slow controller where we slow down the input rate to match the slow measurement rate.

Then we have a conventional slow observer given by

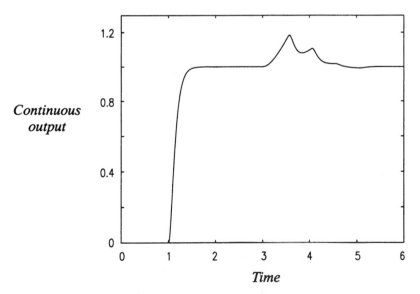

*Figure 12.3.2 : Step response with multirate control – see Example 12.3.1.*

$$\delta_m \begin{bmatrix} \hat{x}[m] \\ \hat{d}[m] \end{bmatrix} = \bar{A}_\delta \begin{bmatrix} \hat{x}[m] \\ \hat{d}[m] \end{bmatrix} + \bar{B}_\delta u[m] + \bar{J}_\delta \Big[ y[m] - C\hat{x}[m] \Big]$$

where $\bar{B}_\delta = \begin{bmatrix} 0.4773 \\ 0.1548 \\ 0 \end{bmatrix}$.

We choose the closed-loop poles to be both at –2 (deadbeat for the slow rate). Now we use the slow sampled discrete version of the system where

$$A_{\delta s} = \begin{bmatrix} -1.7415 & -0.9546 \\ 0.4773 & -0.3096 \end{bmatrix} \qquad B_{\delta s} = \begin{bmatrix} 0.4773 \\ 0.1548 \end{bmatrix}$$

Using Ackermann's formula again we find the feedback gain $K_s$ is now

$$K_{\delta s} = \begin{bmatrix} 2.1235 & 6.0412 \end{bmatrix}$$

and the control law will be

$$u[m] = G_s r[m] - K_{\delta s} \hat{x}[m] - \hat{d}[m]$$

where $G_s = 4.0206$.

The result of implementing this controller on the same continuous system with, again, a unit step input and a unit step input disturbance at $t = 3$ is shown in Figure 12.3.3. For comparison we have redrawn the response of the multirate controller. The improvement with the multirate is clearly quite significant.

△△△

We see from the above example that the use of the multirate controller is significantly better than the slow controller. This is because the input is based on state-estimates clocked at the fast rate even though the output is sampled more slowly.

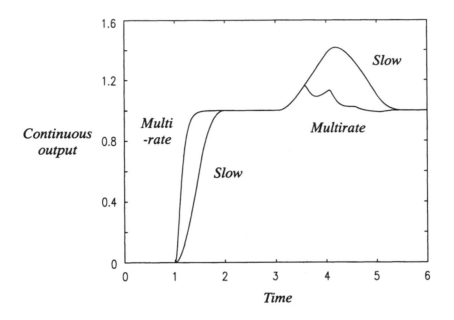

*Figure 12.3.3 : Step response with slow and multirate*
*control – see Example 12.3.1.*

## 12.4  Fast Output Sampling, Slow Input Sampling

Next we consider the converse case of fast output sampling and slow input action. The approach here is, in principle, quite similar to that used in the previous sections. Consider the configuration in Figure 12.4.1. A fast rate observer is being driven by the fast data arriving from the system. The fast state estimates are then sampled at the slow rate to provide the slow rate control law. We assume that the slow input period $\Delta_i = N\Delta$, $\Delta$ being the fast output period.

Consider again the continuous-time system given in equations (12.3.1) – (12.3.2). The fast observer equation here will be

$$\delta_k \begin{bmatrix} \hat{x}[k] \\ \hat{d}[k] \end{bmatrix} = \bar{A}_\delta \begin{bmatrix} \hat{x}[k] \\ \hat{d}[k] \end{bmatrix} + \bar{J}_\delta[y[k] - C\hat{x}[k]] + \bar{B}_\delta \bar{u}[k] \qquad (12.4.1)$$

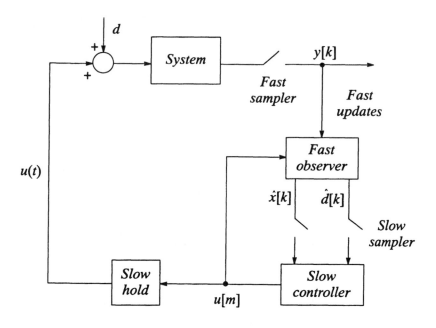

*Figure 12.4.1 : Multirate control with fast output and slow input sampling.*

where $\tilde{A}_\delta$ and $\tilde{B}_\delta$ are as in equations (12.3.4), (12.3.5) and (12.3.7), $\tilde{u}[k] =$ $u[m]$ for $mN \le k < (m+1)N$. However, it is important to note that $\tilde{J}_\delta$ here is based on the fast observer design : $\tilde{J}_\delta$ is chosen so that the matrix $(\tilde{A}_\delta - \tilde{J}_\delta[C \ 0])$ has desired the eigenvalues. We assume that $([C, 0], \ \tilde{A}_\delta)$ is a detectable pair.

To complete the controller design we express the control law as:

$$u[m] = Gr[m] - K_\delta \hat{x}[mN] - \hat{d}[m] \qquad (12.4.2)$$

where $G$ and $K_\delta$ are determined as in the previous section (see equations (12.3.9) $-(12.3.13)$) with the important distinction that the design is for the *slow* sampled equivalent of the continuous system; namely, $\overline{G}_p(\gamma)$ in equation (12.3.9) is the transfer function of the slow sampled system.

**Example 12.4.1** :  Consider the continuous-time system given by

$$\dot{x} = Ax + B(u + d)$$

$$y = Cx$$

where

$$A = \begin{bmatrix} -5 & -8 & -6 \\ 1 & 0 & 0 \\ 0 & 1 & 0 \end{bmatrix} \quad B = \begin{bmatrix} 1 \\ 0 \\ 0 \end{bmatrix} \quad C = \begin{bmatrix} 1 & 7 & 10 \end{bmatrix}$$

The system output is sampled at $\Delta = 0.05$ sec. However the control is constrained to work at a slower rate $\Delta_i = N\Delta$, where $N = 4$.

(a) Design a multirate controller so that the observer is deadbeat (at the slow rate) and $\bar{B}_e(\gamma) = \gamma + 3$ (see equation (12.3.11)).

(b) Test the controller of (a) with unit step input, a unit step disturbance and white output measurement noise.

(c) Modify the design of (a) by choosing a different observer and repeat (b).

(d) Assume the slow input constraint is removed. Redesign a fast input controller so that for the fast sampled system $\bar{B}_e(\gamma) = \gamma + 8$, (note that the choice of –8 here is equivalent to the choice of –2 at the slow rate). Compare the performance with (c).

**Solution**

(a) Following the methodology outlined in this section we calculate

$$\bar{A} = \begin{bmatrix} A & B \\ 0 & 0 \end{bmatrix} = \begin{bmatrix} -5 & -8 & -6 & 1 \\ 1 & 0 & 0 & 0 \\ 0 & 1 & 0 & 0 \\ 0 & 0 & 0 & 0 \end{bmatrix}$$

$$\bar{B} = \begin{bmatrix} B \\ 0 \end{bmatrix} = \begin{bmatrix} 1 \\ 0 \\ 0 \\ 0 \end{bmatrix}$$

and we get

$$\tilde{A}_\delta = \begin{bmatrix} -4.5955 & -7.1926 & -5.2910 & 0.8818 \\ 0.8818 & -0.1864 & -0.1380 & 0.0230 \\ 0.0230 & 0.9968 & -0.0023 & 0.0004 \\ 0 & 0 & 0 & 0 \end{bmatrix}$$

$$\tilde{B}_\delta = \begin{bmatrix} 0.8818 \\ 0.0230 \\ 0.0004 \\ 0 \end{bmatrix}$$

Using Lemma 8.6.1 (Ackermann's formula) to calculate $\tilde{J}_\delta$ for the deadbeat observer we get

$$\tilde{J}_\delta = \begin{bmatrix} 48980 \\ -23206 \\ 11354 \\ 18113 \end{bmatrix}$$

The slow sampled numerator is calculated to be

$$\bar{B}(\gamma) = 1.1522\gamma^2 + 5.631\gamma + 6.1567$$

with zeros at $-3.2358$ and $-1.6514$. We choose

$$\bar{B}_s(\gamma) = (\gamma + 3.2358)(\gamma + 1.6514)$$

and the desired closed loop denominator will be

$$\bar{B}_s(\gamma)\bar{B}_e(\gamma) = \gamma^3 + 7.8872\gamma^2 + 20.0049\gamma + 16.0304$$

To calculate the corresponding $K_\delta$ we use again Ackermann's formula (*for the slow sampled system*) and get

$$K_\delta = \begin{bmatrix} 3.7394 & 18.3276 & 20.0372 \end{bmatrix}$$

$$G = \frac{3}{1.1522} = 2.6037$$

(b) In Figure 12.4.2 we show the results of simulating the above control system with a unit step input at $t = 1$ and a unit step input disturbance (a) once at 4, and (b) once at 4.05. We note the very significant difference in the response to the input disturbance. When the input disturbance hits between the slow sample points the control has to wait to the next sample point before it can take any corrective action. Meanwhile, as we see in the figure the disturbance will have very significant impact on the output.

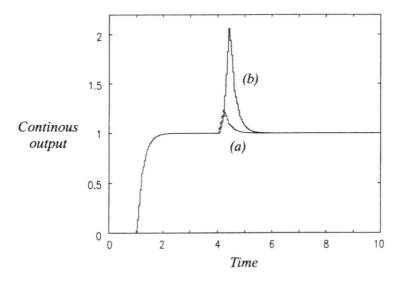

Figure 12.4.2 : *Multirate controller with fast observer and slow controller.*

Next we applied white measurement noise at the output with a variance of $10^{-4}$. In Figure 12.4.3 we see that the noise effect is disastrous on the system output. This is as expected with a deadbeat observer which will be very sensitive to noise due to the rapid observer action.

(c) In view of the problem with measurement noise found above, we modify the observer to have its eigenvalues at –8. We recalculate $J_\delta$ to obtain

$$\bar{J}_\delta = \begin{bmatrix} 605.7648 \\ -289.2554 \\ 144.6239 \\ 463.7028 \end{bmatrix}$$

Using this observer with the same control law as in (b) we see in Figure 12.4.4 similar performance as far as the step response

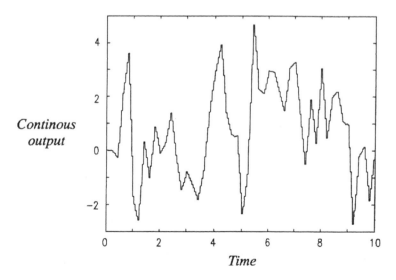

Continous output

Time

*Figure 12.4.3 : Effect of output measurement noise on controller with deadbeat observer.*

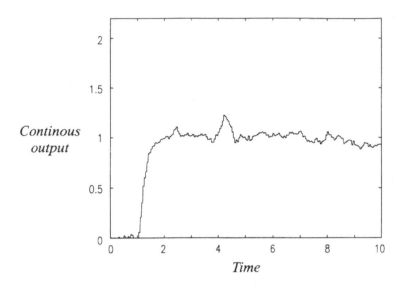

Continous
output

*Figure 12.4.4 : Response of multirate controller having fast rate
observer but with modified pole locations.*

and input disturbance rejection, but the output measurement
noise has considerably less impact on the output.

(d)  In the fast sampled system we have

$$\bar{B}(\gamma) = 1.0468\gamma^2 + 6.633\gamma + 8.8332$$

with zeros at $-4.4332$ and $-1.9035$. So here we choose the
desired closed loop denominator

$$\bar{B}_s(\gamma)\bar{B}_e(\gamma) = (\gamma + 4.4332)(\gamma + 1.9035)(\gamma + 8)$$

$$= \gamma^3 + 14.3367\gamma^2 + 59.1325\gamma + 67.5093$$

The required feedback gain $K_\delta$ is calculated as

$$K_\delta = \begin{bmatrix} 9.3692 & 54.904 & 70.4264 \end{bmatrix}$$

and $G = \dfrac{8}{1.0468} = 7.6426$.

The result of using this fast rate controller (with the observer of (c)) is shown in Figure 12.4.5. We applied again the same measurement noise and the results as shown in Figure 12.4.6. Clearly the result is better than that found in part (c) (Figure 12.4.4).

A clear conclusion from the comparison of (c) and (d) is that a slow input cannot be preferable to a fast one. Thus one would use the multirate control only if the slow input is a system constraint and not a control designer choice.

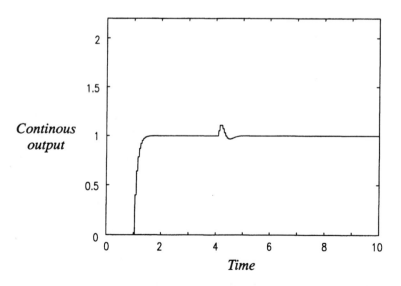

*Figure 12.4.5 : Response of controller where both observer and input changes are clocked at the fast rate.*

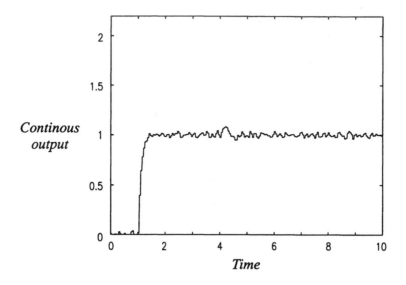

*Figure 12.4.6 :  Effect of measurement noise on controller*
*where both observer and input changes*
*are clocked at the fast rate.*

## 12.5   Further Reading and Discussion

Further reading on the topic of multirate control is contained in Meyer (1990a), Ravi *et. al* (1990), Dahleh *et al.* (1992), Zhang (1992), Albertos (1990), Meyer (1990b), Araki and Yamamoto (1986), Colaneri (1992), Er and Anderson (1991), Er and Anderson (1992), Araki and Hagiwara (1986), Feliu *et al.* (1990), Hagiwara and Araki (1988), Zhang (1993), Shor and Perkins (1993).

In many of these papers, fast output sampling is implicitly used to generate a deadbeat state observer which is a special case of the discussion in Section 12.4.

Finally, it should be remarked that multi-output systems having different sampling rates can be considered as a special case of a periodic system provided the sampling periods are integer multiples of one common period. This topic will be taken up in Chapter 13.

## 12.6 Problems

12.1 (a) Derive equation (12.2.4) from (12.2.3) and the definition of impulse sampled signals.

(b) Derive an expression relating the continuous state estimation error

$$e = \begin{bmatrix} x \\ d \end{bmatrix} - \begin{bmatrix} \hat{x} \\ \hat{d} \end{bmatrix}$$

for $m\Delta_0 \le t < (m+1)\Delta_0$ in terms of its value at $t = m\Delta_0$.

Use the expression to conclude that since $\lim_{m \to \infty} e[m] = 0$, so does

$$\lim_{m \to \infty} e(t) = 0.$$

12.2 Derive equation (12.3.3) with $B_2 = B_1 = B$, $C_1 = C$, $C_2 = 0$ and $\Delta_0 = N\Delta$.

12.3 Consider the continuous time system given by (12.3.1) – (12.3.2). Its output is sampled with period $\Delta_0 = N\Delta$. Two observers are considered: a slow rate observer given by equation (12.2.4) which generates $\hat{x}_1[m]$, $\hat{d}_1[m]$ and a fast one given by equation (12.3.3) which generates $\hat{x}_2[k]$, $\hat{d}_2[k]$. Let

$$e_1[m] = \begin{bmatrix} x(mN\Delta) \\ d(mN\Delta) \end{bmatrix} - \begin{bmatrix} \hat{x}_1[m] \\ \hat{d}_1[m] \end{bmatrix} \quad \text{and} \quad e_2[k] = \begin{bmatrix} x(k\Delta) \\ d(k\Delta) \end{bmatrix} - \begin{bmatrix} \hat{x}_2[k] \\ \hat{d}_2[k] \end{bmatrix}$$

Are $e_1[m]$ and $e_2[mN]$ equal in general? Assume both observers have zero initial conditions. In case your answer is negative evaluate which observer will provide a better estimate at $t = mN\Delta$.

12.4 Consider the following discrete time system in shift form

$$x[k+1] = A_s x[k] + B_s u[k]$$
$$y[k] = C_s x[k]$$

(a)  Suppose $\bar{y}[m] = y_{dec}[m]$ $(= y[mN])$ and $\bar{u}[m] = u_{dec}[m]$ $(= y[mN])$
and assume $u[k] = \bar{u}[m]$ for $mN \leq k < (m+1)N$. Find $\tilde{A}_s$, $\tilde{B}_s$ and
$\tilde{C}_s$ so that $\bar{u}[m]$ and $\bar{y}[m]$ are related through

$$\bar{x}[m+1] = \tilde{A}_s\bar{x}[m] + \tilde{B}_s\bar{u}[m]$$
$$\bar{y}[m] = \tilde{C}_s\bar{x}[m]$$

and show that $\bar{x}[m] = x_{dec}[m]$ $(= x[mN])$.

(b)  Let the following be the deadbeat observer for the given system

$$\hat{x}[k+1] = A_s\hat{x}[k] + B_su[k] + J_s\big[y[k] - C_s\hat{x}[k]\big]$$

and let $\tilde{K}_s$ be chosen so that the matrix $\left(\tilde{A}_s - \tilde{B}_s\tilde{K}_s\right)$ has desired ei-
gen values. Show that, if we choose the control law
$\bar{u}[m] = -\tilde{K}_s\hat{x}_{dec}[m]$ $\left(= -\tilde{K}_s\hat{x}[mN]\right)$ and if $N \geq \dim(x[k])$, $\bar{u}[m]$
also satisfies

$$\bar{u}[m+1] = M\bar{u}[m] - \sum_{i=1}^{N} H_iy[mN + i + 1]$$

(This is actually the multirate controller proposed in Hagiwara and
Araki (1988). Thus we see that this controller combines a deadbeat
fast rate observer with a slow rate control law).

12.5  Let

$$\dot{x} = Ax = Bu$$
$$y = Cx$$

be a linear time-invariant single input system with three outputs. Let the
first entry of $y$ be sampled every $\Delta$, the second every $2\Delta$ and the third
every $3\Delta$. Find the discrete equivalent of this system and show that it is
periodic.

# Chapter 13

# Optimal Control of Periodic Systems

## 13.1  Introduction

In Chapter 11 we studied periodic control of linear time-invariant systems. In this chapter we study control problems for systems which are themselves periodic in nature.

There are numerous examples of periodic control systems including:

(i)  engine control systems, where the periodicity is defined by the combustion cycle;

(ii)  biological systems, e.g. control of blood sugar level where there is a daily cycle; and

(iii)  electric power consumption that exhibits strong daily and seasonal variations.

It is also possible to obtain a periodic system from a time-invariant system as a result of using a constrained sampling strategy at the system output. For example, say a linear time-invariant continuous-time system is sampled at times $\{t_i\} = 0, 2, 3, 5, 6, 8, 9, \dots$ . Including the sampling as part of the system results in a periodic system. Another example of a similar nature arises in a multiple-output, linear, time-invariant system where each output is sampled with a different sampling interval. If these intervals are integer multiples of some basic interval the overall result can be viewed as a periodic system.

## 13.2   Control of Linear Periodic Systems

The problem we are concerned with in this section is the design of a discrete controller for the system (7.2.1) - (7.2.2). The principle will be similar to that seen with time-invariant systems : we design a state feedback controller and an observer and use the estimated state rather than the true state in the controller.

It is a straightforward exercise to show that, as in the linear time-invariant case, here too the controller design and the observer design do not affect each other. The concept commonly referred to as the 'separation principle' applies here as well. The reader is asked to provide the details in Problem 13.1.

We will begin with state feedback control design and later combine this with an observer to estimate the umeasured states.

To illustrate the design of *periodic* controllers for *periodic* systems we will adopt the linear quadratic optimization approach described in Section 9.8.

Thus, let us assume then that we want to find the control of the form

$$u(t) = u[k] \quad \text{for} \quad k\Delta \leq t < (k+1)\Delta \tag{13.2.1}$$

which will minimize the continous cost function

$$J = x(t_1)^T \Sigma_1 x(t_1) + \int_{t_0}^{t_1} \left( x(t)^T Q_c x(t) + u(t)^T R_c u(t) \right) dt \tag{13.2.2}$$

We assume that the system is both controllable and observable and that $\Delta = T/N$ for some integer $N$.

Following closely the discussion for hybrid optimal control of linear time-invariant systems in Section 9.8, we can write for $k\Delta \leq t < (k+1)\Delta$ that

$$\frac{d}{dt} \begin{bmatrix} x(t) \\ u(t) \end{bmatrix} = \overline{A}(t) \begin{bmatrix} x(t) \\ u(t) \end{bmatrix} \tag{13.2.3}$$

where

$$\overline{A}(t) = \begin{bmatrix} A(t) & B(t) \\ 0 & 0 \end{bmatrix} \tag{13.2.4}$$

Then for $k\Delta \le t < (k+1)\Delta$ we have

$$\begin{bmatrix} x(t) \\ u(t) \end{bmatrix} = \overline{\Phi}(t, k\Delta) \begin{bmatrix} x(k\Delta) \\ u(k\Delta) \end{bmatrix} \tag{13.2.5}$$

where $\overline{\Phi}(t, \tau)$ is the transition matrix corresponding to $\overline{A}(t)$.

Substituting (13.2.5) into (13.2.2) we get

$$J = x\left[\frac{t_1}{\Delta}\right]^T \Sigma_1 x\left[\frac{t_1}{\Delta}\right] + \sum_{k=\left(\frac{t_0}{\Delta}\right)}^{\left(\frac{t_1}{\Delta}\right)-1} [x^T[k], u^T[k]] \begin{bmatrix} Q[k] & S^T[k] \\ S[k] & R[k] \end{bmatrix} \begin{bmatrix} x[k] \\ u[k] \end{bmatrix} \tag{13.2.6}$$

where

$$x[k] = x(k\Delta)$$

$$u[k] = u(k\Delta)$$

and

$$\begin{bmatrix} Q[k] & S[k] \\ S^T[k] & R[k] \end{bmatrix} = \int_{k\Delta}^{(k+1)\Delta} \overline{\Phi}^T(t, k\Delta) \begin{bmatrix} Q_c & 0 \\ 0 & R_c \end{bmatrix} \overline{\Phi}(t, k\Delta) \, dt \tag{13.2.7}$$

Also, from (7.2.1) we have

$$\delta x[k] = A_\delta[k]x[k] + B_\delta[k]u[k] \tag{13.2.8}$$

where

$$A_\delta[k] = \frac{1}{\Delta}\left(\Phi((k+1)\Delta, k\Delta) - I\right) \tag{13.2.9}$$

$$B_\delta[k] = \frac{1}{\Delta} \int\limits_{k\Delta}^{(k+1)\Delta} \Phi\big((k+1)\Delta, \tau\big)B(\tau)\ d\tau \qquad (13.2.10)$$

We recognize (13.2.6) to (13.2.8) as a standard discrete-time quadratic optimal control problem which is solvable by Riccati equation methods. However, since $A_\delta[k+N] = A_\delta[k]$ and $B_\delta[k+N] = B_\delta[k]$,

$$\begin{bmatrix} Q[k+N] & S^T[k+N] \\ S[k+N] & R[k+N] \end{bmatrix} = \begin{bmatrix} Q[k] & S^T[k] \\ S[k] & R[k] \end{bmatrix} \qquad (13.2.11)$$

Thus, we see that we actually have a *periodic* LQ problem (see Problem 13.2) which has some interesting special properties.

First, we note that by a simple redefinition of the control $u[k]$ as suggested in Chapter 8 (see Problem 8.22), the criterion in equation (13.2.6) can be rewritten in terms of a block diagonal weighting matrix. Hence, for simplicity we will take the criterion to have the the form

$$J = x[k_1]^T \Sigma_1 x[k_1] + \sum_{k=k_0}^{k_1-1} \big(x^T[k]Q[k]x[k] + u[k]^T R[k]u[k]\big) \qquad (13.2.12)$$

where we denote $k_1 = t_1/\Delta$ and $k_0 = t_0/\Delta$ .

Also, for our discussion here it will be simpler to consider the shift operator version of equation (13.2.8):

$$x[k+1] = A_s[k]x[k] + B_s[k]u[k] \qquad (13.2.13)$$

where

$$A_s[k] = \Delta A_\delta[k] + I \qquad (13.2.14)$$

$$B_s[k] = \Delta B_\delta[k] \qquad (13.2.15)$$

The solution to the above LQ problem has been developed in Chapter 8 and is given by

$$u[k] = -L[k+1]x[k]$$ (13.2.16)

where

$$L[k+1] = \left(R[k] + B_s[k]^T \Sigma[k+1]B_s[k]\right)^{-1} B_s[k]^T \Sigma[k+1]A_s[k]$$ (13.2.17)

and $\Sigma[k+1]$ is the backward solution of the following difference Riccati equation

$$\Sigma[k] = Q[k] + A_s[k]^T \Sigma[k+1]A_s[k] - A_s[k]^T \Sigma[k+1]B_s[k]$$
$$\cdot \left(R[k] + B_s[k]^T \Sigma[k+1]B_s[k]\right)^{-1} B_s[k]^T \Sigma[k+1]A_s[k]$$ (13.2.18)

with the final condition $\Sigma[k_1] = \Sigma_1$. Furthermore, the optimal value of the performance index is equal to $x[k_0]^T \Sigma[k_0]x[k_0]$.

In Section 7.3 we introduced the raised form of a linear discrete-time periodic system. We will next show that the raised form of (13.2.13) provides a means for simplifying the solution of the time-varying Riccati equation (13.2.18). Recall that a raised form of (13.2.13) is given by

$$x_R[m+1] = A_R x_R[m] + B_R u_R[m]$$ (13.2.19)

where

$$\left. \begin{array}{l} x_R[m] = x[mN + k_0] \\[2mm] u_R[m] = \left[u[mN + k_0]^T, \quad \ldots, \quad u[mN + N - 1 + k_0]^T\right]^T \end{array} \right\}$$ (13.2.20)

$$\left. \begin{array}{l} A_R = A_s[k_0 + N - 1]A_s[k_0 + N - 2] \ \ldots \ A_s[k_0] \\[2mm] B_R = \left[A_s[k_0 + N - 1] \ \ldots \ A_s[k_0 + 1]B_s[k_0], \right. \\[2mm] \left. \qquad \ldots, \ A_s[k_0 + N - 1]B_s[k_0 + N - 2], B_s[k_0 + N - 1]\right] \end{array} \right\}$$ (13.2.21)

We now show that the periodic control problem outlined above can be converted into a time-*invariant* problem by use of the raising mechanism.

**Lemma 13.2.1** Assuming $k_1 - k_0 = m_1 N$, the periodic LQ problem defined in equation (13.2.13) and the criterion in (13.2.12), is equivalent to the *stationary* LQ problem defined in equation (13.2.20) and the auxiliary criterion

$$J_R = x_R[m_1]^T \Sigma_1 x_R[m_1] + \sum_{m=0}^{m_1-1} [x_R[m]^T, u_R[m]^T] \begin{bmatrix} Q_R & S_R^T \\ S_R & R_R \end{bmatrix} \begin{bmatrix} x_R[m] \\ u_R[m] \end{bmatrix} \quad (13.2.22)$$

in the sense that $u[k]$ which minimizes (13.2.12) and $u_R[m]$ which minimizes (13.2.22) are related through (13.2.20) where

$$Q_R = E_R^T \begin{bmatrix} Q[k_0] & & 0 \\ & \ddots & \\ 0 & & Q[k_0 + N - 1] \end{bmatrix} E_R \quad (13.2.23)$$

$$S_R = G_R^T \begin{bmatrix} Q[k_0] & & 0 \\ & \ddots & \\ 0 & & Q[k_0 + N - 1] \end{bmatrix} \quad (13.2.24)$$

$$R_R = \begin{bmatrix} R[k_0] & & 0 \\ & \ddots & \\ 0 & & R[k_0 + N - 1] \end{bmatrix}$$

$$+ G_R \begin{bmatrix} Q[k_0] & & 0 \\ & \ddots & \\ 0 & & Q[k_0 + N - 1] \end{bmatrix} G_R \quad (13.2.25)$$

$$E_R = \begin{bmatrix} I \\ A_s[k_0] \\ A_s[k_0 + 1]A_s[k_0] \\ \cdot \\ \cdot \\ A_s[k_0 + N - 2] \ \cdots \ A_s[k_0] \end{bmatrix}$$

$$
G_R = \begin{bmatrix}
0 & 0 & 0 \\
B_s[k_0] & 0 & 0 \\
A_s[k_0 + 1]B_s[k_0] & B_s[k_0 + 1] & 0 \\
\cdot & \cdot & \cdot \\
\cdot & \cdot & \cdot \cdots \cdot \\
\cdot & \cdot & \cdot \\
A_s[k_0 + N - 2] \ \cdots \ B_s[k_0] & A_s[k_0 + N - 3] \ \cdots \ B_s[k_0 + 1] & 0
\end{bmatrix}
$$

$$(13.2.26)$$

**Proof :** Let us use (13.2.20) to rewrite (13.2.12) as

$$
J = x_R[m_1]^T \Sigma_1 x_R[m_1] + \sum_{m=0}^{m_1-1} \sum_{l=k_0}^{k_0+N-1} x[mN + l]^T Q[mN + l]x[mN + l]
$$

$$
+ u[mN + l]^T R[mN + l]u[mN + l] \tag{13.2.27}
$$

Since $Q[k]$ and $R[k]$ are $N$-periodic we get from (13.2.21), (13.2.23) – (13.2.26):

$$
\sum_{l=k_0}^{k_0+N-1} x[mN + l]^T Q[mN + l]x[mN + l] + u[mN + l]^T R[mN + l]u[mN + l]
$$

$$
= \begin{bmatrix} x_R[m]^T u_R[m]^T \end{bmatrix} \begin{bmatrix} Q_R & S_R^T \\ S_R & R_R \end{bmatrix} \begin{bmatrix} x_R[m] \\ u_R[m] \end{bmatrix}
$$

so that $J = J_R$ and the two LQ problems are equivalent.

⌃⌃⌃

Lemma 13.2.1 simplifies the solution to our original periodic LQ problem. The solution to the raised LQ problem is given by

$$
\boxed{u_R[m] = -L_R[m + 1]x_R[m]} \tag{13.2.28}
$$

and

$$\Sigma_R[m] = \Sigma[k_0 + mN] \qquad (13.2.29)$$

where

$$L_R[m+1] = \left(R_R + B_R^T \Sigma_R[m+1] B_R\right)^{-1} \left(B_R^T \Sigma_R[m+1] A_R + S_R\right) \qquad (13.2.30)$$

and $\Sigma_R[m]$ is the backward solution of the following difference Riccati equation with *time-invariant* matrices $Q_R$, $A_R$ etc,

$$\begin{aligned} \Sigma_R[m] &= Q_R + A_R^T \Sigma_R[m+1] A_R - \left(A_R^T \Sigma_R[m+1] B_R + S_R^T\right) \\ &\quad \left(R_R + B_R^T \Sigma_R[m+1] B_R^r\right)^{-1} \left(B_R^T \Sigma_R[m+1] A_R + S_R\right) \end{aligned} \qquad (13.2.31)$$

For the infinite horizon case (13.2.31) becomes an algebraic Riccati equation. Subject to the usual restrictions (e.g. stabilizability and detectability) it is known that the solution of (13.2.31) converges to the solution of the algebraic equation and that the resulting *steady-state* feedback system is stable. This immediately implies, through (13.2.29), that the solution of the original Riccati equation (13.2.18) converges to a "steady state" *periodic* solution which also ensures closed-loop stability.

Moreover, given the solution $\Sigma_R$ to the algebraic raised Riccati equation, the periodic solution to the original Riccati equation is obtained by going through one cycle of the Riccati difference equation (13.2.18) using the final condition

$$\Sigma[k_0 + N] = \Sigma_R \qquad (13.2.32)$$

We illustrate the above by a simple example.

**Example 13.3.1** : Consider the continuous-time system

$$\dot{y}(t) = \cos(\pi t)\, y(t) + \exp\left[\frac{1}{\pi} \sin(\pi t)\right] u(t)$$

This system is to be controlled by a digital controller clocked at $\Delta = 0.4$ sec, so that

$$J = \int_0^\infty \left( y^2(t) + \alpha u^2(t) \right) dt$$

is minimized. Find the optimal control for $\alpha = 1$ and $\alpha = 0.1$.

**Solution** To find the equivalent discrete optimization criterion we have

$$\overline{A}(t) = \begin{bmatrix} \cos(\pi t) & \exp\left[\dfrac{1}{\pi}\sin(\pi t)\right] \\ 0 & 0 \end{bmatrix}$$

and

$$\overline{\Phi}(t,\tau) = \begin{bmatrix} \exp\left[\dfrac{1}{\pi}(\sin(\pi t) - \sin(\pi\tau))\right] & (t-\tau)\exp\left[\dfrac{1}{\pi}\sin(\pi t)\right] \\ 0 & 1 \end{bmatrix} \qquad (13.2.33)$$

So

$$\begin{bmatrix} Q[k] & S[k] \\ S[k] & R[k] \end{bmatrix} = \int_{k\Delta}^{k\Delta+1} \overline{\Phi}^T(t,k\Delta) \begin{bmatrix} 1 & 0 \\ 0 & \alpha \end{bmatrix} \overline{\Phi}(t,k\Delta) dt$$

Using numerical integration the following values result:

$$Q[k] = \int_{k\Delta}^{(k+1)\Delta} \exp\left[\dfrac{2}{\pi}(\sin(\pi t) - \sin(\pi k\Delta))\right] dt$$

$$= 0.5768, \ 0.3858, \ 0.282, \ 0.3311, \ 0.525$$

for $k = 0, 1, 2, 3, 4$

$$S[k] = \int_{k\Delta}^{(k+1)\Delta} (t - k\Delta) \exp\left[\frac{2}{\pi}\sin(\pi t) - \frac{1}{\pi}\sin(\pi k\Delta)\right] dt$$

$$= 0.127, \; 0.1004, \; 0.0594, \; 0.0527, \; 0.0857$$

for $k = 0, \; 1, \; 2, \; 3, \; 4$

$$R[k] = 0.4 \quad \text{for} \quad \alpha = 1$$

$$R[k] = 0.04 \quad \text{for} \quad \alpha = 0.1$$

We define

$$\tilde{u}[k] = u + R^{-1}[k]S[k]y[k] \tag{13.2.34}$$

and will find first $\tilde{u}[k]$. The corresponding change in $\tilde{Q}[k]$ will be

$$\tilde{Q}[k] = Q[k] - R^{-1}[k]S^2[k]$$

Thus

$$\tilde{Q}[k] = 0.5365, \; 0.3606, \; 0.2732, \; 0.3241, \; 0.5067$$

$$\text{for} \begin{cases} \alpha = 1 \\ k = 0, \; 1, \; 2, \; 3, \; 4 \end{cases}$$

and

$$\tilde{Q}[k] = 0.1734, \; 0.1336, \; 0.1938, \; 0.2617, \; 0.3413$$

$$\text{for} \begin{cases} \alpha = 0.1 \\ k = 0, \; 1, \; 2, \; 3, \; 4 \end{cases}$$

From (13.2.33)

$$A_s[k] = \exp\left[\frac{1}{\pi}(\sin(\pi(k+1)\Delta) - \sin(\pi k\Delta)\right]$$

$$= 1.3535,\ 0.8908,\ 0.6878,\ 0.8908,\ 1.3533$$

$$\text{for } k = 0,\ 1,\ 2,\ 3,\ 4$$

$$B_s[k] = \Delta\exp\left[\frac{1}{\pi}\sin(\pi k\Delta)\right]$$

$$= 0.4,\ 0.5414,\ 0.4823,\ 0.3317,\ 0.2955$$

$$\text{for } k = 0,\ 1,\ 2,\ 3,\ 4$$

We then obtain the raised system as

$$A_R = 1$$

$$B_R = [0.2955\quad 0.4490\quad 0.5815\quad 0.4490\quad 0.2955]$$

$$E_R = \begin{bmatrix} 1.0000 \\ 1.3535 \\ 1.2057 \\ 0.8294 \\ 0.7388 \end{bmatrix}$$

$$G_R = \begin{bmatrix} 0 & 0 & 0 & 0 & 0 \\ 0.4 & 0 & 0 & 0 & 0 \\ 0.3563 & 0.5414 & 0 & 0 & 0 \\ 0.2451 & 0.3724 & 0.4823 & 0 & 0 \\ 0.2183 & 0.3317 & 0.4296 & 0.3371 & 0 \end{bmatrix}$$

$$Q_R = 2.0937 \quad \text{for } \alpha = 1$$

$$Q_R = 1.0664 \quad \text{for } \alpha = 0.1$$

$$S_R = \begin{bmatrix} 0.4602 \\ 0.4026 \\ 0.2905 \\ 0.1242 \\ 0 \end{bmatrix} \quad \text{for} \quad \alpha = 1$$

$$S_R = \begin{bmatrix} 0.2639 \\ 0.2910 \\ 0.2130 \\ 0.0837 \\ 0 \end{bmatrix} \quad \text{for} \quad \alpha = 0.1$$

$$R_R = \begin{bmatrix} 0.536 & 0.119 & 0.0858 & 0.0367 & 0 \\ 0.119 & 0.5808 & 0.1304 & 0.0558 & 0 \\ 0.0858 & 0.1304 & 0.5689 & 0.0722 & 0 \\ 0.0367 & 0.0558 & 0.0722 & 0.4558 & 0 \\ 0 & 0 & 0 & 0 & 0.4 \end{bmatrix} \quad \text{for} \quad \alpha = 1$$

$$R_R = \begin{bmatrix} 0.118 & 0.086 & 0.063 & 0.0247 & 0 \\ 0.086 & 0.1707 & 0.0957 & 0.0376 & 0 \\ 0.063 & 0.0957 & 0.1639 & 0.0487 & 0 \\ 0.0247 & 0.0376 & 0.0487 & 0.0776 & 0 \\ 0 & 0 & 0 & 0 & 0.04 \end{bmatrix} \quad \text{for} \quad \alpha = 0.1$$

Solving the raised (steady state) algebraic Riccati equation we get

$$\Sigma_R = 1.5523 \quad \text{for} \quad \alpha = 1$$

$$\Sigma_R = 0.3792 \quad \text{for} \quad \alpha = 0.1$$

and using the solutions above to generate the periodic solutions for the original Riccati equation we obtain

$\Sigma[k+1]$ = 0.6934, 0.6054, 1.1873, 1.5523, 1.531

$$\text{for} \begin{cases} \alpha = 1 \\ k = 0, \ 1, \ 2, \ 3, \ 4 \end{cases}$$

$\Sigma[k+1]$ = 0.2038, 0.2511, 0.4090, 0.3792, 0.3791

$$\text{for} \begin{cases} \alpha = 0.1 \\ k = 0, \ 1, \ 2, \ 3, \ 4 \end{cases}$$

To calculate the corresponding feedback gains we recall the modification we made in (13.2.34):

$$L[k+1] = R^{-1}[k]S[k] + \left(R[k] + B_R[k]^T\Sigma[k+1]B_R[k]\right)^{-1}B_R[k]^T\Sigma[k+1]A_R[k]$$

Thus

$$L[k+1] = 1.0523, \ 0.7567, \ 0.7309, \ 0.9353, \ 1.3617$$

$$\text{for} \begin{cases} \alpha = 1 \\ k = 0, \ 1, \ 2, \ 3, \ 4 \end{cases}$$

$$L[k+1] = 3.4307, \ 2.7665, \ 1.7583, \ 1.5705, \ 2.4931$$

$$\text{for} \begin{cases} \alpha = 0.1 \\ k = 0, \ 1, \ 2, \ 3, \ 4 \end{cases}$$

According to (13.2.34) we finally have

$$u[k] = -L[k+1]y[k] - R[k]^{-1}S[k]y[k]$$

The above control systems was simulated with $y(0) = 1$ and the results are shown in Figure 13.3.1. As expected the control system designed for $\alpha = 0.1$ converges considerably faster than when $\alpha = 1$ is used. The price is larger feedback gains as we see above.

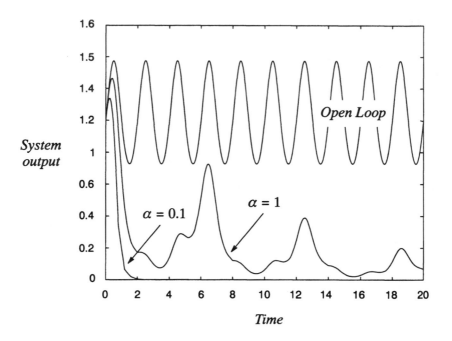

*Figure 13.2.1 : Responses for Example 13.3.1.*

## 13.3   Control Based on State Estimate Feedback

Finally, we show how state estimates can be used in place of the true states in the state variable feedback controllers described in Section 13.2. We illustrate by a simple example.

**Example 13.3.1 :**

(a)  Given the continuous-time system

$$\dot{x} = Ax + Bu$$

with

$$A = \begin{bmatrix} 0 & 0 \\ 1 & 0 \end{bmatrix}, \quad B = \begin{bmatrix} 1 \\ 0 \end{bmatrix}$$

each state is sampled at different instances:

$x_1(t)$ at $\{t_i\}$ = 0, 0.1, 0.3, 0.4, 0.6, 0.7 ....

$x_2(t)$ at $\{t_i\}$ = 0.2, 0.5, 0.8, 1.1, ....

It is desired to design a digital controller clocked at $\Delta = 0.1$ so that

$$J_c = \int_0^\infty \left( x^T Q_c x + u^T R_c u \right) dt$$

is minimized with

$$Q_c = C_1^T C_1$$
$$C_1 = [1, 1]$$
$$R_c = 0.1$$

(b) Simulate the system with $x[0] = \begin{bmatrix} 1 \\ 1 \end{bmatrix}$ and $\hat{x}[0] = \begin{bmatrix} 0 \\ 0 \end{bmatrix}$.

## Solution

(a) Using the separation principle we first design a state-feedback digital controller to minimize the criterion and then an observer to generate a state estimate.

*Controller design:*

Assuming we have data on the two states at sampling period $\Delta = 0.1$, then we can follow the approach in Chapter 9:

$$\overline{A} = \begin{bmatrix} A & B \\ 0 & 0 \end{bmatrix}$$

$$\begin{bmatrix} Q & S \\ S^T & R \end{bmatrix} = \int_0^\Delta e^{\overline{A}^T \tau} \begin{bmatrix} Q_c & 0 \\ 0 & R_c \end{bmatrix} e^{\overline{A}\tau} \, d\tau$$

$$= \begin{bmatrix} 0.1103 & 0.1050 & 0.0055 \\ 0.1050 & 0.1 & 0.0052 \\ 0.0055 & 0.0052 & 0.0104 \end{bmatrix}$$

Solving the corresponding discrete algebraic Riccati equation with

$$A_s = \begin{bmatrix} 1 & 0 \\ 0.1 & 1 \end{bmatrix}, \quad B_s = \begin{bmatrix} 0.1 \\ 0.005 \end{bmatrix}$$

we get the solution of the algebraic Riccati equation

$$\Sigma = Q + A_s^T \Sigma A_s - \left(A_s^T \Sigma B_s + S^T\right)\left(R + B_s^T \Sigma B_s\right)^{-1}\left(B_s^T \Sigma A_s + S\right)$$

$$\Sigma = \begin{bmatrix} 0.4064 & 0.3184 \\ 0.3184 & 0.2797 \end{bmatrix}$$

with the corresponding feedback gain:

$$L = \left(R + B_s^T \Sigma B_s\right)^{-1} \left(B_s^T \Sigma A_s + S\right)$$

$$= [3.4625 \quad 2.604]$$

*Observer design:*

The sampled version of the system is

$$x[k+1] = A_s x[k] + B_s u[k]$$

and the measurement pattern can be viewed as

$$y[k] = C[k]x[k]$$

where

$$C[k] = \begin{cases} [0, 1] & \text{for} \quad k = 3m + 2 \\ [1, 0] & \text{otherwise} \end{cases}$$

For the design of the observer via Kalman filter theory, we will assume

$$Q_s = \begin{bmatrix} 1 & 0 \\ 0 & 1 \end{bmatrix}, \quad R_s = 1$$

and the corresponding periodic discrete Riccati equation (see Chapter 8) is then

$$P[k+1] = Q_s + A_s P[k] A_s^T$$
$$- A_s P[k] C[k]^T \left( R_s + C[k]P[k]C[k]^T \right)^{-1} C[k]P[k]A_s^T$$

We make use of the raised system to solve the above equation. The raised system here will be

$$A_R = \begin{bmatrix} 1 & 0 \\ 0.3 & 1 \end{bmatrix}$$

$$C_R = \begin{bmatrix} 1 & 0 \\ 1 & 0 \\ 0.2 & 1 \end{bmatrix}$$

$$E_R = \begin{bmatrix} 1 & 0 & 1 & 0 & 1 & 0 \\ 0.2 & 1 & 0.1 & 1 & 0 & 1 \end{bmatrix}$$

$$G_R = \begin{bmatrix} 0 & 0 & 0 & 0 & 0 & 0 \\ 1 & 0 & 0 & 0 & 0 & 0 \\ 0.1 & 1 & 0 & 1 & 0 & 0 \end{bmatrix}$$

and then

$$QR = \begin{bmatrix} 3 & 0.3 \\ 0.3 & 3.05 \end{bmatrix}$$

$$RR = \begin{bmatrix} 1 & 0 & 0 \\ 0 & 2 & 0.1 \\ 0 & 0.1 & 3.01 \end{bmatrix}$$

$$SR = \begin{bmatrix} 0 & 1 & 0.1 \\ 0 & 0.2 & 2.02 \end{bmatrix}$$

Solving the algebraic Riccati equation for the raised system we get

$$PR = \begin{bmatrix} 2.6305 & 0.1855 \\ 0.1855 & 1.8138 \end{bmatrix}$$

Substituting $P[0] = PR$ in the periodic Riccati equation we get

$$P[1] = \begin{bmatrix} 1.7246 & 0.1236 \\ 0.1236 & 2.8218 \end{bmatrix}$$

$$P[2] = \begin{bmatrix} 1.633 & 0.1086 \\ 0.1086 & 3.8316 \end{bmatrix}$$

with the corresponding Kalman gains

$$H[k] = A_s P[k] C[k]^T \left( R_s + C[k] P[k] C[k]^T \right)^{-1}$$

$$H[0] = \begin{bmatrix} 0.7246 \\ 0.1236 \end{bmatrix}$$

$$H[1] = \begin{bmatrix} 0.633 \\ 0.1086 \end{bmatrix}$$

$$H[2] = \begin{bmatrix} 0.0225 \\ 0.7953 \end{bmatrix}$$

So the controller will consist of the periodic observer:

$$\hat{x}[k+1] = A_s\hat{x}[k] + B_su[k] + H[k]\big(y[k] - C[k]\hat{x}[k]\big)$$

and the estimated feedback clocked at period $\Delta$ :

$$u[k] = -L\hat{x}[k]$$

(b) The control system was simulated. In Figure 13.3.1 we show the system output

$$y(t) = x_1(t) + x_2(t) = C_1x(t)$$

with $x(0) = \begin{bmatrix} 1 \\ 1 \end{bmatrix}$ and $\hat{x}(0) = \begin{bmatrix} 0 \\ 0 \end{bmatrix}$.

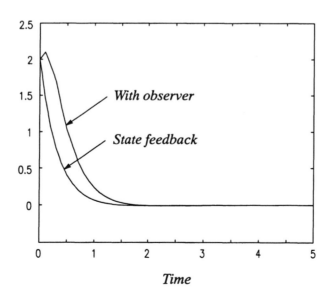

*Figure 13.3.1 : Regulated response from Example 13.3.1.*

Note that the responses are slightly different due to the transient in the state estimator produced by the erroneous initial condition $\hat{x}[0] \neq x[0]$.

Figure 13.3.2 compares the closed loop response with input noise of variance $10^4$ added to $\{u[k]\}$. Note that the system with the observer performs slightly worse than when true state feedback is used as expected.

## 13.4  Further Reading and Discussion

For general background on periodic control problems, see the following papers which relate periodic and discrete time-invariant systems : Colanerim and Longhi (1995), Flamm (1991), Lin and King (1993), Van Dooren and Sreedhar (1994) and Misra (1996). Falmm (1993) shows how one cam compute the stability margins of periodic systems with respect to periodic perturbations of such

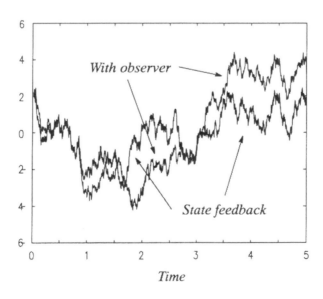

*Figure 13.3.2 :  Response of system in Example 13.3.1 to noise.*

systems. The results are shown to be the same as when computed in the time-invariant representation of the system.

We have given details of the use of linear quadratic optimization to solve the problem of periodic control of periodic system. The associated Riccati equation theory is discussed in more detail in Bittanti, Laub and Willems (1991) and the references therein, especially Bittanti, Colanoi and de Nicolao (1991).

Most of the properties follow directly from the time-invariant case by using the idea of time domain raising.

## 13.5 Problems

13.1 Extend the separation principle to time-varying linear systems.

13.2 Verify (13.2.11), and show that $A_\delta[k + N] = A_\delta[k]$ and $B_\delta[k + N] = B_\delta[k]$ in equation (13.2.8).

13.3 Verify that the system in (13.2.8) and the cost function in (13.2.11) lead to a periodic problem in which all matrices are periodic.

13.4 Consider the system in equations (7.2.1) – (7.2.2). Let $H(t)$ be such that

$$\dot{x}(t) = (A(t) - H(t)C(t))x(t)$$

is asymptotically stable. Define the observer

$$\dot{\hat{x}}(t) = A(t)\hat{x}(t) = H(t)\left(y(t) - C(t)\dot{x}(t)\right) + B(t)u(t)$$

Assuming that $B(t)$ is bounded, show that the system (7.2.1) - (7.2.2) with the control $u(t) = -F(t)\hat{x}(t)$ will asymptotically approach in its behaviour the control $u^*(t) = -F(t)x(t)$ for any bounded $F(t)$.

13.5 Extend Problem 13.2 to general periodic discrete-time stochastic systems.

# Bibliography

[1]     Agarwal, R.C., and C.S. Burrus (1975), 'New recursive digital filter structures having very low sensitivity and roundoff noise', *IEEE Trans. on Circuits and Systems*, Vol. CAS-22, No.12, pp. 921-927.

[2]     Albertos, P. (1990), 'Block multirate input-output model for sampled-data control systems', *IEEE Trans. on Automatic Control*, Vol.AC-35, No.9, pp. 1085-1088.

[3]     Anderson, B.D.O. and J. Moore (1979), *Optimal Filtering*, Prentice-Hall Englewood Cliffs, N.J.

[4]     Antoulas, A.C. (1991), *Mathematical System Theory – The Influence of R.E. Kalman*, Springer-Verlag, Berlin.

[5]     Araki, M. and T. Hagiwara (1986), 'Pole assignment by multirate sample-data output feedback', *Int. Journal of Control*, Vol.44, No.6,  pp. 1661-1673.

[6]     Araki, M. and K. Yamamoto (1986), 'Multivariable multirate sampled-data systems : state-space description, transfer characteristics, and Nyquist criterion', *IEEE Trans. on Automatic Control*, Vol.AC-31, No.2, pp. 145-154.

[7]     Araki, M. and Y. Ito (1992a), 'Frequency response of sampled data systems I : Open loop considerations', Technical Report 92-04, Department of Electrical Engineering II, Kyoto University.

[8]     Araki, M. and Y. Ito (1992b), 'Frequency response of sampled data systems II : Closed loop considerations', Technical Report 92-05, Department of Electrical Engineering II, Kyoto University.

[9]     Araki, M. and Y. Ito (1992c), 'On frequency response of sampled-data control systems', *SICE Symposium on Control Theory*, May 27-29, 1992, Kariya.

[10]    Araki, M. (1993), 'Recent developments in digital control theory', Plenary address, *Proc. IFAC World Congress*, Sydney Australia, July 1993.

[11]    Åström, K.J. (1970), *Introduction to Stochastic Control Theory*, Academic Press, New York.

[12]    Åström, K.J., P. Hagander and J. Sternby (1984), 'Zeros of sampled systems', *Automatica*, Vol.20, pp. 21-39.

[13]    Åström, K.J. and B. Wittenmark (1984), *Computer Controlled Systems : Theory and Design*, Prentice-Hall International.

[14]    Bai, E.-W. and S. Dasgupta (1989), 'On digital control of continuous-time systems using generalized hold functions', *Proc. 28th CDC*, Tampa, Florida, December 13-15, pp. 1241-1246.

[15]    Bamieh, B., M.A. Dahleh and J.B. Pearson (1993), 'Minimization of the $L^\infty$ - induced norm for sampled-data systems', *IEEE Trans. on Automatic Control*, Vol.38, No.5, pp. 717-732.

[16]    Bamieh, B.A. and J.B. Pearson, Jr. (1992a), 'A general framework for linear periodic systems with application to $H^\infty$ sampled-data control', *IEEE Trans. on Automatic Control*, Vol.37, No.4, pp. 1418-1435.

[17]    Bamieh, B.A. and J.B. Pearson (1992b), 'The $H_2$ problem for sampled-data systems', *Systems and Control Letters*, Vol.19, No.1, pp. 1-12.

[18]    Barker, R.H. (1952), 'The pulse transfer function and its applications to sampling servosystems', *Proc. IEE*, 99, part IV, pp. 302-317.

[19]    Benedetto, J.J. (1992), 'Irregular sampling and frames', In Chiu (1992).

[20]    Bhattacharya, R.N. and E.C. Waymire (1990), *Stochastic Processes with Applications*, Wiley and Sons, Inc., New York.

[21]    Bittanti, S., A.J. Laub and J.C. Willems (1991), *The Riccati Equation* Springer-Verlag, Berlin.

[22]  Bittanti, S., P. Colaneri and G. DeNicolao (1991), 'The periodic Riccati equation', in Bittanti, Laub and Willems (1991).

[23]  Bracewell, R.N. (1986) *The Fourier Transform and its Applications* McGraw-Hill International.

[24]  Butzer, P.L., W. Splettstösser and R. Stens (1988), 'The sampling theorem and linear prediction in signal analysis', *Jber. d. Dt. Math-Vérein* 90, 1, 70.

[25]  Butzer, P.L. and R.L. Stens (1992), 'Sampling Theory for not necessarily band-limited functions : A historical overview', *SIAM Review*, Vol.34, No.1, pp. 40-53.

[26]  Caines, P.E. (1988), *Linear Stochastic Systems*, John Wiley and Sons, Inc., New York.

[27]  Chammas, A.B. and C.T. Leondes (1979), 'On the finite time control of linear systems by piecewise constant output feedback', *Int. Journal Contol*, Vol.30, No.2, pp. 227-234.

[28]  Chan, S.W., G.C. Goodwin and K.S. Sin (1984), 'Convergence properties of the Riccati difference equation in optimal filtering of nonstabilizable systems', *IEEE Trans. on Automatic Control*, Vol.AC-29, No.2, pp. 110-118.

[29]  Chapellat, H. and M. Dahleh (1992) 'Analysis of time-varying control strategies for optimal disturbance rejection and robustness', *IEEE Trans. on Automatic Control*, Vol.37, No.11 pp. 1734-1745.

[30]  Chen, C.T. (1970), *Linear System Theory and Design*, CBS College Publishing, Holt, Rinehart and Winston, New York.

[31]  Chen, T. and B.A. Francis (1991), '$H_2$ -optimal sampled-data control', *IEEE Trans. on Automatic Control*, Vol.36, No.4, pp. 387-397.

[32]  Chen, T. and B. A. Francis (1995), *Optimal Sampled-data Control Systems*, Springer Verlag, Berlin.

[33]  Chui, C.K. (1992), *An Introduction to Wavelets*, Academic Press, Boston, U.S.A.

[34]  Colaneri P., R. Scattolini and N. Schiavoni (1992), 'LQG optimal control of multirate sampled-data systems', *IEEE Trans. on Automatic Control*, Vol.37, No.5, pp. 675-682.

[35]  Colanerim P. and S. Longhi (1995), 'The realization problem for linear periodic systems', *Automatica*, Vol.31, No.5, pp. 775-779.

[36]  Cox, D.R. and H.D. Miller (1965), *The Theory of Stochastic Processes*, Chapman and Hall Ltd., London.

[37]  Crochiere, R.E. and L.R. Rabiner, *Multirate Digital Signal Processing*, Englewood Cliffs, NJ: Prentice Hall, 1983.

[38]  Dahleh, M.A., P.G. Voulgaris and L.S. Valavani (1992), 'Optimal and robust controllers for periodic and multirate systems', *IEEE Trans. on Automatic Control*, Vol.37, No.1, pp. 90-99.

[39]  Das, S.K. and P.K. Rajagopalan (1992), 'Periodic discrete-time systems : Stability analysis and robust control using zero placement', *IEEE Trans. on Automatic Control*, Vol.37, No.3, pp. 374-378.

[40]  Davis, J.H. (1972) 'Stability conditions from spectral theory : Discrete systems with periodic feedback', *SIAM Journal of Control*, Vol.10 pp. 1-13.

[41]  Desoer, C.A., R.-W. Liu, J. Murray and R. Saeks (1980), 'Feedback system design : The fractional representation approach to analysis and synthesis', *IEEE Trans. on Automatic Control*, Vol. AC-25, No.3, pp. 399-412.

[42]  de Souza, C.E. and G.C. Goodwin (1984), 'Intersample variances in discrete minimum variance control', *IEEE Trans. on Automatic Control*, Vol.AC-29, No.8, pp. 759-761.

[43]  de Souza, C.E., M.R. Gevers and G.C. Goodwin (1986), 'Riccati equations in optimal filtering of nonstabilizable systems having singular state transition matrices', *IEEE Trans. on Automatic Control*, Vol.AC-31, No.9, pp. 831-838.

[44]  Doob, J.L. (1953), *Stochastic Processes*, John Wiley and Sons, New York.

[45]  Doyle, J.C. and G. Stein (1979), 'Robustness with observers', *IEEE Trans. Automatic Control*, Vol. AC-24, No.4, pp. 607-611.

[46]  Doyle, J.C., K. Glover, P.P. Khargonekar and B.A. Francis (1989), 'State-space solutions to standard $H_2$ and $H_\infty$ control problems', *IEEE Trans. on Automatic Control*, Vol.34, No.8, pp. 831-847.

[47]  Dullerud, G.E. (1996), *Control of Uncertain Sampled-data Systems*, Birkhäuser, Boston.

[48]  Dullerud, G.E. and B.A. Francis, (1992) '$L_1$ analysis and design of sampled-data systems', *IEEE Trans. on Automatic Control*, Vol.37, No.4, pp. 436-446.

[49]  Dym, H. and H.P. McKean (1972), *Fourier Series and Integrals*, Academic Press, New York.

[50]  Edwards, R.E. (1980), *Fourier Series - A Modern Introduction*, Vol. 1 and 2, Springer-Verlag, New York.

[51]  Er, M.J. and B.D.O. Anderson (1991), 'Practical issues in multirate output controllers', *Int. Journal of Control*, Vol.53, No.5, pp. 1005-1020.

[52]  Er, M.J. and B.D.O. Anderson (1992), 'Performance study of multi-rate output controllers under noise disturbances', *Int. Journal of Control*, Vol.56, No.3, pp. 531-545.

[53]  Feliu, V., J.A. Cerrada and C. Cerrada (1990), 'A method to design multi-rate controllers for plants sampled at a low rate', *IEEE Trans. on Automatic Control*, Vol.35, No.1, pp. 57-60.

[54]  Feuer, A. (1993) 'Periodic control of LTI systems : A critical point of view', *Proc. IFAC World Congress*, Sydney, July 1993.

[55]  Feuer, A. and G.C. Goodwin (1993), 'Generalized sample-hold functions : Analysis of robustness, sensitivity and intersample difficulties', Tech. Report EE9141, University of Newcastle, also *IEEE Trans. Automatic Control*, Vol.39, No.2, May 1994, pp.1042-1047.

[56]  Feuer, A. and G.C. Goodwin (1995) 'Sampled data systems' in *CRC Handbook of Control*, Edited by B. Levine.

[57]    Flamm, D.S. (1991), 'A new shift-invariant representation for periodic linear systems', *Systems and Control Letters*, Vol.17, pp. 9-14.

[58]    Flamm, D.S. (1993), 'Single-loop stability margins for multirate and periodic control systems', *IEEE Trans. on Automatic Control*, Vol.38, No.8, pp. 1232-1236.

[59]    Forsyth,. W. and R.M. Goodall (1991), *Digital Control*, MacMillan Education Ltd., London.

[60]    Franklin, G.F., J.D. Powell and M.L. Workman (1990), *Digital Control of Dynamic Systems*, Addison-Wesley Publishing Company Ltd., New York.

[61]    Francis, B.A. and T.T. Georgiou (1988), 'Stability theory for linear time-invariant plants with periodic digital controllers', *IEEE Trans. on Automatic Control*, Vol.33, No.9, pp. 820-832.

[62]    Gelb, A. (1974) *Applied Optimal Estimation*, The Analytic Sciences Corporation.

[63]    Gevers, M. and G. Li (1993), *Parameterization in Control, Estimation and Filtering Problems*, Springer-Verlag, London.

[64]    Goodall, R.M. and D.S. Brown (1985), 'High-speed digital controllers using an 8-bit microprocessor', *Software & Microsystems*, Vol.4, No. 5 & 6, pp. 109-116.

[65]    Goodall, R.M. (1990), 'The delay operator $z^{-1}$ – Inappropriate for use in recursive digital filter', *Trans. Institute of Meas. and Control*, Vol.12, No.5, pp. 246-250.

[66]    Goodwin, C.G. and K.S. Sin (1984) - *Adaptive Filtering, Prediction and Control*, Prentice-Hall Inc., Englewood Cliffs, NJ.

[67]    Goodwin, G.C. and A. Feuer (1992), 'Linear periodic control: A frequency domain viewpoint', *Systems & Control Letters*, Vol.19, pp.379-390.

[68]    Goodwin, G.C., R.H. Middleton and H.V. Poor (1992), 'High-speed digital signal processing and control', *Proceedings of IEEE*, Vol.80, No.2, pp. 240-259.

[69]   Goodwin, G.C. and M.E. Salgado (1993), 'Frequency domain sensitivity functions for continuous time systems under sampled data control', Tech.Report EE9163, The University of Newcastle also *Automatica*, Vol.30, No.8, 1994, pp.1263-1270.

[70]   Hagiwara, T. and M. Araki (1988), 'Design of a stable state feedback controller based on the multirate sampling of the plant output', *IEEE Trans. on Automatic Control*, Vol.33, No.9, pp. 812-819.

[71]   Haykin, S. 1986, *Adaptive Filter Theory*, Prentice-Hall.

[72]   Hansen, E.R. (1975), *A Table of Series and Products*, Prentice-Hall Inc. Englewood Cliffs, N.J.

[73]   Higgins, J. (1977), *Completeness and Basic Properties of Sets of Special Functions*, Cambridge University Press.

[74]   Honig, M.L. and D.L.G. Messerschmitt 1984, *Adaptive Filters – Structures, Algorithms and Applications*, Kluwer-Academic Publishers, 1984.

[75]   Hori, N., K. Kanai and P. Nikiforuk (1990), 'Robustly stable, discrete-time, adaptive flight control using the Euler operator', *Proc. IFAC World Congress*, Tallin, 1990.

[76]   Hurewicz, W. (1947), 'Filters and servo systems with pulsed data' in *Theory of Servo Mechanism* ed. H.M. James, N.B. Nichols and R.S. Philips, New York, McGraw Hill.

[77]   Hyslop, G.L., H. Schöttler and T.-J. Tarn (1992), 'Descent algorithms for optimal periodic output feedback control', *IEEE Trans. on Automatic Control*, Vol.37, No.12, pp. 1893-1904.

[78]   Itakura, F. and S. Saito (1971), 'Digital filtering techniques for speech analysis and synthesis', *Proc. 7th Int. Conf. Acoust.* Budapest, Vol.25-C-1, pp. 261-264.

[79]   James, H.M., N.B. Nichols, R.S. Phillips (1947), *Theory of Servo Mechanisms*, McGraw Hill, New York.

[80]   Jerri, A.H. (1977), 'The Shannon sampling theorem - its various extensions and applications : A tutorial review'. *Proceedings of the IEEE*, Vol. 65, No.11, pp. 1565-1596.

[81]    Jury, E.I. and J. Tschauner (1971), 'On the Inners formulation of stability condition within the shifted unit circle', *Regelungstechnik*, Vol. 7, pp. 299-300.

[82]    Kabamba, P and Hara, S. (1990), 'On computinging the induced norm of sampled data feedback systems', *Proc. American Control Conf.*, San Diego, CA., May 23-25, pp. 319-320.

[83]    Kabamba P.T. (1987), 'Control of linear systems using generalized sampled-data hold functions', *IEEE Trans. on Automatic Control*, Vol.32, No.9, pp. 772-783.

[84]    Kailath, T. (1974), 'A view of three decades of linear filtering theory', *IEEE Trans. on Inform. Theory*, Vol IT. 20, No.2, pp. 146-181.

[85]    Kanai, K., Y. Kishimoto, N. Hori and P. Nikiforuk (1990), 'Adaptive flight control of CCV aircraft with limiting zeros', AIAA Guidance, Navigation and Control Conference, August 1990, Portland, Oregan.

[86]    Katz, P. (1981), *Digital Control using Microprocessors*, Prentice-Hall, Englewood Cliffs, N.J.

[87]    Kawake, T., Y. Yamamura and K. Kanai (1990), 'Robust controller for a servo positioning system of an automobile', *Proc. of the 29th CDC*, Nonolulu, Hawaii, December 5-7, pp. 2170-2175.

[88]    Keller, J.P. and B.D.O. Anderson (1992), 'A new approach to the discretization of continuous-time controllers', *IEEE Trans. on Automatic Control*, Vol.37, No.2, pp. 214-223.

[89]    Kennedy, R.A. and R.J. Evans (1990), 'Digital redesign of a continuous controller based on closed loop performance', *Proc. of 29th CDC*, Honolulu, Hawaii, pp. 1898-1901.

[90]    Khargonekar, P.P., K. Poolla and A. Tannenbaum (1985), 'Robust control of linear time-invariant plants using periodic compensation', *IEEE Trans. on Automatic Control*, Vol. AC-30, No.11, pp. 1088-1096.

[91]    Kitamori, T. (1983), 'Fusion of continuous and the discrete control theory', *Journal Society Instrumentation and Control Engineers*, (in Japanese), Vol.22, No.7, pp. 599-605.

[92]  Körner, T. (1988), *Fourier Analysis*, Cambridge University Press, Cambridge, England.

[93]  Kranc, G.M. (1957), 'Input-output analysis of multirate feedback systems', *IRE Trans. Automatic Control*, Vol.3, pp. 21-28.

[94]  Kuo, B.C. (1992), *Digital Control Systems*, Saunders College Publishing, HBJ Orlando, Florida.

[95]  Lagrange, J.L. (1759), 'Sur l'intégration d'une équation différentiale a différences finies, qui contient la théorie des suites récurrentes', *Miscell. Taurin.*, Vol.I, pp. 33-42.

[96]  Lagrange, J.L. (1792–1793), 'Mémoire sur la méthod d'interpolation', *Nouv. Mém. Acad. Sci.* Berlin, Vol. XLV, pp. 276-288.

[97]  Laplace, P.S. (1820), Théorie analytique des probabliliés, *Revue et augmentée par l'Auteur.* (3rd Ed.) Paris.

[98]  Lee, S., S.M. Meerkov and T. Runolfsson (1987), 'Vibrational feedback control : zero placement capabilities', *IEEE Trans. on Automatic Control*, Vol.32 No.7, pp. 604-611.

[99]  Lévy, B., T. Kailath, L. Ljung and M. Morf (1979), 'Fast time-invariant implementations for linear least-squares smoothing filters', *IEEE Trans. on Automatic Control*, Vol. AC-24, No.5, pp. 770-774.

[100]  Li, G. and M. Gevers (1990), 'Comparative study of finite wordlength effects in shift and delta operator parameterizations', Tech. Report, Universite Catholique de Louvain, Centre for Systems Engineering and Applied Mechanics.

[101]  Lin, C.A. and C.W. King (1993), 'Minimal periodic realizations of transfer matrices', *IEEE Trans. on Automatic Control*, Vol.38, No.3, pp. 462-466.

[102]  Linnemann, A. (1992), '$L_\infty$ -induced optimal performance in sampled-data systems', *Systems & Control Letters*, Vol.18, pp. 265-275.

[103]  Mayne, W.H., 'Common reflection point horizontal data stacking techniques', *Geophys.*, Vol.27, pp. 927-938, December 1962.

[104] Meyer, D.G. (1990a), 'A parameterization of stabilizing controllers for multirate sampled-data systems', *IEEE Trans. on Automatic Control*, Vol.35, No.2, pp. 233-236.

[105] Meyer, D.G. (1990b), 'A new class of shift-varying operators, their shift-invariant equivalents, and multirate digital systems', *IEEE Trans. on Automatic Control.*, Vol.35, pp. 429-433.

[106] Meyer, R.A. and C.S. Burrus, 'A unified analysis of multirate and periodically time-varying digital filters', *IEEE Trans. Circuits Systems*, Vol. CAS-22, pp. 162-168, March 1975.

[107] Middleton, R.H. and G.C. Goodwin (1986), 'Improved finite word length characteristics in digital control using delta operators', *IEEE Trans. on Automatic Control*, Vol. AC-31, No.11, pp. 1015-1021.

[108] Middleton, R.H. and G.C. Goodwin (1990), *Digital Control and Estimation : A Unified Approach*, Prentice-Hall, Englewood Cliffs, NJ.

[109] Middleton, R.H. and J. Freudenberg (1993), 'Zeros and non-pathological sampling for generalized sampled-data hold functions', Technical Report EE9318, Department of Electrical and Computer Engineering, The University of Newcastle, Australia also *International Journal of Control*, Vol.61, No.6, June 1995, pp.1387-1421.

[110] Mintzer, F. and B. Liu, 'Aliasing error in the design of multirate filters', *IEEE Trans. Acoustics, Speech and Signal Processing*, Vol. ASSP-26, pp. 76-88, February 1978.

[111] Miscra, P. (1996), 'Time-invariant representation of discrete periodic systems', *Automatica*, Vol.32, No.2, pp. 267-272.

[112] Moore, K.L., S.P. Bhattacharyya and M. Dahleh (1989), 'Arbitrary pole and zero assignment with n-delay input control using stable controllers', *Proc. 28th CDC*, Tampa, Florida, December 13-15, pp. 1253-1258.

[113] Mori, T., P.N. Nikiforuk, M.M. Gupta and N. Hori (1987), 'A class of discrete time models for a continuous time system', *Proc. 1987 American Control Conference*, Minneapolis, MN. June 10-12, pp. 953-957.

[114] Neuman, C.P. (1988), 'Properties of the delta operator model of dynamic physical systems', Technical Report SMC 088-12-1258, Electrical Engineering, Carnegie Mellon University.

[115] Oppenheim, A.V., A.S. Willsky and I.T. Young (1983), *Signals and Systems*, Prentice-Hall International.

[116] Oppenheim, A.V. and R.W. Shafer (1989), *Discrete-time Signal Processing*, Prentice-Hall, New Jersey.

[117] Orlandi, G. and G. Martinelli (1984), 'Low-sensitivity recursive digital filters obtained via the delay replacement', *IEEE Trans. on Circuits and Systems*, Vol. CAS-31, No.7, pp. 654-657.

[118] Osburn, S.L. and D.S. Bernstein (1993), 'An exact treatment of the achievable closed loop $H_2$ performance of sampled data controllers : From continuous time to open loop' *Proc. IEEE CDC Conference* San Antonia, Texas, pp.325-330 (also to appear Automatica).

[119] Peterka, V. (1986), 'Control of uncertain processes: Applied theory and algorithms', *Kybernetika*, Vol.22, pp. 1-102.

[120] Premaratne, K. and E.I. Jury (1992b), 'Tabular method for determining root distribution of delta-operator formulated real polynomials' Tech. Report, Department of Electrical Engineering, University of Miami.

[121] Premaratne, K. and E.I. Jury (1992), 'On the application of polynomial array method to discrete-time system stability', Technical Report, Department of Electrical Engineering, University of Miami.

[122] Pridham, R.G. and R.A. Mucci, 'Digital interpolation beamforming for low-pass and bandpass signals', *Proc. IEEE*, Vol.67, pp. 904-919, June 1979.

[123] Ptolemy (1992), Available by anonymous ftp from ptolemy.berkeley.edu.

[124] Ragazzini, J.R. and L.A. Zadeh (1952), 'The analysis of sampled-data systems' *AIEE Trans.*, Vol. 71, part II, pp. 225-234.

[125] Rattan, K.S. (1984), 'Digitalization of existing continuous control systems', *IEEE Trans. on Automatic Control*, Vol.AC-29, Vol.3, pp. 282-285.

[126] Ravi R., P.P. Khargonekar, K.D. Minto and C.N. Nett (1990), 'Controller parameterization for time-varying multirate plants', *IEEE Trans. on Automatic Control*, Vol.35, No.11, pp. 1259-1262.

[127] Rosenuasser, Ye. N. (1994), *Linear Theory of Digital Control in Continuous Time*, (in Russian), Nauka, Moscow.

[128] Rosenuasser, Ye. N. (1995a), 'Optimal synthesis of sampled-data systems in $L_2$-space' *2nd Russia - Swedish Control Conference*, August 29-31, Saint Petersburg, Russia pp. 73-76.

[129] Rosenuasser, Ye. N. (1995b), 'Mathematical description and analysis of multivariable sampled-data systems in continuous time : Part 1 – Parametric transfer function and weight function of multivariable sampled data systems', *Automation and Remote Control*, Vol.56, No.4, pp. 526-540.

[130] Rosenuasser, Ye. N. (1995c), 'Mathematical description and analysis of multivariable sampled-data systems in continuous time. Part 2 Analysis of sampled-data systems under deterministic and stochastic disturbances', *Automation and Remote Control*, Vol.55, pp.684-697.

[131] Rosenuasser, Ye. N., Y. Yu. Polyakou and B.P. Lampe (1996), 'Frequency method for $H_2$-optimization of time-delayed sampled-data systems', preprint.

[132] Salgado, M.E., R.H. Middleton and G.C. Goodwin (1986), 'Connection between continuous and discrete Riccati equations with applications to Kalman filtering', *Proc. IEE*, Part D, Vol.135, No.1, pp. 28-34.

[133] Scheibner, D.J. and T.W. Parks, 'Slowness aliasing in the discrete randon transform : A multirate system approach to beamforming', *IEEE Trans. Acoustics, Speech and Signal Processing*, Vol. ASSP-32, pp. 1160-1165, December, 1984.

[134] Shamma, J.S. and M.A. Dahleh (1991) 'Time-varying versus time-invariant compensation for rejection of persistent bounded disturbances and robust stabilization', *IEEE Trans. on Automatic Control*, Vol. 36, No.7, pp. 838-847.

[135] Shenoy, R.G., D. Burnside and T.W. Parks (1994), 'Linear periodic systems and multirate filter design', *IEEE Trans. Signal Processing*, Vol.42, No.2, September, pp. 2242-2256.

[136] Shor, M.H. and W.R. Perkins (1993), 'Sampled-data decentralized controller design', *Proc. IFAC World Congress*, Sydney, July 1993.

[137] Sivashankar, N. and P.P. Khargonekar (1991a), '$L_\infty$ -induced norm of sampled-data systems', *Proc. American Control Conf.*, June 26-28, Boston, MA, pp. 167-172.

[138] Sivashankar, N. and P.P. Khargonekar (1992b), 'Robust stability and performance of sampled-data systems', *Proc. 30th CDC* Brighton, U.K. December 11-13, pp. 881-886.

[139] Smith, M.J.T. and T.P. Barnwell, 'A new filter bank theory for time-frequency representation', *IEEE Trans. Acoustics, Speech and Signal Processing*, Vol. ASSP-35, pp. 314-327, March 1987.

[140] Soh, C.B. (1991), 'Robust stability of discrete-time systems using delta operators', *IEEE Trans. on Automatic Control*, Vol.36, Vol.3 pp. 377-380.

[141] Stirling, J. (1730), *Methodus Differentialis : Sive Tractatus de Summatione et Interpolatione Serierum Infinitarum*, London.

[142] Stuart, R.D. (1961), *An Introduction to Fourier Analysis*, Methuen & Co. Ltd. London.

[143] Toivonen, H.T. (1990), 'Sampled data control of continuous-time system with an $H_\infty$ optimality criterion', Rep.90-1 Department of Chemical Engineering, Abo Akademic, Finland.

[144] Tschauner, J. (1963), *Introduction a la Theorie des Systemes Echantillon*, Dunod : Paris.

[145] Tsypkin, Y.Z. (1949), (1950), 'Theory of discontinuous control', *Automati Telemekh*, 3 (1949), 5 (1949), 5 (1950).

[146] Vaidyanathan, P.P., 'Multirate digital filters, filter banks polyphase networks, and applications : A tutorial', *Proc. IEEE*, Vol.78, pp. 56-93, January 1990.

[147] Vaidyanathan, P.P., *Multirate Systems and Filterbanks*, Englewood Cliffs, NJ: Prentice Hall, 1993.

[148] Van Dooren, P. and J. Sreedhar (1994), 'When is a periodic discrete-time system equivalent to a time-invariant one?', *Linear Algebra and its Application*, Vol. 212/213, pp. 1220-1225.

[149]  Vetterli, M. (1987), 'A theory of multirate fulter banks', *IEEE Trans. Acoustics, Speech and Signal Processing*, Vol. ASSP-35, No.3, March, pp. 356-372.

[150]  Vetterli, M. (1989), 'Invertibility of linear periodically time-varying filters', *IEEE Trans. on Circuits and Systems*, Vol.36, No.1, January, pp. 148-150.

[151]  Weller, S.R., A. Feuer, G.C. Goodwin and H.V. Poor (1993), 'Interrelations between continuous and discrete lattice filter structures', to appear, *IEEE Trans. on Circuits and Systems*.

[152]  Widrow, B. and S.D. Stearns, 1985, *Adaptive Signal Processing*, Prentice-Hall.

[153]  Williamson, D. (1988), 'Delay replacement in direct form structures', *IEEE Transactions on Acoustics, Speech and Signal Processing*, Vol.36, No.4, pp. 453-460.

[154]  Williamson, D. (1991), *Digital Control and Implementation – Finite Wordlength Considerations*, Prentice-Hall International.

[155]  Yamamoto, Y. (1990), 'New approach to sampled-data control systems – A function space method', *Proc., 29th CDC*, December 5-7, Honolulu, HI, pp. 1882-1887.

[156]  Yamamoto, Y. (1994), 'Frequency response and its computation for sampled data systems', *Systems and Networks : Mathematical Theory and Applications, Mathematical Research 79, Proc. MTNS-93*, Regensburg, Germany, U. Helmke, R. Mennicken and J. Saurer, Edo Berlin, Academic Verlag, 1994, pp. 573-574.

[157]  Yamamoto, Y. and P.P. Khargonekar (1996), 'Frequency response of sampled data systems', *IEEE Trans. on Automatic Control*, Vol.41, No.2, pp. 166-177.

[158]  Yan, W.Y., B.D.O. Anderson and R.R. Bitmead, (1991), 'On the gain margin improvement using dynamic compensation based on generalized sampled-data hold functions', Technical Report, ANU.

[159]  Youla, D.C., J.J. Bongiorno and H.A. Jabr (1976), 'Modern Wiener-Hopf design of optimal controllers', *IEEE Trans. on Automatic Control*, Vol.AC-21, pp. 3-13 and 319-338.

[160] Zadeh, L.A. (1950), 'Frequency analysis of variable networks', *Proc. IRE*, March 1950, pp.291-299.

[161] Zhang, C. (1992), 'A dual rate digital compensator for zero assignment', *Systems & Control Letters*, Vol.19, No.3, pp. 225-232.

[162] Zhang, C. (1993), 'Unification of a class of periodic and multirate controllers', Technical Report, University of Melbourne, Australia.

# Index

# Systems & Control: Foundations & Applications

*Founding Editor*
Christopher I. Byrnes
School of Engineering and Applied Science
Washington University
Campus P.O. 1040
One Brookings Drive
St. Louis, MO 63130-4899
U.S.A.

*Systems & Control: Foundations & Applications* publishes research monographs and advanced graduate texts dealing with areas of current research in all areas of systems and control theory and its applications to a wide variety of scientific disciplines.

We encourage the preparation of manuscripts in TEX, preferably in Plain or AMS TEX—LaTeX is also acceptable—for delivery as camera-ready hard copy which leads to rapid publication, or on a diskette that can interface with laser printers or typesetters.

Proposals should be sent directly to the editor or to: Birkhäuser Boston, 675 Massachusetts Avenue, Cambridge, MA 02139, U.S.A.

Estimation Techniques for Distributed Parameter Systems
*H.T. Banks and K. Kunisch*

Set-Valued Analysis
*Jean-Pierre Aubin and Hélène Frankowska*

Weak Convergence Methods and Singularly Perturbed
Stochastic Control and Filtering Problems
*Harold J. Kushner*

Methods of Algebraic Geometry in Control Theory: Part I
Scalar Linear Systems and Affine Algebraic Geometry
*Peter Falb*

$H^\infty$-Optimal Control and Related Minimax Design Problems
*Tamer Başar and Pierre Bernhard*

Identification and Stochastic Adaptive Control
*Han-Fu Chen and Lei Guo*

Viability Theory
*Jean-Pierre Aubin*

Representation and Control of Infinite Dimensional Systems, Vol. I
A. Bensoussan, G. Da Prato, M. C. Delfour and S. K. Mitter

Representation and Control of Infinite Dimensional Systems, Vol. II
A. Bensoussan, G. Da Prato, M. C. Delfour and S. K. Mitter

Mathematical Control Theory: An Introduction
Jerzy Zabczyk

$H_\infty$-Control for Distributed Parameter Systems: A State-Space Approach
Bert van Keulen

Disease Dynamics
Alexander Asachenkov, Guri Marchuk, Ronald Mohler, Serge Zuev

Theory of Chattering Control with Applications to Astronautics,
Robotics, Economics, and Engineering
Michail I. Zelikin and Vladimir F. Borisov

Modeling, Analysis and Control of Dynamic Elastic
Multi-Link Structures
J. E. Lagnese, Günter Leugering, E. J. P. G. Schmidt

First Order Representations of Linear Systems
Margreet Kuijper

Hierarchical Decision Making in Stochastic Manufacturing Systems
Suresh P. Sethi and Qing Zhang

Optimal Control Theory for Infinite Dimensional Systems
Xunjing Li and Jiongmin Yong

Generalized Solutions of First-Order PDEs: The Dynamical
Optimization Process
Andreĭ I. Subbotin

Finite Horizon $H_\infty$ and Related Control Problems
M. B. Subrahmanyam

CPSIA information can be obtained at www.ICGtesting.com
Printed in the USA
LVOW102134171012

303372LV00002B/52/P